Lecture Notes in Mathematics

Edited by A. Dold and B. Eckmann

809



Igor Gumowski
Christian Mira

# Recurrences and Discrete Dynamic Systems

Springer-Verlag
Berlin Heidelberg New York 1980

**Authors**
Igor Gumowski
Christian Mira
U.E.R. de Mathématiques, Université Paul Sabatier,
118, Route de Narbonne
31077 Toulouse
France

AMS Subject Classifications (1980): Primary: 58-02
Secondary: 26A18, 34C15, 34C35, 39A15, 58F05, 58F10, 58F15, 58F20, 58F99

ISBN 3-540-10017-2 Springer-Verlag Berlin Heidelberg New York
ISBN 0-387-10017-2 Springer-Verlag New York Heidelberg Berlin



Printing and binding: Beltz Offsetdruck, Hemsbach/Bergstr.
2141/3140-543210

# P R E F A C E

The idea to study recurrences took root in 1949 during an informal lecture of P. Montel. A discussion about possible types of fixed points in 1959 led to a lasting collaboration with C. Mira. The analytical insight increased more rapidly after an invariant curve germ, based on a Lattès series, was successfully continued by means of numerical computations. The material described in this monograph constitutes a synopsis of the slowly accumulated particular results. Frequent discussions with R. Thibault, R. Clerc, Ch. Hartmann, J. Couot, C. Gillot, O. Rössler and G. Targonski produced substantial improvments.

During the last few years recurrences appeared in several fields of applied science and they have become a major research topic of the interdisciplinary Dynamic Systems Research Group of Toulouse University. The creation of such a group would have been impossible without the continued support of J.C. Martin, President of Toulouse University.

The text was typed by Mrs. C. Grima and many of the figures were drawn by G. Roussel. The preparation of the typescript was encouraged by Ph. Leturcq. All contributions are gratefully acknowledged.

Toulouse, March 1980. I.G.

CONTENTS

Chapter IV - Stochasticity in almost conservative recurrences



Chapter V - Stochasticity in strongly non-conservative recurrences

# INTRODUCTION AND STATEMENT OF THE PROBLEM

The content of this monograph is intended to be easily accessible to readers from various disciplines, such as mathematics, physics, biology, biometry, medical biometry, ecology, etc., who face time-evolving phenomena, or as the latter are concisely called : dynamic systems. For this reason, the simplest possible terminology has been adopted, in order not to add artificial vocabulary-hurdles to the intrinsic difficulty of the subject. Ars-pro-artis generalisations, so popular in contemporary mathematics, have been deliberately sacrified in order not to obscure the dominating internal mechanisms of the evolution processes and thus preserve a phenomenological transparency. The emphasis on "what happens" instead of on symbols appears to be a valid motivation of all efforts to understand nature (as well as mathematics). The resulting formulation of problems may therefore appear primitive, sometimes even simplistic, and the terminology old-fashioned. Abstractly inclined readers will no doubt find occasions to feel irritated ; the authors accept in advance full blame for these distractions from the subject-matter in question. Experience in dynamic systems teaches, however, that apparently simple problems turn out to have complicated solutions, whose description happens to encroach on the "influence spheres" of several distinct mathematical disciplines, each having already a frozen professional vocabulary. One way out of this linguistic predicament appears to be the pursuit of maximal simplicity, and thus the search for a maximal interdisciplinary common part ; which is the exposition method adopted by the authors. The monograph is restricted to the study of discrete dynamic dystems, taken from a "natural" context, expressed in the form of one or two-dimensional real-valued point-mappings, or according to old French usage, recurrences.

Recurrences occur in many branches of mathematics, ranging from number theory to functional equations (c.f. [M 10]). They appear also independently as natural descriptions of evolution phenomena in physics, biology, etc. (c.f. [M 2]). Individual results on recurrences are therefore scattered in different types of publications, and are generally expressed in a widely varying vocalubary. The purpose of this monograph is to provide a systematic and unified treatment of presently known and physically, biologically, etc. relevant properties of first and second order real-valued recurrences. Several new results are published here for the first time. For conciseness, all constants, parameters, variables and functions are assumed to be real-valued, unless the contrary is explicitly stated.

The conventional specialized symbols used to express this property will be omitted. Functional spaces will be treated in the same manner ; in most cases only a verbal specification will be given. Readers in need of constantly repeated specialized symbols will experience no difficulty in supplying their own.

Autonomous first order recurrences of the form

$$x_{n+1} = f(x_n, c), \qquad n = 0, \pm 1, \pm 2, \ldots \tag{0-1}$$

where c is a parameter and f(x, c) a single-valued smooth function of both x and c, constitute one of the oldest mathematical notions (examples : relation between two successive numbers of a sequence ; Hurwitz's definition of the logarithm in terms of the limit of $x_n$, $n \to \infty$ when $x_{n+1} = \sqrt{x_n}$, $x_o > 0$). A rather fundamental role is played by recurrences in the works of Poincaré, although at first sight it appears that they are introduced there merely as an artifice for the study of trajectories of dynamic systems, defined by a system of ordinary differential equations, where the independent variable represents time. The advantage obtained by means of this artifice consists in the reduction of the dimensionality of the problem by one unit, which is accomplished by the elimination of one dependent variable via a suitable "surface of section". As an illustration of the procedure consider a particular example [A 5] possessing explicit expressions in terms of elementary functions at every stage (this is completely untypical in a non linear context).

Let the dynamic system be of order two :

$$\ddot{x} + 2b\dot{x} + x = 0, \ \dot{x} < 0 \ ; \ = a, \ \dot{x} = 0 \ ; \ = 1, \ \dot{x} > 0 \ ; \ t \geqslant 0 , \tag{0-2}$$

where $0 < b \ll 1$, $0 < a < 1$, and $x = x(t)$, $\dot{x}(t) = y(t)$ are continuous functions. It is required to determine the nature of the solutions of (0-2) in the phase plane x, y ; and in particular to ascertain the possible existence of periodic solutions, which to quote Poincaré, constitute the main breach in the natural defences (i.e. in the intrinsic complexity) of non linear dynamic systems.

Since the dynamic system (0-2) is linear for $y < 0$ and $y > 0$, the corresponding component-solutions are :

$$\begin{aligned} &y < 0: \ x_-(t) = e^{-bt}(\bar{x} \cos \omega t + \bar{z} \sin \omega t), \ \bar{z} = \frac{1}{\omega}(\bar{y} + b\bar{x}), \\ &y > 0: \ x_+(t) = 1 + x_-(t), \ \omega^2 = 1 - b^2, \end{aligned} \tag{0-2'}$$

where $\bar{x}$, $\bar{y}$ is either $x_-(0)$, $y_-(0)$ or $x_+(0)$, $y_+(0)$. The assumed continuity of x(t) and $\dot{x}(t)$ implies that all phase-plane trajectories G(x, y) = const. of (0-2), defined parametrically by (0-2'), are continuous, cross the x-axis, and are not tangent to the latter. Thus the x-axis, or a part of it, is an appropriate (one-dimensional) surface of section. Fix the initial point ($\bar{x}$, $\bar{y}$), $\bar{y} \neq 0$, and let $0 < x_o \neq a$ be the first x-axis crossing point of a trajectory of (0-2) passing through ($\bar{x}$, $\bar{y}$). It follows from (0-2') that the next positive x-axis crossing occurs at

$$x_1 = 1 + e^{-c} + x_o e^{-2c}, \ c = \pi b/\omega \ ,$$

and, in general, two successive positive x-axis crossing points of G(x, y) = const. are related by the linear recurrence

$$x_{n+1} = 1 + e^{-c} + x_n e^{-2c}, \ x_n > 0, \ n = 0, 1, 2, \ldots \ . \tag{0-3}$$

By construction, the <u>first</u> order recurrence (0-3) contains the same amount of information about the function G(x, y) as the <u>second</u> order differential equation (0-2). If (0-3) is to be useful, however, it is necessary to know how to find its relevant solution. Because of the simplicity of both (0-2) and (0-3), this solution can be found by trial and error :

$$x_n = x_o e^{-2cn} + (1 - e^{-2cn}) / (1 - e^{-c}) \tag{0-4}$$

without any need of a preliminary abstract analysis characterizing its nature, and the functional space in which it is located. A known explicit solution is, of course, more informative than an abstract theorem affirming its existence, uniqueness and continuity, but unfortunately in the general context of non linear recurrences such a favourable situation occurs only exceptionally. From an examination of $x_n$ in (0-4) as a function of n, it is easily seen that the dynamic system (0-2) admits a unique periodic solution, whose closed phase plane trajectory $G_\infty$ crosses the x-axis at

$$x_\infty = \lim_{n\to\infty} x_n = 1/(1 - e^{-c}) \tag{0-5}$$

The uniqueness of $G_\infty$ results from the fact that $x_\infty$ is independent of $x_o$. All other trajectories G are spirals which approach $G_\infty$ asymptotically. The periodic solution is therefore asymptotically stable (in the sense of Liapunov) and its influence domain is the whole phase plane x, y, i.e. it is reached from arbitrary $\bar{x}$, $\bar{y}$. The point ($x_o \neq a$, 0), excluded in the determination of (0-3), is an unstable constant solution of (0-2).

It is known that in certain cases (example : problem of the climate, [L 9]), the trajectory structure of a differential equation is described by a recurrence whose order is <u>two</u> units lower. The reduction of dimensionality by one or more units is obviously not a general property of an intrinsic structural equivalence between a differential equation and a recurrence. Consider in fact the first order autonomous differential equation :

$$\dot{x} = g(x, c),\ x = x(t),\ t > 0,\ x(0) = x_o, \tag{0-6}$$

where $x_o$, c are parameters and g is a single-valued continuously differentiable function of its arguments. Let

$$x(t) = H(x_o, t, c) \tag{0-7}$$

be the general solution of (0-6). Replacing x(t) by the finite difference $(x(t+h) - x(t)) / h$, $h > 0$ and letting $t_n = nh$, $x_n = x(t_n)$, equation (0-6) turns into the recurrence

$$x_{n+1} = x_n + hg(x_n, c) \tag{0-8}$$

which is of the <u>same</u> order as the differential equation (0-6). For $h \ll 1$ the solution structure of (0-6) and (0-8) is known to be equivalent (Euler's method of discretization, but this equivalence does not persist for larger h, [G 9], p. 46).

The two examples (0-2), (0-3) and (0-6), (0-8), raise the following question : how is it possible to decide whether for a specific f the recurrence (0-1) is structurally equivalent to a differential equation of order one, two or higher ? A meaningful answer to this question is impossible unless the notions "solution of a recurrence" and "structural equivalence", used above in a rather self-evident fashion, have been given unambiguous definitions. The detailed nature of these definitions appears to have far-reaching consequences with respect to what constitutes a characteristic property of a recurrence, because such a characterization is to be consistent with the properties of associated differential equations, regardless of a possible difference of order. Moreover, since there exists a very strong link (c.f. [K 6]) between the recurrence (0-1), the functional iterates

(0-9) $$f_{n+1}(x, c) = f(f_n(x, c)),\ f_1(x, c) = f(x, c),\ f_o(x, c) = x,$$

$n = 0, 1, 2, \ldots$ and some functional equations, like for example

(0-10)
$$w(f(x, c)) = s\, w(x) \quad \text{(Schröder)}$$
$$w(f(x, c)) = w(x) \quad \text{(automorphic functions)}$$
$$w(f(x, c)) = (w(x))^m \quad \text{(Böttcher)}$$
$$w(f(x, c) = w(x) + a \quad \text{(Abel)}$$
$$\sum_i u_i\, w(v_i) = w(x) \quad \text{(Perron-Frobenius, special case)}$$

where m, s, a, are parameters, $u_i = u_i(f(x, c))$, $v_i = v_i(f(x, c)$ known functions of f, and f is the same as in (0-1), any characteristic property of a recurrence sheds considerable light on the properties of iterations and functional equations. It should be stressed at this point that x is a discrete dependent variable in (0-1), whereas in (0-9), (0-10), it is usually a continuous independent one.

Due to the assumed discreteness of x, the recurrence (0-1) is not equivalent to a nominally identical difference equation. In a difference equation x is assumed to be a continuous variable. This apparently minor distinction has major consequences. For example, when (0-1) is considered as a recurrence, a completely and unambiguously defined initial data (i.e. Cauchy) problem is formulated for $n > 0$ by specifying an isolated initial value $x_o$ of x, whereas when (0-1) is considered as a difference equation, it is necessary to specify x on a continuum of values (for example, on the interval $x_o < x < x_1$). In the first case the solution is a sequence of real numbers $\{x_n\}$, $n = 0, 1, 2, \ldots$, and in the second a rather complicated mathematical entity (a functional of the initial function). In the case of a linear recurrence (with respect to $x_n$), this theoretically fundamental distinction dissolves into practical insignificance (see for example eq. (0-3) and (0-4)), but for non linear f no deeper analysis of (0-1) is possible without it.

Even a cursory examination of the literature on non linear recurrences shows that their properties are extremely complex. The simplest possible "generic" example of (0-1) :

(0-11) $$x_{n+1} = x_n^2 + c, \qquad -2 \leq c \leq \frac{1}{4} \quad ,$$

has been studied intensely during at least two decades [M 11]-[M 15], but its basic solution structure has been identified only recently. This recurrence, examined in detail in Chapter I, constitutes therefore a yardstick with which to measure the efficiency of arguments and the amount of progress made on other first order recurrences.

Keeping in mind the fact that recurrences appear fundamentally as natural descriptions of observed evolution phenomena, because most measurements of time-evolving variables (except for short-period continuous "analog"-recordings) are discrete, and only incidentally via differential or functional equations (which from a fundamental physical point of view are more far-fetched abstractions), it is possible to refer to them as dynamic systems in their own right. Whenever a distinction is required between a dynamic system expressed in terms of differential equations and one in terms of recurrences, the former will be called continuous and the latter discrete. In these designations the adjectives continuous and discrete refer only to the independent variables t and n, respectively. The dynamic system point of view presents the advantage of giving access to an efficient and widely known terminology, which should be able to cover at least in part the variety of situations expected to arise in recurrences. In fact, there exists a hard-core part of dynamic system terminology, which possesses a physically transparent phenomenological meaning, and which has withstood the wear of prolonged theoretical as well as experimental usage. All efforts will be made to stick to this "naturally selected" part.

The complex nature of non linear dynamic processes requires some comments on what is to be understood by the notion "relevant solution" of a recurrence. A formal, or perhaps better a formalistic definition is easy enough. Similarly to a continuous dynamic system, the recurrence (0-1) is considered as an implicit definition of a function

(0-12) $$x_n = F(x_o, n, c),$$

which leads to an identity in n, c, and $x_o$ after insertion into (0-1). While n takes only integer values, the parameters $x_o$, c are essentially continuous. Since $x_o$ plays in F and H, eq. (0-7), an analoguous role, it is natural to define (0-12) as the general solution of (0-1). The definition (0-12) is theoretically quite satisfactory, but "operationally" (i.e. practically) it is almost useless, because in general the function F is unknown. The physical, biological, etc. information content of (0-12) is therefore very low. In fact, since the function F is known to be in general extremely complicated (in all non-contrived cases it cannot be expressed explicitly in terms of known elementary and transcendental functions), it is illusory to study F by means of, say, series expansions, integral representations, etc. Such particular expressions of F possess at most a local significance and cannot be used to test whether all globally relevant properties of (0-1) have been obtained or not.

Passing to an opposite extreme, motivated by the existence of numerical computation facilities, the expression (0-12) can be thought of as being merely a shorthand notation for a set of qualitatively different (real number) sequences $\{x_n\}$, $n = 0, \pm 1, \pm 2, \ldots$, generated by a suitable chosen set of initial values $x_o$. This definition is fully operational for $n > 0$, because having fixed the set of $x_o$, the sequences $\{x_n\}$, $n \geq 0$, called for brevity sequences of consequents of the $x_o$, are unambiguously defined and can be straightforwardly computed for any fixed $x_o$ and c, The definition of (0-12) in terms of the $\{x_n\}$ is therefore widely used. A "graphical" display in phase space of the points $\{x_n\}$, $n = 0, 1, 2, \ldots$ is called a discrete half-trajectory of (0-1), and whenever it is permissible to omit the qualification $n = 0, 1, 2, \ldots$, simply a trajectory of (0-1). The numerical and graphical knowledge of a finite number of discrete half-trajectories of (0-1) has obviously a very low theoretical information content. In fact, there is no simple way of knowing whether a given finite, or even enumerable set of $\{x_n\}$, $n = 0, 1, 2, \ldots$ is qualitatively exhaustive, i.e. whether it contains a sufficient number of relevant "samples" permitting to establish all characteristic properties of the function F in (0-12). The doubt about qualitative exhaustivity is reinforced by the observation that the "complementary" discrete half-trajectories $\{x_n\}$, $n = 0, -1, -2, \ldots$, called for brevity sequences of antecedents of $x_o$, are rarely, if ever examined. The reason for the practical avoidance of the $\{x_n\}$, $n < 0$ lies in the operational difficulty of inverting (0-1), i.e. in finding the $x_{n-1}$ corresponding to a known $x_n$. It is obvious that simple single-valued smooth functions $f(x_n, c)$ may have complicated multi-valued and not necessarily smooth inverse functions $f^{-1}(x_n, c)$. The computational determination of antecedents requires thus the use of (real-valued) root-finding algorithms. As a rule, the non-uniqueness of $f^{-1}$ triggers a complex branching process, whose existence undermines seriously the presumed practical usefullness of the operational definition of (0-12).

A third definition of the solution of (0-1), complementary to the analytical one, is essentially indirect. It is based on the notion of a set of singularities of the function F in (0-12). Both F and its singularities are assumed to be defined implicitly by the function f in (0-1). Following an idea introduced by Poincaré in connection with continuous dynamic systems, a meaningful characterization of F consists in the identification of its singularities, and in the description of the behaviour of the latter as $x_o$ and c vary. As in the case of continuous dynamic systems, any change of the singularities, or of their properties, is called a bifurcation. The function F in (0-12) is said to be known, i.e. fully characterized, when all its singularities, and all bifurcations of the latter in the admissible range of $x_o$ and c, have been described. This characterization of F is similar to that of a meroporphic function by means of poles and zeros in the complex plane.

The indirect definition is used extensively in this monograph, the implicitly defined singularity structure serving as a conceptual skeletton to which all properties of a recurrence are related. Such an ordering of otherwise isolated particular properties (microscopic, macroscopic or collective ones) discloses many of the intrinsic interrelationships of the latter. Once the decision has been taken to use the singularity structure as a key tool, the remaining basic problem consists in discovering what constitutes a relevant, and possibly a complete set of singularities of F. For firts order recurrences with continuous and at least piecewise differentiable f, a substantial set of fundamental, i.e. building block-like, singularities is already known (c.f. chapter I). The problem of completeness is still open, because the number of all possible qualitatively distinct singularities has not yet been determined. Accumulations of "elementary" singularities occur frequently, giving rise to composite singularities ; the number of the latter is not necessarily finite, which gives rise to new accumulations and thus to "higher-order" composite singularities, etc. Some singularities are due to the form of f in (0-1), i.e. they exist even if f is a polynomial in $x_n$ ; others are due to the limited smoothness of f, i.e. they exist only when f lacks a sufficient number of continuous derivatives, but is otherwise arbitrarily close to a polynomial (in the sense of some norm consistent with the limited smoothness). In the case of a specific recurrence, i.e. with a given f in (0-1), the singularities of F in (0-12) can only be determined one by one, or at most set by set, and then ordered into sequences of similar elements. If these sequences converge (generally non uniformly), their limits may constitute additional qualitatively distinct singularities. In order to avoid unnecessarily awkward sentences in what follows, it is understood that whenever some parameters or functions are mentioned, they belong to their respective admissible spaces. All instances to the contrary are explicitly mentioned.

Similarly to the case of continuous dynamic systems, a value or a set of values of $x_o$ is called a singularity of the recurrence (0-1), if and only if, for this value or set the function F in (0-12) describes a stationary state, or a constitutuent part of the latter. A stationary state is a dynamic system-term for an invariant manifold of (0-1), which in some sense, to be specified in each case, is independent of n. A necessary, but not always a sufficient condition for an $x_o$ of (0-1) to be singular, is the violation of uniqueness of F at $x_o$. Since the recurrence (0-1) contains a single dependent variable, its singularities are either zero- or one-dimensional, i.e. they consist either of points or of segments of the x-axis. The simplest point-singularity $x_n = \bar{x}$ is given by an isolated finite root of the "algebraic" equation :

(0-13) $$f(x, c) - x = 0 \quad .$$

Eq. (0-13) may possess, of course, more than one root. No loss of generality occurs by considering one isolated root at a time. The corresponding stationary state is : $x_n \equiv \bar{x} = \bar{x}(c)$, which is invariant with respect to (0-1) and independent of n in an obvious manner. It is also a point of non-uniqueness of F. A constant stationary state of (0-1) constitutes a fixed point of the mapping, defined by f, of the x-axis onto itself. The successive iterates $x_{n+k}$, k = 1, 2, ..., of $x_n$ are called for conciseness consequents of $x_n$ of rank k, with respect to the recurrence (0-1). The statement about the rank is usually omitted unless necessity dictates otherwise. All consequents of a fixed point $\bar{x}$ coincide, of course, with $\bar{x}$, but this is not necessarily so for all antecedents. Suppose that $\bar{x}$ admits a non-zero neighbourhood $X_\varepsilon$ : $-\varepsilon_1 < x_n - \bar{x} < \varepsilon$, $0 < \varepsilon_1$, $0 < \varepsilon$, free from other singularities of (0-1). The existence of such an $X_\varepsilon$ is not always guaranteed a priori, except when the root $\bar{x}$ is known to be isolated, but it is useful to make the existence assumption provisionally, subject to an a posteriori confirmation. Consider a point $x_\varepsilon \neq \bar{x}$ inside $X_\varepsilon$, and the set of its consequents $x_{\varepsilon n}$, n = 1, 2, ... Three cases are possible as n increases indefinitely :

a) the (Euclidian) distance $d_n$ between $x_{\varepsilon n}$ and $\bar{x}$ diminishes and approaches zero,
b) $d_n$ increases till one of the $x_{\varepsilon n}$ reaches the boundary of $X_\varepsilon$ or leaves $X_\varepsilon$ entirely,
c) $d_n$ remains strictly bounded below and above, and all $x_{\varepsilon n}$ remain inside $X_\varepsilon$.

In the first case the fixed point $\bar{x}$ is said to be attractive, in the second repulsive, and in the third neutral. In the terminology of dynamic systems the equivalent statement is : the constant stationary state (or static equilibrium) $\bar{x}$ is asymptotically stable, unstable, and (simply or indifferently) stable, respectively. Stability is understood to be in the sense of Liapunov, extended to discrete dynamic systems, except when specified otherwise. The singularity $x = x_\infty$, eq. (0-5) of the recurrence (0-3) is an example of an attractive fixed point, whose $X_\varepsilon$ is the whole positive x-axis, minus the point $x_n = a$, excluded from (0-3) - (0-4) by construction. Since (0-3) is a linear recurrence, the singularity $x_\infty$ is of course unique (when the singularity at infinity is omitted).

Isolated roots of eq. (0-13) are not the only possible point-singularities of the recurrence (0-1). In fact, no information is added to (0-1) by determining successively k > 1 consequents of $x_n$. Eliminating the "intermediate" variables $x_{n+1}, \ldots, x_{n+k-1}$, it is possible to construct from (0-1) the iterated recurrence

(0-14) $$x_{n+k} = f_k(x_n, c), \quad k > 1, \; f_1 = f, \; n = 0, 1, 2, \ldots$$

where the function $f_k$ is unambiguously defined and single-valued. The construction of (0-14) from (0-1) may be tedious, but it is straightforward. The inverse problem of constructing f from the sole knowledge of $f_k$ is not straightforward at all ; its general solution is still unknown. It is related to the more general problem of determining fractional iterates of a given function ; for example, determining the "half-iterate" f(x) of f(f(x)) = g(x) when g is given. The recurrence (0-14) is not different in principle from the recurrence (0-1), and it may also admit isolated point-singularities, defined by the roots of the algebraic equation

(0-15) $$f_k(x, c) - x = 0, \quad k = 2, 3, \ldots$$

The case k = 1 is of no interest because eq. (0-15) is then the same as eq. (0-3) . Traditionally a root x of (0-15) is said to be a cycle (or a periodic point) of (0-1), provided x is not simultaneously a root of (0-15) when k is replaced by one of its divisors, unity included. Every x defines k - 1 > 0 distinct consequents, obtained by means of (0-1), which are also roots of (0-15). The k-th consequent of any root coincides with itself. The set of k points forming a cycle (of order k) is invariant with respect to the iterated recurrence (0-15) in the same way as a fixed point (a root of (0-13)) is invariant with respect to (0-1). In a physical, biological, etc., content, when a fixed point of the recurrence (0-1) describes a periodic solution ("main" resonance) of the corresponding continuous dynamic system (example : (0-5), (0-3) and (0-2), respectively), a cycle describes a subharmonic periodic solution (subharmonic resonance or frequency division). A fixed point or a cycle of a fractional iterate, which is not simultaneously a fixed point or a cycle for k = integer, describes a harmonic or fractionally periodic solution (a harmonic resonance or frequency multiplication, and a combination resonance or rational frequency conversion, respectively).

The situation of a cycle with regard to neighbouring points, or equivalently, its stability, is therefore the same as that of a fixed point : a cycle (if isolated) is either stable, asymptotically stable or unstable. The points of a cycle represent a stationary state of (0-1), independent of n, modulo-k. A cycle constitutes the simplest possible non-constant stationary state of the recurrence (0-1). Constant stationary states of a continuous dynamic system like (0-6), are defined by real roots x of $g(x, c) = 0$. With the assumed smoothness of g, eq. (0-6) has no continuous periodic solutions.

Unless the form of f in (0-1) is severely restricted (to smooth monotonic functions, for example), the number of different cycles of (0-1) is not finite, and accumulations of point-singularities are possible inside finite parts of the x-axis (cf. [S 6]). This is so for the quadratic recurrence (0-11) ; its cycles and bifurcations as a function of c are described in Chapter 1. Recurrences of form (0-1) possess also invariant segments, provided f(x, c) has at least one local extremum. Two examples of an invariant segment $X_i : x < x_n < \bar{x}_e$ are shown in Fig. 0-1, where x, $\bar{x}$, are unstable fixed points, $x_e$ is the abscissa of the extremum, and $\bar{x}_e$ a consequent of $x_e$. Fig. 0-1 constitutes an illustration of a rather well known geometrical method of analyzing real-valued one-dimensional recurrences. From an inspection of Fig. 0-1, it is obviously that, except for a point set $X_p$ of zero measure, the consequents of any internal point $x_n$ of $X_i$ will remain inside $X_i$, without any possibility of escaping or settling down as n increases. The segment $X_i$ does not contain any stable point singularity ; it contains cycles, but these are all unstable. The excluded set $X_p$ consists of points (and antecedents) of the unstable cycles and of the unstable fixed point x. In the case of the quadratic recurrence (0-14), the situation of Fig. 0-1-a occurs for several values of c. One example is : k = 1, c = -2, x = 2, $x_e = 0$, $\bar{x}_e = f(x_e, c) = -2$, $X_i : -2 < x_n < 2$. An invariant segment like $X_i$ represents a "complex","disorderly",

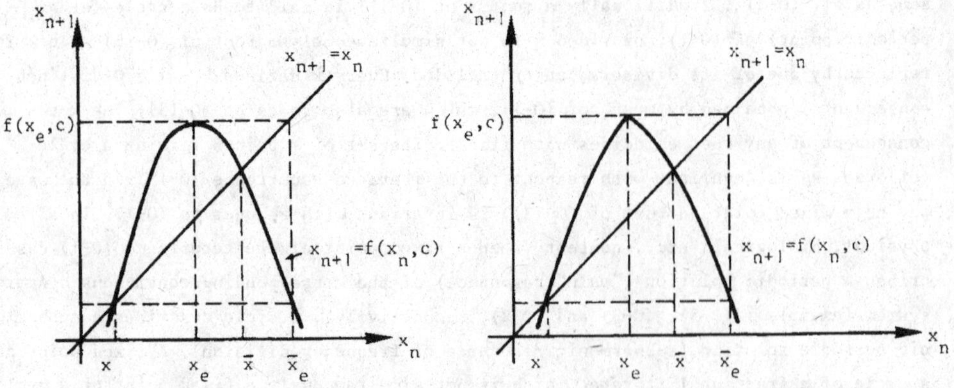


a) f and $\frac{\partial f}{\partial x_n}$ continuous

b) f continuous
$\frac{\partial f}{\partial x_n}$ piecewise continuous

Fig. 0-1. One type of invariant segment of the recurrence (0-1), c = constant.

"chaotic" or "stochastic" stationary state. If $X_1$ is attractive with respect to neighbouring external points, then it is also called a "strange" attractor, but of course all strange attractors need not be of this very special type. The adjectives complex, disorderly, chaotic, stochastic and strange are used synonimously to express the fact that within the stationary state the behaviour of consequents of an individual initial "state" $x_o$ is "intuitively" (i.e. without the knowledge of (0-1)) unpredictable : the function $F(x_o, n, c)$, eq. (0-12), possesses some "random" features. This randomness of F is routinely exploited in digital computers for (pseudo-) random number generation. The best known recurrence used for this purpose is "Lehmer's congruential algorithm" :

(0-16)

$$x_{n+1} = f(x_n, a, b, m) = \text{mod}_m(a+bx_n), \; n = 0, 1, 2, \ldots$$

where a, b, m, are suitably chosen positive constants (due to a finite computer word length they are all integers). In contrast to (0-1), the function f in (0-16) is not continuous with respect to $x_n$.

In a systematic study of functions H(t), defined implicitly by continuous dynamic systems, Birkhoff and Andronov (see p. 108-109 of [A 4]) have proposed a classification of stationary states in the order of increasing complexity (decreasing orderliness), each class containing the preceding one : I) constant, II) periodic, III) quasiperiodic, i.e. $H(t) = \sum_m a_m \cos(b_m t + c_m)$, max $m < \infty$, all angular frequences $b_m$ mutually incommensurable, IV) almost periodic, i.e. H(t) the same as in class III), except $m \to \infty$, V) recurrent and stable in the sense of Poisson, VI) recurrent and unstable in the sense of Poisson, and if a presently popular limit-class is added, VII) (pseudo-) random. If the same classification is used for the stationary states of an autonomous recurrence of order one, then a correct description of an invariant segment requires a function F, eq. (0-12), whose Birkhoff's class is not less than III). The problem of determining the exact Birkhoff class of a specific F is not only still open, but the study of several particular (generic) recurrences suggests that Birkhoff's classification is too coarse, and probably incomplete. This conjecture is reinforced by the study of stationary states of second order autonomous recurrences (cf. Chapters III-V).

Consider a continuous dynamic system of form (0-6), where x and g are no longer scalars but m-vectors, $m = 1, 2, \ldots$ If the functions g are single-valued and sufficiently smooth (without any substantial loss of generality : analytic), then it is well known that a stationary state belonging to Birkhoff's class II) is impossible unless $m > 1$ ; a state belonging to class III), or eventually class IV), is impossible unless $m > 2$. Since an autonomous recurrence of order one may possess a stationary state of class III), or higher, functions defined implicitly by recurrences are intrinsically more complex than those defined by differential equations. Hence, a global structural equivalence between a recurrence and a differential equation can exist only when the stationary states of both belong to the same maximal Birkhoff class. This is the case for the example described by eq. (0-2) and (0-3). A local structural equivalence may exist under less stringent conditions. This is so in all "computationally stable" discretization schemes currently used in digital computers, the simplest example of which is the recurrence (0-8). It is well known that a global structural equivalence between (0-6) and (0-8) is in general (excluding linear equations and those reducible to linear ones) impossible, no matter how small $h > 0$ is chosen.

The main reason for the greater complexity of functions defined by recurrences is the limited invertibility of the latter, already refered to briefly in connection with the definition of F in (0-12) in terms of discrete half-trajectories. In fact, when the function g in (0-6) is single-valued, then the function H in (0-7) can be determined in principle for $t > 0$ and $t < 0$ with equal facility. An analogous situation exists only exceptionally in the case of the recurrence (0-1), as was also mentioned before.

The single-valuedness of f for $n = 0, 1, 2, \ldots$ does not imply anything about the existence and uniqueness of $x_n$ for $n = -1, -2, \ldots$, because the inverse recurrence

(0-17) $$x_n = f_{-1}(x_{n+1}, c), \qquad n = -1, -2, \ldots$$

where $f_1(f_{-1}(x)) = f_{-1}(f_1(x)) = x$, may not exist, or it may be multi-valued. For example the inverse recurrence of (0-11) is

(0-18) $$x_n = \pm \sqrt{x_{n+1} - c}$$

and there are no real $x_n$ for $x_{n+1} < c$, while for $x_{n+1} > c$ the $x_n$ are double-valued. In analogy with the rank of consequents, the $x_{n+m}$, $m < 0$, of $x_n$ are called its antecedents of rank $k = -m$. The determination of antecedents of rank $k > 1$ is much more difficult than that of consequents, because the iterated inverse recurrence

$$x_n = f_{-k}(x_{n+k}, c),\ f_{-(k+1)} = f_{-1}(f_{-k}),\ k > 0,\ n = -1, -2, \ldots \tag{0-19}$$

involves in general more complicated functions than the directly iterated recurrence (0-14). For this reason, eq. (0-19) is only exceptionally used for the determination of cycles, although all information contained in the algebraic equation (0-15) is also contained in $f_{-k}(x, c) - x = 0$. Contrarily to the antecedent half-trajectories $\{x_n\}$, $n < 0$, there is no fundamental need to determine the position of cycles via the inverse recurrence (0-19), but this is not always so for other properties of cycles. For example, a study of global stability of a cycle requires both (0-14) and (0-19). The additional complexity of (0-19) is generated primarily, but not exclusively by the roots $x = x_c$ of the Jacobian determinant of (0-1) :

$$J = f'(x, c) = \frac{\partial}{\partial x} f(x, c) = 0 \tag{0-20}$$

and the consequents of $x_c$ up to a certain rank k. The consequents of the points $x_c$ have been called by Julia and Fatou [J 2], [F 1] critical points of (0-1). Critical points play an obvious role in the branching of inverse iterates, and a less obvious, but an equally important one, in the determination of the number of asymptotically stable fixed points and cycles with finite coordinates, and in the appearance of certain bifurcations leading to composite singularities, like the invariant segments shown in Fig. 0-1.

When a recurrence possesses a stochastic stationary state, or as it sometimes said, when it involves "complex dynamics", then some implicit regularity or order (in the dictionary sense) may become discernable by going over to a suitably chosen "collective" representation of the chaotically evolving "microscopic" $x_n$. In principle the transition from microscopic to collective, or "macroscopic" variables can be carried out in various ways, but up to now only two practical approaches were used ; one inspired by ergodic theory and the other by statistical physics. In the first, the objective is to determine some physically, biologically, etc., meaningful invariant quantities, the simplest of which appear to be invariant densities (related to some invariant measures), and the topological entropy. Invariant measures are easier to determine than the corresponding invariant densities, but the latter contain more "dynamic" information, provided they are not too degenerate (i.e. provided they are not more singular than, say Dirac measures). An invariant density $w(x)$ of a recurrence like (0-1) is usually sought in the form of a solution of the Perron-Frobenius functional equation

$$w(x, c) = \frac{d}{dx}\int_{s_o}^{s} w(y, c)\, dy, \quad s_o \leqslant s \leqslant s_1 \qquad x \in X_i\ , \tag{0-21}$$

where $s_o$, $s_1$ are constants, $s = s(x, c) = f^{-1}(x, c)$, and $X_i$ is an x-axis segment of "no escape" for the $\{x_n\}$, $n \geqslant 0$ (but not necessarily for the $\{x_n\}$, $n < 0$). One example of a no-escape segment $X_i$ is the invariant segment shown in Fig. 0-1. If the operations of (Lebesgue-) integration and differentiation are carried out analytically, which presupposes that the hurdles presented by the complexity of the antecedents s(x, c), and the possible degeneracy of w(x, c) have been successfully negotiated, then eq. (0-21) reduces to the purely functional form given in (0-10). Even if the smoothness of w(x, c) with respect to x turns out to be quite satisfactory, the smoothness with respect to c is generally very poor. As a rule, w(x, c) is at most continuous with respect to c, without being differentiable, and its study involves the most abstruse part of Lebesgue's integration theory. In order to salvage from this overequipped mathematical framework a physically, biologically, etc., relevant part, an equivalent definition of w(x, c) which bypasses (0-21) will be given. Although apparently less general, the alternate definition of w(x, c) is more operational than (0-21), especially as at present there are no efficient methods of solving (0-21), and the number of (analytically) known solutions is rather small. An iterative method, based on the convergence of numerically computed Cesaro-sums of approximants, has been tested successfully on a few piecewise linear recurrences (symmetrically and unsymmetrically "cut-hat" functions f), and the results so obtained have been extended to smoother recurrences, and in particular to (0-11), by means of the so called conjugate transformations. Since conjugate transformations are prone to pathologies (i.e. they are often singular), the extended results are unreliable without an a posteriori justification (cf. Chapter I). It should be noted that the functional iterates (0-9), and some of the other functional equations in (0-10) constitute also instances of a collective representation of the solution (0-12) of (0-1). The well known difficulties of obtaining non-trivial global solutions of the functional equations (0-9) and (0-10) can clearly be attributed to the existence of complex stationary states of the associated recurrence (0-1). The known particular solutions of (0-9) and (0-10) are either quite special or local in character (cf. [K 6]), i.e. their domain of definition coincide with an <u>orderly</u> part of the phase space of (0-1). More specifically, the validity of these local solutions is limited intrinsically to a rather small neighbourhood (a part of the immediate influence domain) of an isolated attractive fixed point or attractive cycle of (0-1). Whether equation (0-9), and most of the equations (0-10), possess smooth dynamically meaningful solutions when their x-domain covers several distinct cycles, or a stochastic stationary state, is still open to debate (cf. Chapter VI).

The topological entropy h(c) of the solution (0-12) of (0-1) can be defined in several, partially or fully equivalent ways : abstractly in terms of covers, partitions or measures of an interval $X_i$ of no escape, dynamically in terms of the number of cycles contained inside $X_i$, and operationally in terms of successive (pseudo-) eigenvalues of a single "generating" half trajectory $\{x_n\}$, $x \geqslant 0$ (see Chapter I). In all definitions some form of logarithmic averaging is involved. It is tempting to assume that the knowledge of h(c) for a given $f(x_n, c)$ in (0-1) permits a description of the amount of "disorderliness" of its solution $F(x_o, n, c)$ : a larger h(c) corresponding to more disorder, or to a stronger stochasticity. This would be so if h(c) > 0 implied that (0-1) does not admit inside $X_i$ any attractive fixed points or cycles (of a finite order). Unfortunately, no such inference can be made (see Chapter I). A quantitative relationship between the values of h(c) and the corresponding Birkhoff classes of $F(x_o, n, c)$ is unknown. Like the invariant density w(x, c), the topological entropy h(c) has poor smoothness properties with respect to c : as a rule, a mere continuity without differentiability. Moreover, h(c) has in principle a much smaller dynamic information content than w(x, c), because by construction it obliterates all differences between stable and unstable singularities. h(c) is a purely topological invariant of the singularity structure, without any relationship to the physically, biologically, etc. all-important macroscopic "one-way" evolution : past → present → future. An analytical determination of h(c) has been successful only for very special f's in (0-1). An extension to topologically equivalent cases is possible by means of (non-singular) transformations.

Some attempts have been made to characterize stochastic stationary states (or strange attractors contained therein) in terms of fractal objects, but this approach is still in a rather exploratory stage [R 3-4]. It is therefore too early to assess the dynamic information contributed by the study of fractals.

In the statistical physics approach the total admissible phase space of the discrete dynamic system is subvided rather arbitrarily into a finite number of macroscopic "phase cells". All discrete trajectory points belonging to the same phase cell are assumed to possess the same properties with respect to a given type of averaging, regardless of the singularity structure of the recurrence in question. The locally averaged phase points (or phase states) are expected to exhibit a fixed (statistical) distribution, or to give rise to some other orderly pattern (for example, one analogous to that of an ideal gas in or near a state of thermodynamic equilibrium). The results obtained by this approach are, so far, rather disappointing (even in the opinion of some advocates, c.f. [C 3]). The dynamic ineffectiveness of the direct transposition of averaging methods inspired by statistical physics becomes quite plausible after it is recognized that the stochastic stationary states of low-order recurrences describe evolution processes which have nothing in common with those of N-particle systems, N >> 1, when the interactions of the latter result in a state close to thermodynamic equilibrium. For instance, the stochastic stationary state depicted in Fig. 0-1 describes an essentially irreversible and strongly dissipative process, intrinsically different from thermodynamic equilibrium.

Instead of starting from equilibrium thermodynamics it is also possible to start from the notion of randomness in order to obtain some statistical insight into complex stationary states of discrete dynamic systems. More specifically, it is possible to compare the discrete half-trajectories of (0-1) to those of Lehmer's recurrence

(0-16). When (0-16) is used to generate (pseudo-) random number sequences $\{x_n\}$, $n \geq 0$ (for example, in a practical implementation of the Monte-Carlo method), the parameters in (0-16) are chosen so that the $x_n$ satisfy a certain number of statistical tests. The most important of these tests are uniform distribution of the $x_n$ inside a fixed (say, unity) interval, and a lack of correlation between $x_n$ and $x_{n+k}$, $k = 1, 2, \ldots$ . When sequences $\{x_n\}$, $n \geq 0$, generated for example by (0-11) with $c = -2$ (more precisely $c = -1,999\ldots$) are subjected to the same tests, the results are invariably negative, i.e. compared to (0-16) the dynamics of (0-11) are quite regular [G 7]. There is no uniform distribution and no progressive decrease of correlation. On the contrary, a constant saturation level of correlations is reached for relatively low values of k. The same conclusion holds for various cut "hat" functions with one extremum. The standard randomness tests are very efficient in showing the presence of regularity, but they are quite unwieldy in quantifying the latter. More specifically, they do not provide any obvious scale of measuring the amount of order (or disorder), and thus do not provide a relationship to the values of topological entropy or to Birkhoff's complexity classes.

Consider now an autonomous second order recurrence

(0-22)
$$y_{n+1} = f(x_n, y_n, c),\ x_{n+1} = g(x_n, y_n, c),\ n = 0, \pm 1, \pm 2, \ldots$$

where, as before, the f, g, are assumed to be sufficiently smooth functions of the variables $x_n$, $y_n$ and of the parameter c. The recurrence (0-22) may be interpreted as a point-mapping of the (phase) plane $x_n$, $y_n$ onto itself. Let the functions defined implicitly by (0-22) be

(0-23)
$$y_n = F(x_o, y_o, n, c),\ \ x_n = G(x_o, y_o, n, c)\ ,$$

where $(x_o, y_o)$ is an initial point in the phase plane. The functions F, G, in (0-24) admit, of course, all the singularities of the general solution (0-12) of (0-1), in addition to some of their own, generated by the increased dimensionality. Two-dimensional singularities are thus possible, at least in principle. The point-singularities of (0-21) are given, analoguously to (0-13), by roots x, y of the algebraic equations

(0-24)
$$f_k(x, y, c) - y = 0,\ g_k(x, y, c) - x = 0,\ k > 0\ ,$$

where the $f_k$, $g_k$ are defined by the (direct) iterated recurrence

(0-25)
$$y_{n+k} = f_k(x_n, y_n, c),\ x_{n+k} = g_k(x_n, y_n, c),\ n = 0, 1, 2, \ldots$$

with $f_1 = f$, $g_1 = g$, and $k > 0$ fixed. The roots of (0-24) correspond to fixed points ($k = 1$) and cycles ($k > 1$) of (0-22). Similarly to the recurrence (0-1), the fixed points and cycles of (0-22) are either stable, asymptotically stable or unstable. The higher dimensionality of (0-22) permits, however, a more varied relation of point-singularities with regard to neighbouring points. In order to characterize this relation the notion of an invariant curve is required. A curve described in the phase plane by $H(x_n, y_n) = C$, $C$ = constant, is called an invariant curve of (0-22), provided the function H satisfies the functional equation

$$H(g(x, y, c), f(x, y, c)) = H(x, y, c) . \tag{0-26}$$

The constant C is determined, for example, by means of an initial condition $(x_o, y_o)$. A similar definition of an invariant curve applies to the iterated recurrence (0-25), and to the "simple" and iterated recurrences of antecedents.

Compared to the first-order recurrence (0-1) a new fundamental difficulty arises in the study of (0-22), because very little is known about functional equations of type (0-26). Except in very special cases (defined locally, see Chapter II), no existence, uniqueness and continuity (or differentiability) theorems are available. Although the solutions of (0-26) constitute a fundamental tool for the unraveling of the properties of (0-22), in the most crucial cases they can only be constructed approximately, subject to an independent confirmation of validity, the latter replacing the corresponding unavailable existence, uniqueness and continuity theorems. The whole approach can be described as a process of symbiotic improvements : a little more understanding of a concrete recurrence (0-22) permits the construction of a better approximation of a particular solution of (0-26), and a more precisely known invariant curve increases the understanding of (0-22), and so forth. The sequence of successive improvements is initiated by the following known properties :

a) some point-singularities are the same in both discrete and continuous dynamic systems, i.e. the invariant curves near a fixed point (or a point of a cycle) of (0-22) are of the same qualitative type as the phase plane trajectories of an "analoguous" differential equation near a constant solution (position of static equilibrium),

b) convergent or asymptotic series expansions are valid at least locally in "regular" cases, i.e. a quantitative description of small segments of invariant curves, whose qualitative type is already known, can be obtained either analytically or numerically (for example, via roots of algebraic equations, when the coefficients of an invariant curve series are known analytically),

c) in some "singular" cases the same qualitative situation exists in a recurrence and is an analoguous differential equation, i.e. the problem is not immediately inextricable when both a) and b) fail. Whenever the invariant-curve structure near a point singularity has been identified, a concise terminology can be introduced. For convenience the same names are given to point-singularities of recurrences and of analoguous differential equations, because the notions invariant curve and (continuous) trajectory are dynamically equivalent.

A complete characterization of a set of point-singularities of (0-22) requires obviously the knowledge of both local and global solutions of (0-26). Unfortunately, there are no efficient systematic methods of solving the functional equation

(0-26) globally, except when the recurrence (0-22) is linear :

$$y_{n+1} = cx_n + dy_n, \quad x_{n+1} = ax_n + by_n , \tag{0-27}$$

a, b, c, d, being constants. In the case of (0-27), the equations of invariant curves can be expressed explicitly in terms of elementary functions. The non-degenerate singularity (0, 0) of (0-27) turns out to be of four possible familiar types : centre, saddle, focus and node, depending on the values of a, b, c, d, although the number of distinct sub-types is larger than in the case of two analoguous linear differential equations. The focus and node are either asymptotically stable or unstable. In the first case, they are approached by consequents of an initial point, and in the second by antecedents. By means of a continuity argument it is easy to see that the invariant curves near a focus, node and saddle are qualitatively unaffected by smooth non linear perturbations, because linear terms dominate sufficiently close to the point-singularity. This is not so for the centre and for some degenerate forms of (0, 0), because non linear perturbations dominate then the linear terms in (0-27). The situation is entirely parallel to that of continuous dynamic systems. In both cases the dominance of non linearities is said to constitute a critical case (in the sense of Liapunov).

When the existence of invariant curves has been established, and the point-singularities of (0-22) turn out to be of the same type as those of second order autonomous differential equations, then the iterated recurrence (0-25) is also valid for fractional and continuous values of k [B 4]. This iterated recurrence constitutes thus a one-parameter continuous group of transformations, k is an additive parameter of that group, and (0-22) possesses under these conditions a (local) structural equivalence to the continuous dynamic system

$$\dot{y}(k) = \bar{f}(x(k), y(k), c), \ \dot{x}(k) = \bar{g}(x(k), y(k), c) ,$$

where the functions $\bar{f}$, $\bar{g}$, are unambiguous defined by the solutions of an equation of type (0-26) [B 4]. The equivalence between (0-22) and (0-28) is similar to that between (0-8) and (0-6). For unrestricted f, g, $\bar{f}$, $\bar{g}$, of comparable smoothness the functions defined implicitly by (0-22) are much more complex than those defined by (0-28). Like in the case of the recurrence (0-1), this property is due to the fact that (0-22) admits a greater variety of singularities than (0-28) : at least two distinct types of stochastic stationary states are not only possible in (0-22), but their occurence is rather the rule than the exception. The first type is simply a two-dimensional counterpart of an essentially one-dimensional stochasticity, the second dimension playing dynamically a "decorative" role, whereas the second type is a salient feature of the increased dimensionality. It appears mathematically straightforward to imbed the characteristic properties of invariant segments, illustrated in Fig. 0-1, into a recurrence of order two. A known example of stochasticity inconsistent with (0-1), regardless of the form of f, are Birkhoff's instability rings (cf. Chapter

III). The richness of the singularity structure of a recurrence is therefore a basic reason for the general non-solvability of the inverse problem : is there an autonomous differential equation of the same order as a "nominally equivalent" recurrence ? If the singularity structure is to be the same, the difference of order is generally at least unity in favour of recurrences.

Some qualifying comments are needed on the notions of "dynamically" meaningful continuity with respect to a parameter and that of dominance of a part of dynamic system over other parts. Without such comments a wide door is opened to semantic misunderstandings, which involve key works like generic problem, generic solution, regular and singular perturbations, inert systems (in French : systèmes grossiers, in transliterated Russian : grubye sistemy), effect of noisy environment, etc. The opportunity for misunderstandings is provided by the coexistence of two intrinsically equivalent but operationally distinct approaches, each with a nominally similar but in reality distinct vocabulary. These two approaches can be roughly describes as follows :

A : As a starting point a set of axioms is chosen (motivation : mathematical convenience), fixing indirectly the conventional meaning of terms like dynamic system, its trajectories, etc... Possible properties of trajectories, and of their singularities, are studied by essentially deductive arguments (example : Morse-Smale systems). If the results accumulated by this approach are confronted with a dynamic system brought in from the "outside", i.e. from physics, biology, etc., then some of the properties of the latter may appear "strange", "structurally unstable", and so on, whereas in their natural content these properties are entirely well behaved (see Chapter I and V).

B : As a starting point concrete dynamic systems describing a physically, biologically, etc. genuine process are chosen (motivation : physical, biological, etc. insight), with an objective of identifying their "typical" or "generic" internal mechanism (examples of such mechanism : resonance, synchronization, phase stability). Because of the diverse natural origins, the same process may have several equivalent "practically close" descriptions, for instance, several mathematical slightly different recurrences, or a recurrence and a differential equation resulting from physically, biologically, etc. equivalent, but mathematically apparently different assumptions. The mathematically specified degree of smoothness of the state variables may change from case to case, and sometimes even during the evolution of the same case. Moreover, physical, biological, etc. parameter values are not sharp ; they represent rather averaged measured values with a more or less known accuracy. It is therefore meaningless to ask whether they are, say, rational or irrational. The only certainty is : the dynamic system is structurally stable with respect to its admissible perturbations (noise included), because otherwise there would be no physically, biologically, etc. relevant mathematical problem to start with. The precise nature of admissible perturbations is not always known in detail, the characterization of admissibility constituting often the main problem.

The method used in the approach B is therefore essentially inductive, and its main feature is the initially provisional assignment of functional spaces governing functions in equations, boundary conditions, and solutions. In short, the relevant working framework is all but axiomatic, because all key limitations have their source outside of mathematics. The intended interdisciplinary scope of this monograph generates a bias in favour of the approach B. This does not mean that its content will consist merely of a collection of disparate problems without any unifying mathematical thread. Quite to the contrary, the authors attempt to start where Hadamard left off : propose an improved mathematical characterization of physically, biologically, etc. meaningful and mathematically "correctly formulated" problems. It is perhaps useful to recall that an abstract mathematical problem is said to be correctly formulated in the sense of Hadamard if its solution exists, is unique, and depends continuously on its "data". In Hadamard's opinion, it is precisely this continuous dependence on data which reflects the experimental origin of the problem (finite accuracy of measurements). A now classical example is the incorrectness of the abstract Cauchy problem for the Laplace equation with (physically) insufficiently smooth initial functions. Consider in fact

$$\text{(0-28)} \qquad \begin{gathered} \Delta u = u_{xx} + u_{yy} = 0, \qquad u = u(x, y) \\ y > 0, \; -\infty < x < +\infty, \quad u(x, 0) = 0, \quad u_y(x, 0) = g(x), \end{gathered}$$

where the subscripts stand for partial differentiation, and $g(x)$ is a continuous but not necessarily a continuously differentiable function. It can be verified by substitution that

$$\text{(0-29)} \qquad g(x) = \frac{\sin nx}{n} \quad \text{implies} \quad u(x, y) = \frac{1}{n^2}(\operatorname{sh} ny)(\sin nx) .$$

Taking the limit $n \to \infty$ in (0-28), the boundary condition $u_y(x, 0) \to 0$ defines a unique solution : $u(x, y) \equiv 0$, whereas the limit solution obtained from (0-29) is entirely different ($\frac{\operatorname{sh} ny}{n^2}$ is not bounded as n increases). The lack of continuity of $u(x, y)$ with respect to the "boundary data" $g(x)$ is a source of chaos (or stochasticity) to an observer of the "effect" of $g(x)$ on $u(x, y)$, as was already pointed out by Poincaré. A similar type of chaos-generating mechanism operated in some discontinuous recurrences, like (0-16) or that related strictly (i.e. without "qualitative equivalence" simplifications) to the Lorenz climate problem [L 9], where a single flap or flutter of one butterfly wing may make a difference in the subsequent evolution. The existence of say, stable micro-climates, illustrates the physical irrelevance of the preceding statement. The incorrectness and chaos of the continuous dynamic system (0-28) should be contrasted with the fact that physicists measure routinely the magnetic and gravitational fields near the earth surface and compute successfully the corresponding field penetrations into the earth interior. The results obtained are reliable, because by phenomenological insight the initial data are smoothed sufficiently before the actual computation via (0-28). This physically motivated smoothing converts an incorrectly formulated problem into a correct one, because the mathematically meaningful but physically too abstract space of g(x) has been changed into a physically realistic one. It can be easily verified that the discontinuity of u(x, y) with respect to initial data disappears when g(x) is, for example, continuously differentiable. It should be noted, however, that the inverse situation may also occur : the physically admissible space is more abstract than that sanctioned by the current mathematical state of the art. As an example it is sufficient to mention the now historical controversy between the users of Heavyside's operational calculus ("too general" to be "well founded") and those of Laplace's transformation ("properly rigorous" but failing to admit $\delta$-"functions").

The redefinition of the admissible functional space for the initial functions in (0-28) can be considered as a regularization of a physically meaningless mathematical pathology. Such a regularization can of course be applied to other problems, not necessarily defined by partial differential equations like (0-28). In particular, it can be applied to the physically, biologically, etc. quite relevant description of dynamically complex microscopic regimes by means of suitably chosen collective variables. In most cases, the density w(x, c) of an invariant measure and the topological entropy h(c) referred to earlier, have insufficient smoothness with respect to a small variations of data to possess a dynamically appreciable information content.

A useful tool in the study of regular problems, i.e. correctly formulated in the sense of Hadamard and having in addition continuous differentiability with respect to the data, is sensitivity analysis [X 1],[G 10]. In fact, the qualitative requirement of Hadamard can be given a quantitative complement by means of partial derivatives in the case of parameter parturbations, and functional derivatives (in the sense of Volterra [V 1-3]) in the case of functional ones. These derivatives are called sensitivity coefficients, even if they happen to be functions or functionals. The knowledge of sensitivity coefficients permits to judge the range of accuracy or of uncertainty of a solution in terms of the corresponding range of the data, and thus to establish the physical, biological, etc. relevance of the mathematical problem under study as well as of the results obtained. If the physical, biological, etc. problem is known to be genuine, then sensitivity analysis permits a discrimination between admissible and inadmissible parameter and "structural" parturbations, regardless of the purely mathematical merits of the latter. In other words, it is physically, biologically, etc. irrelevant whether a perturbation is regular or singular, stable or unstable, etc. in some mathematical sense. Especially in the context of complex dynamics there appears to be no correlation between physical, biological, etc. relevance and an a priori mathematical simplicity (or lack thereof).

A basic working assumption of this monograph is the conjecture that there exists a one-to-one relation between all physically, biologically, etc. operating internal mechanisms and a set of singularities in the sense of Poincaré of the corresponding mathematical dynamic system, regardless of the concrete form of the latter. It is the set of singularities which is "generic" for a given internal mechanism, and not the form of the specific equation of evolution admitting this set. A physical, biological, etc. process under study may, of course, involve more than one internal mechanism, operating alternately, successively or simultaneously, depending on initial (or boundary) conditions and external influences. The latter may be taken into account either directly, via the form of the equations, or indirectly, via the set of admissible perturbations. A correct mathematical description must contain the corresponding singularity sets and allow for the changes implied by the adjectives alternate, successive and simultaneous.

A change in a set of singularities, or just of the nature of one singularity, will be called a bifurcation. The study of bifurcations is therefore a key tool of this monograph. Since in a dynamically complex process non classical bifurcations are to be expected, special attention has to be paid to any circumstantial evidence pointing to their occurence. For this reason, once a singularity of (0-1) or of (0-22) has been found, its properties are studied as a function of a suitably chosen parameter c. More specifically, all known "characteristic qualities" of the singularity are carefully watched for some possibly significant changes (as c is varied "adiabatically"), and whenever feasible, the bifurcational origin and disappearance of the singularity are traced. In this procedure, it becomes necessary to devise special algorithms for the solution of the algebraic equations (0-15), (0-24) and of the functional equation (0-26). Whenever the stock of known functions failed to cope with the task at hand, recourse was taken to accuracy-controlled digital computations. It became thus possible to determine a large number of cycles and of invariant curves. An extensive table of cycles and of types of associated invariant curves permits to order the cycles in various ways, and thus to identify the existence of their characteristic qualifiers. In most cases the numerical results were used merely as a substitute for a set of properties of analytically defined, but so far unstudied transcendental functions (solutions of (0-15), (0-24) and (0-26), and <u>not</u> as a tool of analysis.

For operational and historical reasons, preference is given to dynamic system terms introduced by Poincaré and Liapunov, and by some of their direct successors, like Julia, Fatou, Lattès and Andronov. Thus "domain of influence" is prefered to "basin", "cycle" to "periodic point", "saddle" to (Birkhoff's) "hyperbolic point", "centre" to "elliptic point", "inert" (grossier) to "structurally stable", etc..

The monograph consists of six chapters. The first is devoted to first order recurrences. Since second order recurrences are considerably more complex, their study is divided into four chapters : II : General properties in the absence of stochasticity ; III : Hamiltonian case (the Jacobian determinant J of (0-22) is identically equal to unity ; IV : Almost Hamiltonian case ($J = 1 + \varepsilon h(x_n, y_n)$, $0 < \varepsilon \ll 1$,

$\|h\| < \infty$) ; V : Strongly dissipative case. A more detailed description is given in the table of content.

The references cited are those which actually contain the ideas used in the course of various arguments. The list of references is by no means complete, because it was not materially feasible to trace each idea to its primary originator. In case of doubt, or independent co-discovery, the chronologically oldest reference known to the authors is given. References to papers which are not directly related to the leitmotif of this monograph, but which are otherwise quite important either independently or within an other context, were excluded for reasons of conciseness. The aim of this monograph is to introduce non-specialized readers to the subject of recurrences and discrete dynamic systems ; the monograph pretends in no way to be complete or exhaustive.

CHAPTER I

# SOME PROPERTIES OF FIRST ORDER RECURRENCES

## 1.0 Introduction

Direct descriptions of evolution processes by means of first order recurrences appear rather seldom in theoretical and applied physics. One example from control engineering is a digital servo with feedback, having one time-constant $T_o$ and loop gain k. Let $u = f(t)$ be a smooth time dependent signal. If u is sampled at $t = t_n$ and held constant during $t_n \leq t \leq t_{n+1} = t_n + T$, $n = 0, 1, \ldots$, $T$ = positive constant, then the operation of the servo is described by

$$(1\text{-}1) \qquad x_{n+1} = ax_n + k(1-a)\, f(-x_n), \quad a = \exp(-T/T_o) \; .$$

The technological problem consists in choosing k and $T_o$ so that a maximally stable orderly state is realized and the possibility of any disorderly states, representing a faulty operation, is minimized.

In biology two examples from population dynamics of species with non-overlapping generations are given by [R 2], [B 2-3], [M 2] :

$$(1\text{-}2) \qquad x_{n+1} = x_n \exp\left[c\,(1-x_n/K)\right] ,$$

$$(1\text{-}3) \qquad x_{n+1} = ax_n - bx_n^2 , \quad a = 2 - c, \; b = (c-1)/K \quad ,$$

where $x_n$ designates the population density of the species in the n-th generation, c is the intrinsic growth rate, and K the carrying capacity of the environment. Although the recurrences (1-2) and (1-3) have been formulated independently from each other, (1-3) is a particular case of (1-2) resulting from an approximation of the exponential function. Chaotic regimes are known to occur in both (1-2) and (1-3), but since these recurrences represent very coarse descriptions of the biological situation, their relevance (or lack thereof) depends on the particularities of the species concerned. The recurrence (1-3) has also arisen in a purely mathematical context : iteration of quadratic polynomials [M 13]. Most of this chapter is devoted to its study.

In economics there exist several models leading to first order recurrences (cf. [S 2], [T 2], [G 1]). Since first order recurrences are very primitive mathematical entities, the practical value of the corresponding models is inherently limited (cf. [G 17]). The first example deals with a delayed price evolution. It is a rather common experience that for some products the supply q reacts to the price p only after a time delay r, whereas the price adjusts itself to the current supply almost instantaneously. Assume for simplicity that r is constant and let $t_n = t_o + nr$, $q(t_n) = q_n$, $p(t_n) = p_n$, $n = 0, 1, \ldots$, where $t_o$ is a suitably chosen initial time. If $t_{n+1}$ is made to represent the present, then $q_{n+1}$ depends on the past via $p_n$, and independently on the present via $p_{n+1}$, modified by a psychological parameter c characterizing the relatively slowly evolving purchasing trend. The values of c are related to factors like advertizing, seasonal appeal, sales tax, presence of competition, etc... . Formally :

$$q_{n+1} = f(p_n) \quad \text{and} \quad q_{n+1} = g(p_{n+1}, c) \quad , \tag{1-4}$$

where the non linear functions f and g are not known exactly, their specific forms depend on detailed considerations of current economic thought. The two recurrences in (1-4) constitute a parametric representation of the implicit first order recurrence

$$g(p_{n+1}, c) - f(p_n) = 0 \quad . \tag{1-5}$$

If f and g are continuously differentiable functions admitting a unique real root $p_n = p_{n+1} = \bar{p}$ (static equilibrium state), then for sufficiently small deviations $r_n = p_n - \bar{p}$, eq. (1-5) simplifies into the linear recurrence

$$r_{n+1} = \lambda r_n, \quad \lambda = f'(\bar{p})/g'(\bar{p}, c), \quad ' = \frac{\partial}{\partial p} \quad . \tag{1-6}$$

The equilibrium state $\bar{p}$ is locally asymptotically stable if the slope of the demand curve at $\bar{p}$ is absolutely smaller than that of the supply curve, i.e. if $|\lambda| < 1$. The assumption r = const. is not intrinsically necessary, a functional dependence $r = r(p_n, q_n, p_{n+1}, q_{n+1}, n)$ will do just as well, except that the recurrence analoguous to (1-5) will be of order two, or higher.

The second example deals with the evolution of present spending s in terms of past earnings e and present investment i. Using the same time discretization as in the preceding example, except that a constant r represents now the delay between the availability and spending of the earnings, one obtains $s_{n+1} = f(e_n)$. Assume for simplicity that the interest is constant and relatively low, so that one can speak about the possibility of a steady state. Since the investment depends on earnings, and the earnings are split into spending and investment, i.e. $i_n = g(e_n)$ and $e_n = s_n + i_n$, the evolution of earnings is described by the implicit recurrence

$$g(e_{n+1}) - e_{n+1} + f(e_n) = 0 \quad , \tag{1-7}$$

where for a given economic situation, f and g are fixed non linear functions. If f and g are continuously differentiable at the static equilibrium $e_{n+1} = e_n = \bar{e}$, then locally (1-7) is equivalent to (the variational equation)

$$r_{n+1} = \lambda r_n, \; \lambda = f'(\bar{e})/[1 - g'(\bar{e})], \; ' = \frac{\partial}{\partial e}, \; r_n = e_n - \bar{e} \,. \tag{1-8}$$

An increase of order of (1-7) occurs when the constancy of the delay r is abandonned.

The third example deals with (ex ante) saving s(t) and (ex ante) investment i(t). The simplest relation between s(t) and i(t) is given by the Harrod-Domar model [H 4], [D 2], expressed in terms of discretized time, constant delay and the linear recurrences

(1-9) $$s_{n+1} = ai_n,\ i_{n+1} = b(s_{n+1} - s_n),$$

where a is the marginal propensity to save and b the acceleration coefficient. In an equilibrium economy, $s_{n+1} = i_{n+1}$, i.e.

(1-10) $$s_{n+1} = \lambda s_n, \qquad \lambda = (a+b)/b \ .$$

If $|\lambda|$ is not small compared to unity, the assumptions leading to (1-10) are generally unjustified, and non linear corrections have to be taken into account. A chaotic evolution is in principle possible in all three examples when $|\lambda| > 1$. Whether such a chaos describes some features of contemporary economic evolution is a matter of debate (cf. [C 1]).

Indirect descriptions of evolution processes by means of first order recurrences are quite common in physics, biology, etc. The simplest example from physics is the explicitly solvable recurrence (0-3). An analoguous example from economics deals with business cycles, resulting from the destabilization of a constant steady state and the appearence of a stationary asymptotically stable periodic one (cf. [G 5]). Another well known example from physics is the Lorenz formulation of a meteorological problem [L 9],[L 10], in which the vertical convection of a fluid with a large Prandtl number is examined. If mode interaction is neglected, then from Boussinesq's approximation of the equations of hydrodynamics it is possible to derive three even more approximate non-dimensional ordinary differential equations [S 1]

(1-11) $$\dot{x} = -a(x+y),\ \dot{y} = cx - y - xz,\ \dot{z} = -bz + xy\ ,$$

where x, y, z, designate the intensity of convective motion, the temperature difference between the ascending and descending currents and the deviation of the vertical temperature profile from a linear one, respectively. If (1-11) is integrated numerically in the so called supercritical region (for example, for $a = 10$, $b = 8/3$ and $c = 28 > c_c = 470/19 \approx 24.7$, where $c_c$ is a normalized critical Reynolds number of the fluid), then one obtains on the Poincaré surface of section z = const. trajectory intersection points located on two disjoint curve segments. Typical shapes of these curve segments $y = g(x)$ are shown in Fig. 1-1 (for more details, see [R 5]). It should be noted that the function g(x) is neither single-valued nor continuous. If the existence of the very special curves $y = g(x)$ of Fig. 1-1 is used as an argument to eliminate one dependent variable, then by an essentially heuristic procedure a one-dimensional recurrence $u_{n+1} = f_o(u_n)$ is obtained, where $u_n$ represents successive relative maxima of z. The function $f_o(u_n)$ possesses a single maximum forming an upwards pointing cusp (Fig. 1-2). A qualitatively equivalent approximation to $u_{n+1} = f_o(u_n)$ is believed to be given by the piecewise linear recurrence [L 9]

(1-12) $$u_{n+1} = f(u_n) = 2u_n,\ 0 < u_n \leqslant \frac{1}{2}\ ; \quad = 2(1-u_n),\ \frac{1}{2} \leqslant u_n \leqslant 1,$$

the summit of f replacing the cusp of $f_o$. The function f in (1-12) is often called a

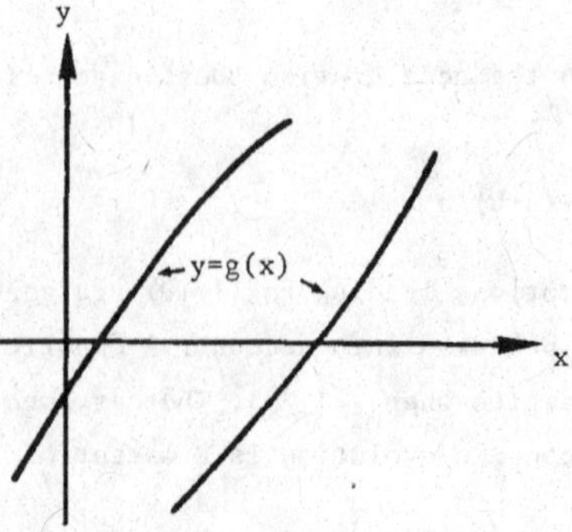


Fig. 1-1

Trajectory intersections of (1-11) on the plane z ≗ const.

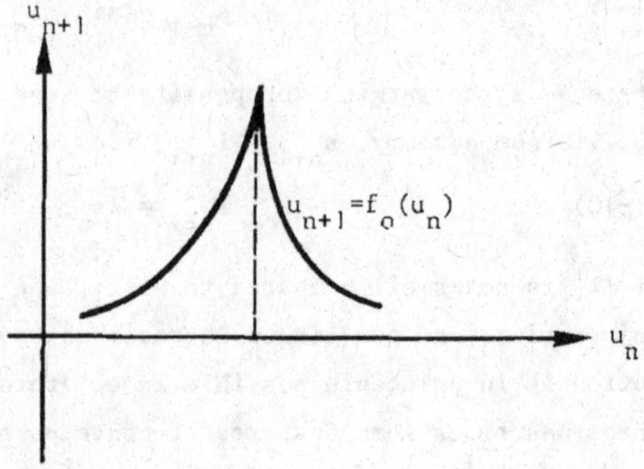


Fig. 1-2

One-dimensional recurrence obtained from (1-11) and the curve segments y = g(x) of Fig. 1-1.

(symmetrically cut) "hat" function, and the recurrence (1-12) a "baker" transformation. Because of its simplicity, the properties of (1-12) have been extensively studied (cf. [G 2], [C 10], [C 11] and the references therein). Unfortunately, in spite of its physical origin, the type of flow implied by (1-12) does not take place in any known physical fluid [M 9]. The negative result is apparently due to the numerous simplifications and approximations introduced between the realistic initial fluid flow equations and the final energy relation (1-12).

No such loss of physical content occurs in the description of a non linear feedback system whose loop gain can take values covering both the amplifying and oscillating operation modes. Although the preceding sentence uses an electronic circuits vocabulary, the same phenomenological situation can be described in biological, medical, ecological or economic terms. The underlying observation is that an (adiabatic) increase of a key parameter c (loop gain of the feedback system, intrinsic growth rate of a population, etc.) can lead to a sequence of qualitatively distinct stationary states : amplifier → almost sinusoïdal oscillator → periodic non-sinusoïdal oscillator → complex oscillator. In the vocabulary of dynamic systems the transition amplifier → almost sinusoïdal oscillator corresponds to the Poincaré bifurcation, first identified physically by Andronov in 1929 : stable constant state → unstable constant state + stable periodic state of limit cycle type. The transition almost sinusoïdal oscillator → periodic non-sinusoïdal oscillator does not produce any new singularity structure, it involves merely a deformation of the limit cycle from an oval (close to an ellipse or a circle) into a more complicated closed curve. As to the last transition, it still defies a complete description, because the notion "complex oscillator" is intrinsically negative, i.e. it says only that the stationary state is neither constant nor periodic, without giving any additional information. Observations of electronic circuits suggest that the complex oscillations in question range from quasi-periodic with a few incommensurable periods (amplitude modulation or weak phase modulation) to apparently completely chaotic ones. It was also observed that what happens inside the complex oscillator state is determined essentially by the shape of the dominant non linearity, i.e. by the internal amplitude or energy limitation, and the number of dominating reactances, i.e. the minimal order of a differential equation providing a qualitatively correct description (see § 4.3, p.20 of [G 9] for an example). The chaotic oscillations are of course of no interest to electronic engineers designing amplifiers or low distortion sinusoïdal oscillators, but they lack no relevance in other contexts. For example, complex oscillators are indispensable for (pseudo-) random function generation in analog computers. An experimental device possessing the full transition range amplifier → complex oscillator is called a "universal" oscillator [M 1].

The simplest type of universal oscillator is described by a third-order continuous dynamic system with one small parameter $\varepsilon > 0$, reducing for $\varepsilon \to 0$ to a qualitatively and quantitatively equivalent first-order recurrence of form (0-1). Consider as an example

$$x' = [az(x-b) - (y-c)(z-d)]/d \,, \quad x = x(t), \quad ' = \frac{d}{dt} \,,$$

(1-13) $$y' = [ayz - (z-d)]/d \,, \quad y = y(t), \quad a > 0, \ b > 1, \ d > 0,$$

$$\varepsilon z' = F(x, y, z) \,, \quad z = z(t) \,,$$

where $F(x, y, z) = 0$ describes in the phase space x, y, z, a surface S with a fold. When $\varepsilon$ is small the motion of a particle on a trajectory of (1-13) is slow on the surface S ; because of this property the surface S is said to be the locus of slow motion (time scale : $\varepsilon^{\alpha}t$, $\alpha > 0$) compared to the fast motion elsewhere in phase space (time scale : $t/\varepsilon^{\beta}$, $\beta > 0$). If S consists of a folded plane, two parts $P_1$, $P_2$ parallel to the x, y-plane and separated by a distance d ($P_1$ : $z = d$, $x > 0$ ; $P_2$ : $z = 0$, $x < 1$) with the edges $z = d$, $x = 0$ and $z = 0$, $x = 1$, respectively, then the third part $P_3$ is a part of the inclined plane $z + dx - d = 0$ (see Fig. 1-3). S separates the phase space of (1-13) into two fast regions : $F < 0$, $z > d$ and $F > 0$, $z < 0$. The fast parts of the trajectories of (1-13) approach normals to the x, y-plane as $\varepsilon \to 0$. When the motion on a fast part of a trajectory attains S, it is slow untill it reaches an edge and becomes fast again. When the sequence of slow and fast motions is trapped inside a finite part of the phase space bounded by a part of S, a chaotic stationary regime results. This happens indeed in the case of (1-13) (see [R 3], [R 4] for other simple examples). If $a > 0$, $b > 1$, then there exists an equilibrium point (b, 0, d) on $P_1$, which is an unstable star-node. On $P_2$ the slow trajectories are parabolas described by $x = \frac{1}{2}(y^2 - \bar{y}^2) + c(y-\bar{y})$, where $\bar{y}$ is the ordinate on the edge $x = 1$, $z = 0$. If the

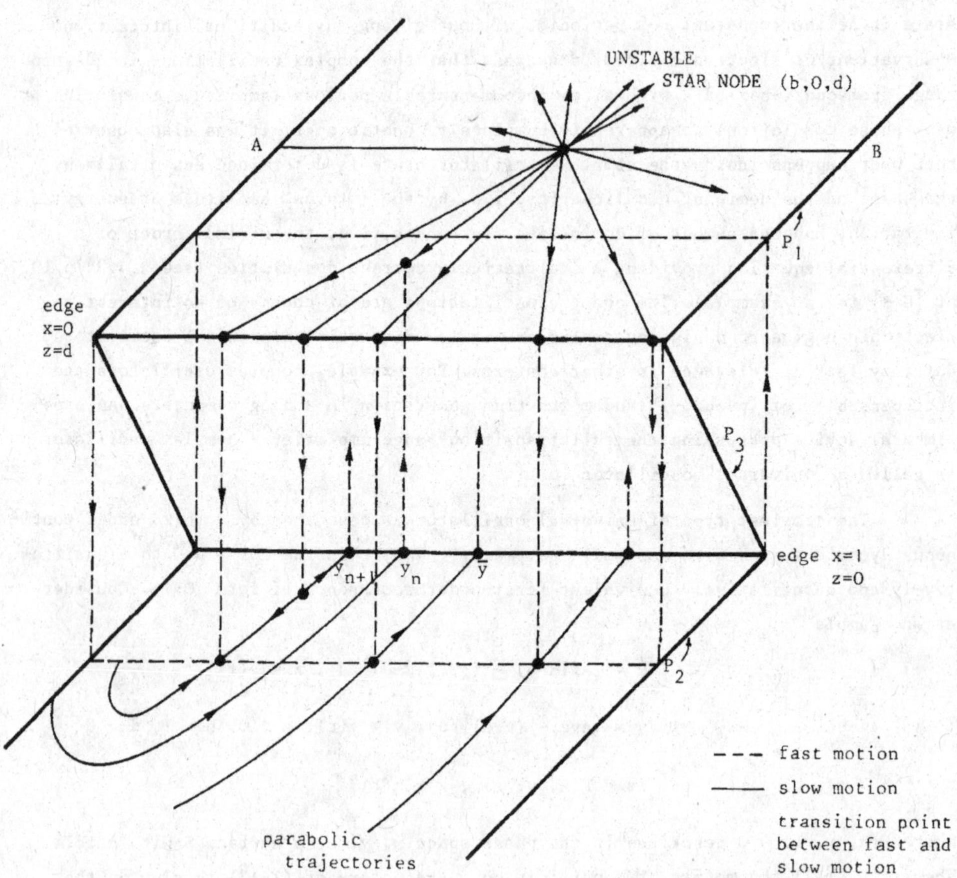


Fig. 3 : A part of the phase space of (1-13) possessing a one-dimensional stationary chaotic regime

initial fast part of a trajectory reaches either $P_2$, or $P_1$ in front of the line AB in Fig. 1-3, then it will move towards the edge, jump rapidly from $P_1$ to $P_2$, or vice-versa, move slowly again till it reaches an edge, and so forth. Either edge constitutes thus a segment without contact for the trajectories of (1-13), and the point mapping of the edge onto itself defines completely the overall trajectory structure. After some elementary operations, it is found that (1-13) is equivalent for $\varepsilon \to 0$ to the first-order recurrence [M 6]

(1-14) $$y_{n+1} = -c + (2+c^2+k^2y_n^2 + 2kcy_n)^{\frac{1}{2}} \quad , \quad k = b/(b-1) \ .$$

For a small but finite $\varepsilon$ it is straightforward to correct (1-14) by means of a standard singular asymptotic expansion, analoguous to those used for boundary layers in fluid dynamics. Except for the method of derivation, the relationship between the

Lorenz problem (1-11), (1-12) and the universal oscillator problem (1-13), (1-14) is completely analoguous. The correspondence between a third-order universal generator (with $\varepsilon = 0$), and a first order recurrence is therefore generic, i.e. it applies to a significant class of problems.

### 1.1 Elementary properties of first order recurrences

Consider the recurrence

$$x_{n+1} = f(x_n, c), \qquad n = 0, \pm 1, \pm 2, \ldots \quad , \tag{0-1}$$

where f is a single-valued smooth function of $x_n$ and c (i.e. continuous and admitting a certain number of continuous derivatives with respect to both $x_n$ and c). The simplest possible singularity of (0-1) is given by a set of points x corresponding to real roots of the algebraic equations

$$x_{n+k} = x_n = \bar{x}, \qquad k = \text{integers}, \qquad \bar{x} = \bar{x}(c) \; , \tag{1-15}$$

defined by (0-1) and the iterated recurrence

$$x_{n+k} = f_k(x_n, c), \qquad f_1 = f \; . \tag{0-14}$$

When k = 1 the singular point x is called a fixed point of (0-1). Since (0-1) is in general non linear, it may possess more than one fixed point. When k > 1, a root x is said to define a cycle (or a periodic solution) of order k of (0-1), provided x is not simultaneously a root of (1-15) when k is replaced by one of its divisors, unity included. Call $x_{n+1}$ the consequent of $x_n$, and $x_{n-1}$ its antecedents. Any root $\bar{x}$ of order k > 1 of (1-15) defines k-1 distinct consequents, which are also roots of (1-15). The cycle of k points so obtained represents dynamically a k-th order sub-harmonic resonance.

Except for very special cases and small values of k, explicit expressions of (0-14) and (1-15), not to mention the roots of the latter, are rarely known, because they would take too much space to write out. Computer experiments using symbolic algebra have shown that the explicitation of (0-14) and (1-15) triggers usually a combinatorial explosion. Recourse to numerical computation becomes therefore unavoidable in almost all concrete cases. This fact may appear merely as a practical complication, but the problem is fraught with its own technical difficulties (see for example [R 1]). There are no foolproof methods of finding numerically even the real roots of moderate degree polynomials ! For most recurrences the roots of (1-15) have to be found laboriously by a combination of ingenious accuracy-controlled methods for which there is no general recepee.

Once a cycle $\bar{x}$ of (0-1) has been determined, one may enquire about its local stability in the sense of Liapunov, i.e. about its relationship with respect to neighbouring half-trajectories. Consider a consequent half-trajectory $\{x_n\}$, n = 0, 1,

2, ..., passing through the initial point $x_o = \bar{x} + \delta$, $0 < |\delta| << 1$. Similarly to the definition used in continuous dynamic systems, $\bar{x}$ is said to be stable with respect to the initial perturbation $\delta$ if for any $\varepsilon > 0$, no matter how small, it is possible to choose $\delta$ so that $|x_{n+m} - \bar{x}| < \varepsilon$ for $n > N$, where m and $N >> 1$ are suitably chosen integers. The choice $m >> 1$ serves to eliminate uncharacteristic "transient" effects. When no such $\delta$ can be found, x is said to be unstable. If in addition to the preceding $\varepsilon$-inequality one has $|x_{n+m} - \bar{x}| \to 0$ as $n \to \infty$, then $\bar{x}$ is said to be asymototically stable. Let $\lambda_k = \lambda_k(c) = f'_k(x, c)$, $' = \frac{\partial}{\partial x_n}$, be unambiguously defined ; this is always possible because of the assumed smoothness of (0-1). The linear recurrence

$$y_{n+k} = \lambda_k\, y_n \tag{1-16}$$

is called the variational equation of (0-1) at $\bar{x}$, and the number $\lambda_k$ its eigenvalue. The corresponding eigenfunction, i.e. general solution of (1-16) is

$$y_{n+k} = C\, \lambda_k^n \quad , \tag{1-17}$$

where C is an arbitrary constant. A cycle of order k is obviously locally asymptotically stable (or locally attractive) when $|\lambda_k| < 1$, and unstable when $|\lambda_k| > 1$. The consequents of $y_n \neq 0$ are all on the same side of the fixed point $y_n = 0$ of (1-17) when $\lambda_k > 0$, and on both sides when $\lambda_k < 0$. In the latter case the cycle $\bar{x}$ of (0-1) is sometimes said to be stable or unstable with reflection. The values $\lambda_k = \pm 1$ are critical in two ways :

a) the root $\bar{x}$ of (1-15) is not necessarily simple, and
b) the stability of $\bar{x}$ cannot be deduced from (1-16), (1-17), non linear terms of (0-14) have to be taken into account. For these reasons $|\lambda_k| = 1$ is said to constitute a critical stability case (in the sense of Liapunov). The values of the $\lambda_k$ are also rarely known explicitly, and again a recourse to numerical computation is unavoidable.

When a cycle is locally asymptotically stable, it is possible to enquire about its influence domain, i.e. about the largest admissible initial perturbation $\delta$, or more exhaustively, about the set $X_i$ of points $x_n$ such that the consequents of all $x_n \in X_i$ approach asymptotically and successively the k points of the cycle $\bar{x}$ as $n \to \infty$. A preferential approach to some points of the cycle, to the detriment of others, is clearly impossible ; the linearization (1-16) would not exist at every point of the cycle. Depending on the form of f in (0-1), the influence domain of a cycle is an isolated point set (antecedents of the points $\bar{x}$), a set of line segments, or both. Contrarily to continuous dynamic systems these line segments may be contiguous or disjoint, i.e. the continuous part of $X_i$ may be singly or multiply connected. A self-explanatory illustration of a multiply connected continuous influence domain $X_i$ is shown in Fig. 1-4. The stability of the fixed point $\bar{x}_2$ is readily established by the Königs-Lemeray graphical staircase construction (point on $x_n$-axis $\to$ point on $x_{n+1} = f(x_n, c)$ $\to$ point on $x_{n+1}$-axis $\to$ transfer to the $x_n$-axix, and so forth).

$\bar{x}_2$ = stable fixed point

$\bar{x}_1$, $\bar{x}_3$ = unstable fixed point

$a_i$, i = 1, 2, ... antecedents of $\bar{x}_1$

$\lim\limits_{i \to \infty} a_i = \bar{x}_3$

$a_1 < x_n < \bar{x}_1$ : $X_i$ = immediate
         influence domain of $\bar{x}_2$

$a_{2m+1} < x_n < a_{2m}$, m = 1, 2, ... : disjoint
        influence domain of $\bar{x}_2$

$\sum$ ▬ = $X_t$-total influence domain of $\bar{x}_2$

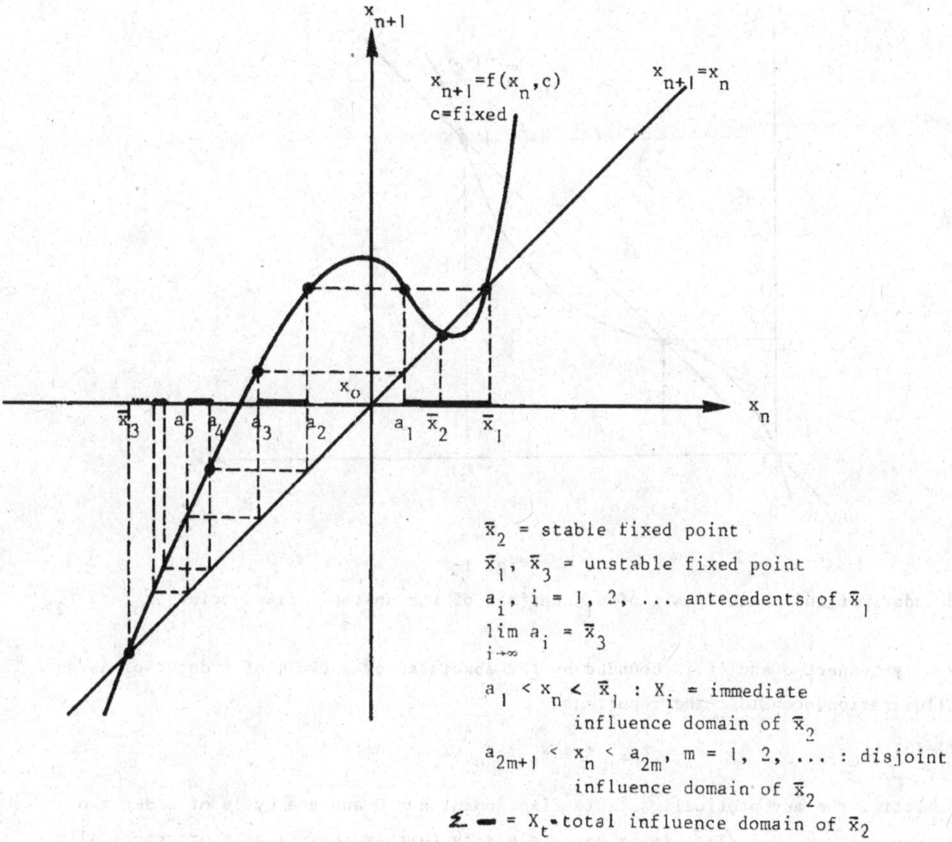

Fig. 1-4 : Disjoint influence domain of a fixed point of $x_{n+1} = f(x_n, c)$.
The complement of $X_t$ inside $\bar{x}_3 < x_n < \bar{x}_1$ belongs to the instability domain,
for example, the consequents of $x_0$, $a_2 < x_0 < a_1$ become unbounded.

In order to study the influence domain of a fixed point or cycle, it is
useful to know something about the nature of stability-boundary points. The simplest
possible case occurs when f in (0-1) is monotonically increasing and cuts the
$x_{n+1} = x_n$ line (example : Fig. 1-5, $\bar{x}_1$ = asymptotically stable fixed point, $\bar{x}_0 = \bar{x}_2$ =
unstable fixed points). Similarly to the case of first-order differential equations
(cf. (0-6), where non-degenerate static equilibrium points are given by isolated real
roots $\bar{x}_i$ of g(x, c) = 0), the constant steady states of (0-1) are alternately asymptotically stable and unstable. The influence domain of an asymptotically stable fixed
point is singly connected and bounded by the closest unstable fixed points. When f
in (0-1) is monotonically decreasing, its graph cuts the $x_{n+1} = x_n$ line only once.
If the resulting fixed point is asymptotically stable, its influence domain is also

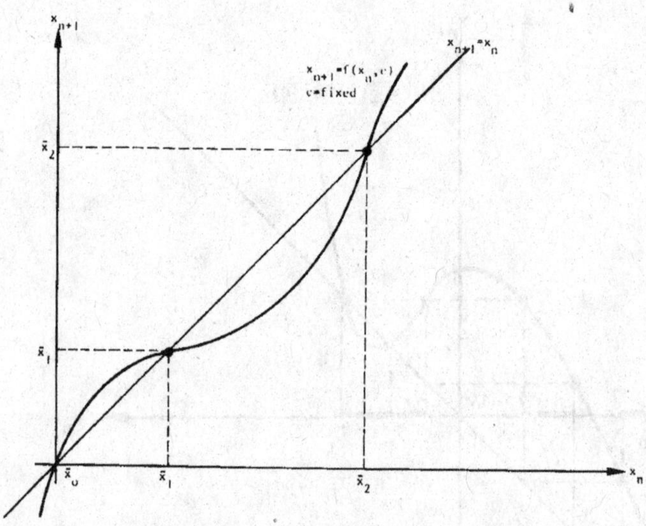


Fig. 1-5

Boundary of influence domain of $\bar{x}_1$ consists of the unstable fixed points $\bar{x}_o$ and $\bar{x}_2$.

singly connected and it is bounded by the abscissae of a cycle of order two. As an illustration, consider the recurrence

$$x_{n+1} = a\,x_n - x_n^3, \qquad |a| < 1, \tag{1-18}$$

admitting the asymptotically stable fixed point $\bar{x} = 0$ and the cycle of order two $\bar{x}_1 = -\sqrt{1+a}$, $\bar{x}_2 = +\sqrt{1+a}$. It is easy to verify (either analytically or graphically via the Königs-Lemeray construction) that the influence domain of $\bar{x} = 0$ is $X_i$ : $\bar{x}_1 < x_n < \bar{x}_2$.

The case of a non-monotonic f in (0-1) is more complicated, because there exist then points at which the slope of the curve $x_{n+1} = f(x_n, c)$ changes sign. Assume first that f is at least twice continuously differentiable with respect to $x_n$ and let $x_{c_i}$, i = 1, 2, ..., m be the (real) roots of $f'(x_m, c) = 0$, $' = \frac{\partial}{\partial x_n}$. The consequents of the $x_{c_i}$ are called critical points (in the sense of Julia-Fatou). There exists a theorem [F 1] which relates the number of critical points of a rational function f(x, c) in the complex plane x = u+iv to the number of aymptotically stable singularities of the recurrence (0-1). These singularities are fixed points or cycles, and by an obvious extension, invariant segments of the type shown in Fig. 0-1a. There is no exact equivalent of this theorem in terms of real variables, except when f(x, c) is at least three times continuously differentiable with respect to x and admits a strictly negative Schwarzian derivative, i.e. when

$$S.f = f'''/f' - \frac{3}{2}(f''/f')^2 < 0, \quad ' = \frac{d}{dx} \; .$$

In such a case the recurrence (0-1) is known to admit only a finite number of asymptotically stable cycles and fixed points (see for example [S 7]). There exists however a conjecture based on an extension of the Julia-Fatou theorem, giving a better estimate : if f(x, c) is sufficiently smooth then for each c the number of stable singularities $m_s$ on the finite part of the $x_n$-axis does not exceed the total number of segments $n_c$ of the curve $x_{n+1} = f(x_n, c)$ which have curvatures of a different sign, i.e. $m_s \leqslant n_c$.

The simplest possible case occurs when $n_c = 1$, and $m_s = 1$, i.e. when f has only one extremum and the curvature of the curve $x_{n+1} = f(x_n, c)$ does not change sign. Let for instance the extremum be a minimum as shown in Fig. 1-6a. The total influence domain $X_t$ of the stable fixed point $\bar{x}$ is singly-connected and consists of the segment $\bar{x}'_1 < x_n < \bar{x}_1$, bounded by the unstable fixed point $\bar{x}_1$ ans its antecedent $\bar{x}'_1$ , respectively. For a different value of c in f, Fig. 1-6a may take the form of Fig. 1-6b, where both the fixed points $\bar{x}$ and $\bar{x}_1$ are unstable. The stable singularity is still unique. It is either a cycle or a finite set of cyclic invariant segments, each similar to that of Fig. 0-1a, when f(x, c) is replaced by a suitably chosen iterate $f_k(x, c)$. The total influence domain of this stable singularity is the "no-escape" interval $X_t : \bar{x}_{1,-1} < x_n < \bar{x}_1$. After a finite number of iterations m, the consequents $x_m$ of any $x_o \in X_t$ move inside a smaller no-escape interval $X_o : x_{c1} < x_n < 0$. When the stable singularity contained inside $X_o$ is a cycle $\bar{x}_i$, i = 1, 2, ..., k, the subset $X_e$ of points of $X_t$, which do not belong to the interior of the total influence domain of $\bar{x}_i$, consists of points of unstable cycles, their antecedents, and of possible accumulation points of the former and latter. The exceptional points $x_n \in X_e$ are boundary points of partial influence domains of the stable cycle $\bar{x}_i$ and accumulation points of these boundary points (elementary example : Fig. 1-4, k = 1, $X_t$ contained inside $\bar{x}_3 < x_n < \bar{x}_1$, $X_e$ : points $a_i$, i = 1, 2, ... and the accumulation $\bar{x}_3$).

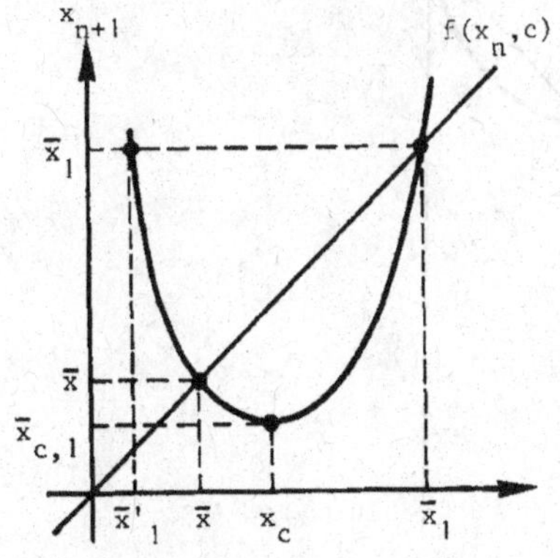


$\bar{x}$ = stable sixed point

$\bar{x}_1$ = unstable fixed point

$\bar{x}'_1$ = antecedent of $\bar{x}_1$

$x_c$ = root of f'(x,c) = 0

$x_{c,1}$ = consequent of $x_c$

$\bar{x}$ has a singly- connected influence somain $X_t : \bar{x}_{1,1} < x_n < \bar{x}_1$

Fig. 1-6a.

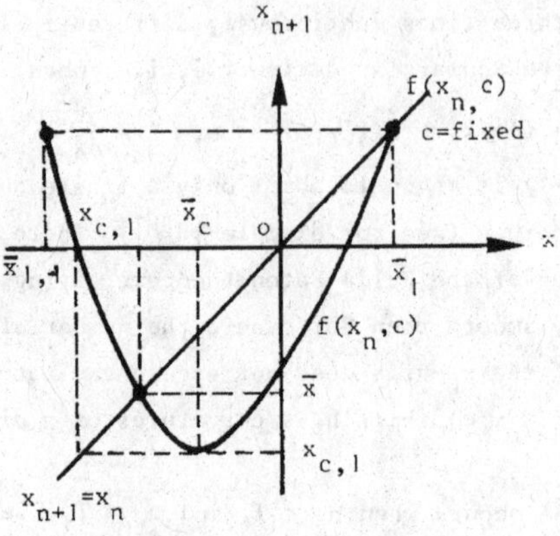


Figure 1-6b : Same as (a), except both $\bar{x}$ and $\bar{x}_1$ = unstable fixed points.
Influence domain $X_i$ of stable singularity : $\bar{x}_{1,-1} < x_n < \bar{x}_1$
Consequents of $x_n \in X_i$, enter $X_o$ : $x_{c,1} < x_n < 0$ after a finite number of iterations and remain inside $X_o$ thereafter.

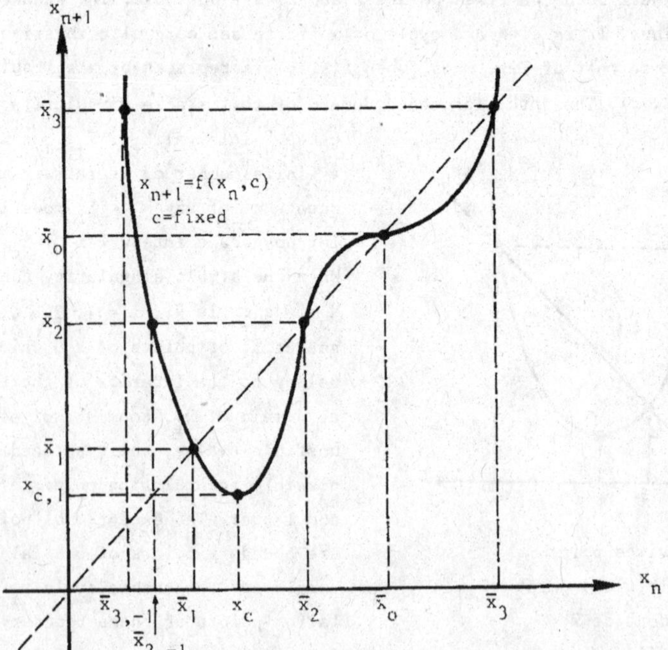


Figure 1-7(a)

Coexistence of an asymototically stable fixed point and of an invariant segment.

$\bar{x}_o$ : stable fixed point

$X$ : $\bar{x}_{2,-1} < x_n < \bar{x}_2$ = invariant segment

$X_t$ : $\bar{x}_{3,-1} < x_n < \bar{x}_3$, composite influence domain

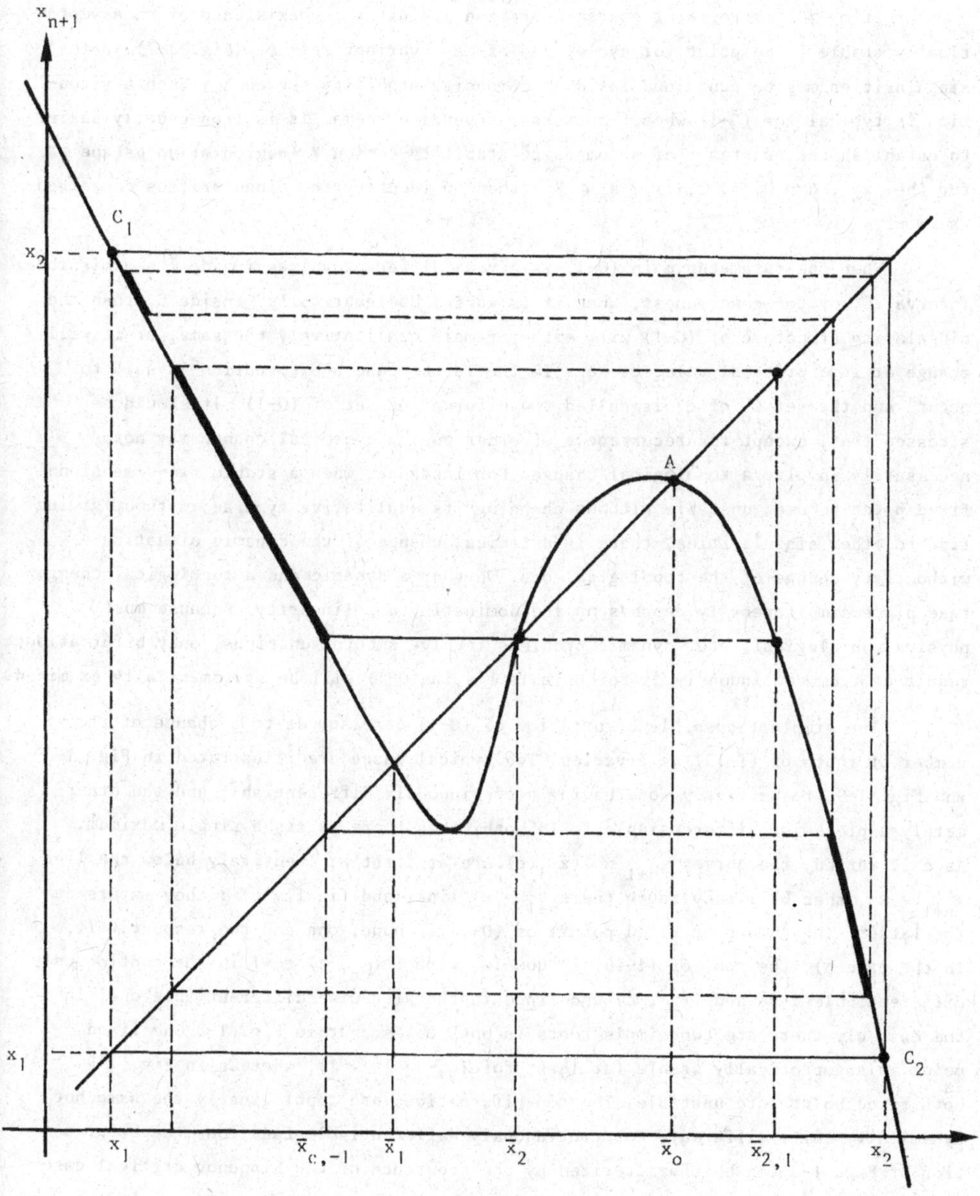


Figure 1-7(b)

Coexistence of an asymptotically stable fixed point and of an invariant segment.

$\bar{x}_o$ : stable fixed point, $x_1$, $x_2$ : points of a cycle of order k=2,
X : $\bar{x}_{c,-1} < x_n < \bar{x}_2$ = invariant segment, its influence domain: ▬.

If $m_s \gg 2$ there is of course no reason against the coexistence of an asymptotically stable fixed point (or cycle) and of an invariant segment (Fig. 1-7). Both singularities may be contained inside a composite stability region $X_c$. Such a situation is typical for (0-1) when f possesses several extrema. It is then usually easier to establish the existence of a composite stability region $X_c$ (region of no escape for the $x_n$, n = 0, 1, 2, ..., $x_o \in X_c$) than to identify the singularities contained therein.

When the parameter c in (0-1) is given different values inside its admissible interval C, or for conciseness, when it is varied "adiabatically" inside C, then the singularity structure of (0-1) will either remain qualitatively the same, or it will change at some critical value $c = c_o$. In the latter case a bifurcation is said to occur, and the set $C_o$ of $c_o$ is called the bifurcation set of (0-1). It should be stressed that, except for recurrences of order one, a dynamical change may not necessarily involve a topological change. For instance, when a stable two-dimensional fixed point becomes unstable without changing its qualitative type and without giving rise to other singularities, there is a radical change of the dynamic situation without any change of the topological one. Whether a dynamic and a topological change take place simultaneously depends on the dominating non linearity. Because most physical, biological, etc. dynamic problems involve smooth functions, only bifurcations resulting from continuously differentiable f's in (0-1) will be systematically examined.

The simplest possible bifurcation of (0-1) corresponds to a change of the number of roots of (1-15) as c varies. Two typical cases are illustrated in Fig. 1-8 and Fig. 1-9, respectively, one having a continuously differentiable and the other merely a piecewise differentiable f. In both cases there exists a single maximum. As c is varied, the curves $x_{n+1} = f(x_n, c)$ are at first a) : entirely below the line $x_{n+1} = x_n$, then b) : they touch the $x_{n+1} = x_n$ line, and finally c) : they intersect the latter. The number of fixed points of (0-1) is none, one and two, respectively. In the case b), the root of (1-15) is double, with $f'(x_n, c) = +1$ in the continuously differentiable case and $f(x_n, c)$ undefined in the piecewise differentiable one. In the case c), there are two simple roots in both cases, but in Fig. 1-8 one fixed point is asymptotically stable (at least for $|c - c_o| \ll 1$), whereas in Fig. 1-9 both fixed points are unstable. The two bifurcations are topologically the same but dynamically quite different. For continuously differentiable functions the bifurcation of Fig. 1-8 can be characterized by the occurence of the Liapunov critical case $\lambda_k = +1$, k = 1 in (1-16), reflecting the fact that the curve $x_{n+1} = f(x_n, c)$ is tangent to the line $x_{n+1} = x_n$. The corresponding bifurcation scheme is

(1-19) double cycle ($\lambda_k = 1$) $\twoheadrightarrow$ { asymptotically stable cycle ($-1 < \lambda_k < +1$) ; unstable cycle ($\lambda_k > 1$) } .

No systematic characterization in terms of an eigenvalue (or of a similar quantity) is available when $f(x_n, c)$ is less smooth. A change of the number of fixed

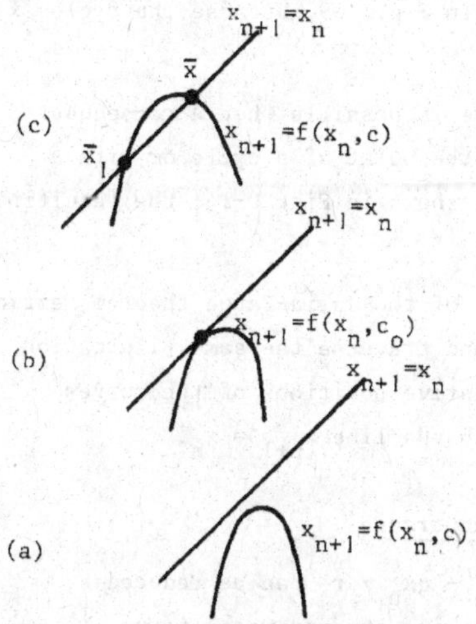


Fig. 1-8 : $f(x_n,c)$ continuously differentiable with respect to $x_n$ and continuous with respect to c at and near $c = c_o$.

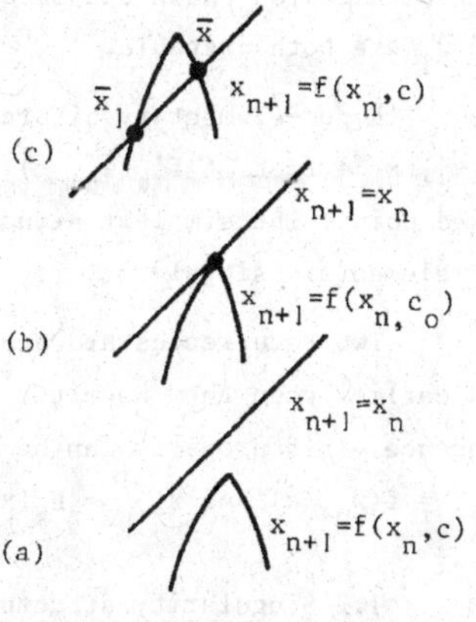


Fig. 1-9 : $f(x_n,c)$ piecewise differentiable with respect to $x_n$, and at least continuous with respect to c at and near $c = c_o$. $f'(x_n,c_o)$ = undefined, $|f'(x_n,c)| > 1$ when $c \neq c_o$.

points or cycles is defined by the algebraic equations (1-15). Unfortunately, there are at present no adequate analytical theorems covering the existence, uniqueness and continuity of (real) roots of these equations, so most concrete bifurcations have to be studied via numerical methods.

Another elementary bifurcation of (0-1) occurs when the stable fixed point $\bar{x}$ in Fig. 1-8 loses its stability when a further variation of c causes an upward movement of the curve $x_{n+1} = f(x_n, c)$. The same situation arises for the points of a cycle and the curve $x_{n+k} = f_k(x_n, c)$. The transition stability → instability is characterized by $\lambda_k = -1$ in (1-16), and is accompanied by the appearance of a bifurcated stable cycle of order 2k. The corresponding bifurcation scheme is

$$\text{(1-20)}\qquad \text{asymptotically stable cycle } (-1 < \lambda_k < 1) \xrightarrow[\lambda_k=-1]{} \begin{cases} \text{asymptotically stable cycle of doubled order } (-1<\lambda_{2k}<+1) \\ \text{unstable cycle } (\lambda_k = -1) \end{cases}$$

The root $\bar{x}$ of (1-15) is simple when $-1 < \lambda_k < +1$, it becomes triple when $\lambda_k = -1$, and is simple again when $\lambda_k < -1$. As can be easily seen by constructing locally the curve $x_{n+2k} = f_{2k}(x_n, c)$ from the curve $x_{n+k} = f_k(x_n, c)$, the bifurcated cycle of doubled order comes into existence with the eigenvalue $\lambda_{2k} = +1$. A continued upward motion of the curve $x_{n+1} = f(x_n, c)$ will therefore yield a bifurcation chain with the successive generation of cycles of order $k = 2^m$, m = 1, 2, ... The limit $m \to \infty$ of this chain may give rise to a non-elementary singularity and to a non-elementary bifurcation. No such bifurcation chain can occur in the case shown in Fig. 1-9, because the cycles $\bar{x}$ and $\bar{x}_1$ are both unstable.

A non-elementary bifurcation of another type is possible when a consequent of $x_c$, $x_c$ = root of $f'_k(x_n, c) = 0$, coincides with the point of a cycle or with a fixed point. The simplest situation of this type is shown in Fig. 0-1 ; the resulting non elementary singularity is an invariant segment.

Two recurrences are equivalent in the sense of the Julia-Fatou theorem mentioned earlier when they have the same singularities and traverse the same bifurcation sequence. This property can be deduced from the relative positions of the curves $x_{n+1} = f(x_n, c)$ and $x_{n+k} = f_k(x_n, c)$ with respect to the line $x_{n+1} = x_n$.

## 1.2 Singularity structure of a quadratic recurrence

The general quadratic recurrence $x_{n+1} = px_n^2 + qx_n + r$ can be reduced by means of a linear transformation to one of the following one-parameter forms

(1-21) $$x_{n+1} = x_n^2 + c\,, \qquad -2 \leq c \leq \frac{1}{4}\,,$$

(1-21a) $$x_{n+1} = ax_n(1 - x_n)\,, \qquad 1 \leq a \leq 4\,.$$

The parameter ranges indicated correspond to the existence of bounded trajectories $\{x_n\}$, n = 0, 1, 2, ... Since the first form ($c = pr - q^2 + q$) has been more extensively studied than the second, it will be prefered in what follows.

The fixed points and the cycle of order two of (1-21) can be found analytically in terms of elementary functions. In fact, a double fixed point $\bar{x} = 1/2$ appears at $c = 1/4$, which splits for $c < 1/4$ into two simple fixed points

(1-22) $$\bar{x} = \frac{1}{2} \pm \sqrt{\frac{1}{4} - c}$$

A cycle of order k = 2, whose coordinates $\bar{x}$ satisfy $x^2 + x + c + 1 = 0$, bifurcates from the initially stable fixed point at $c = -3/4$, $\bar{x} = -1/2$, $\lambda_1 = -1$, the other fixed point remaining unstable. When k = 3, the equation (1-15) can be factored, and after omission of a redundant quadratic factor there results

(1-23) $$x^6 + x^5 + (1+3c)x^4 + (1+2c)x^3 + (1+3c+3c^2)x^2 + (1+c)^2x + c(1+c)^2 + 1 = 0,$$

which can no longer be solved in an elementary way. A double root appears at $c = -1.75$, and it splits into two simple ones as c decreases. Higher order cycles have to be found numerically. Some of their properties can, however, be obtained analytically by means of an indirect argument [M 13]. Consider in fact the k-th iterate of (1-21) written in the form

(1-24) $$x_{n+k} = f_k(x_n, c) = x_n^{2^k} + \ldots + G_k(c),$$

where $G_k(c)$ is a polynomial in c of degree $2^{k-1}$ defined recursively by

$$G_{k+1}(c) = G_k^2(c) + c,\ k > 1,\ G_1(c) = c\,.$$

Myrberg has shown that when a cycle of order k of (1-21) has the eigenvalue $\lambda_k = 0$, then the corresponding value of c is a root of $G_k(c) = 0$. To prove this, let $x_1, x_2, \ldots x_k$ be the abscissae of this cycle. From (1-16) it follows that $\lambda_k = (2x_1).(2x_2)\ldots(2x_k) = 2^k x_1.x_2\ldots x_k$. If one of the points, say $x_i$, is at the minimum $x_n = 0$ of $f_k(x_n, c)$, then $\lambda_k = 0$, and the next point becomes $x_{i+1} = c$. For example, for k = 1, the fixed point x = 0 having $\lambda_2 = 0$ is given by the root c = 0 of $G_1(c) = 0$ ; for k = 2, the point of the cycle x = 0 having $\lambda_2 = 0$ is given by the root c = -1 of $G_2(c) = 0$, etc. Some of the next roots are given in Table 1 ; multiple entries occur for the same value of k because cycles of order k > 3 are non-unique. In fact, the number of cycles $N_k$ of order k increases very rapidly with k, as can be seen from Table 2. It is straightforward but somewhat tedious [M 5] to express $N_k$ and the

| k | c for $\lambda_k = 0$ | k | c for $\lambda_k = 0$ |
|---|---|---|---|
| 3 | -1.754877666 | 6 | -1.476014643 |
| 4 | -1.940799806 | 8 | -1.521317232 |
| 5 | -1.985424253 | 8 | -1.536243271 |
| 5 | -1. 860782522 | 10 | -1.501716839 |
| 5 | -1.625413716 | 10 | -1.447008841 |

Table 1 - Some roots of $G_k(c) = 0$ for the cycle point x = 0 whose $\lambda_k = 0$.

corresponding number $N_c(k)$ of bifurcation values of c in terms of k, because the resulting formulae depend on the decomposition of k into prime factors. The principal expressions are [M 5] : If k is prime, then

$$N_k = (2^k - 2) / k \tag{1-25a}$$

| k | 1 | 2 | 3 | 4 | 5 | 6 | 7 | 8 | 9 | 10 | 11 | 12 | 13 | 14 | 15 | ... | 20 |
|---|---|---|---|---|---|---|---|---|---|---|---|---|---|---|---|---|---|
| $N_k$ | 2 | 1 | 2 | 3 | 6 | 9 | 18 | 30 | 56 | 99 | 186 | 335 | 630 | 1161 | 2182 | ... | 52377 |
| $N_c(k)$ | 1 | 1 | 1 | 2 | 3 | 5 | 9 | 16 | 28 | 51 | 93 | 170 | 315 | 585 | 1091 | ... | 26214 |

Table 2 - Number of cycles $N_k$ of order k and number $N_c(k)$ of the corresponding bifurcation values of c.

If, on the contrary, $k = k_1^{\alpha_1}.\, k_2^{\alpha_2} \ldots k_n^{\alpha_n}$ and the $k_i$, i = 1, 2, ..., n, are primes, ordered so that $k_{i+1} < k_i$, the $\alpha_i$ being integers, the expression of $N_k$ is more complex. Let $\ell_\nu = k/k_\nu$, $\nu = 1, 2, \ldots, n$, $(\ell_\nu, \ell_\mu, \ell_\theta, \ldots \ell_\gamma)$ be the greatest common divisor of $\ell_\mu, \ell_\theta, \ldots, \ell_\nu$, then [T 1]

$$N_k = \left[2^k - \sum_{\nu=1}^{n} 2^{\ell_\nu} + \sum_{\mu=2}^{n} \sum_{\nu=1}^{\mu-1} 2^{(\ell_\nu \ell_\mu)} - \sum_{\theta=3}^{n} \sum_{\mu=2}^{\theta-1} \sum_{\nu=1}^{\mu-1} 2^{(\ell_\nu \ell_\mu \ell_\theta)} + \sum\sum\sum \ldots\right] / k \;. \tag{1-25b}$$

The number of bifurcation values is obviously $N_c(k) = \frac{1}{2} N_k$ when k is odd (only the bifurcation scheme (1-19) is possible), whereas $N_c(k) = \frac{1}{2} (N_k + N_c(k/2))$ when k is even (the bifurcation schemes (1-19) and (1-20) provide contributions).

Three other properties of (1-24) are known [M 12 - 15] :

(1-26) $\begin{cases} \text{If } -2 < c < -1, \text{ then between the zeros } c_{k-2} \text{ and } c_{k-1} \text{ of } G_{k-2}(c) \text{ and} \\ G_{k-1}(c), \text{ respectively, there exists at least one zero } c_k \text{ of } G_k(c). \end{cases}$

(1-27) All zeros of $G_k(c)$ are inside the interval $-2 < c \leq 0$.

Since both the parametric space $C : -2 \leq c \leq 1/4$ and the composite stability domain $X_c : \bar{x}_{1,-1} < x_n < \bar{x}_1$, where $\bar{x}_1 = \frac{1}{2} + \sqrt{1/4-c}$ and $\bar{x}_{1,-1}$ is the antecedent of $\bar{x}_1$, of (1-21) are bounded, accumulations are possible inside C and $X_c$. Moreover, due to the theorem of Julia-Fatou (cited in section 1.1), for every c there exists only one stable singularity of (1-21), i.e. a fixed point, a cycle or an invariant segment. The composite stability domain $X_c$ is therefore either the (logical) sum of the singularity and its influence domain, and the boundary of the latter, or $X_c$ and the singularity coincide. The latter case occurs obviously for $c = -2$.

In order to understand the properties of the general solution (0-12) of (1-21), it is necessary to order the singularities into regular patterns. This ordering can be done in various ways, but the most natural one appears to be related to the corresponding generating bifurcations. In fact, the cycle of lowest order in the chain $k = m.2^i$, m = fixed, i = 0, 2, ... appears via (1-19) and the others via (1-20). For conciseness let → designate the bifurcation process. Some bifurcation values of c are given in Table 3. Similar tables can be computed for other cycle chains with

| m = 1 | | m = 3 | |
|---|---|---|---|
| k → 2k | c | k → 2k | c |
| 0 → 1 | +0.250 | 0 → 3 | -1.75 |
| 1 → 2 | -0.75 | 3 → 6 | -1.7549 |
| 2 → 4 | -1.25 | 6 → 12 | -1.7729 |
| 4 → 8 | -1.3816 | 12 → 24 | -1.7783 |
| 8 → 16 | -1.3969 | 24 → 48 | -1.785093 |
| 16 → 32 | -1.4002 | 48 → 96 | -1.7850990 |
| i → ∞ | -1.401155189 | 96 → 192 | -1.7850991 |
| | | i → ∞ | -1.7850992 |

Table 3 - The bifurcation chains $k = 1.2^i$ and $k = 3.2^i$ of (1-21) .

moderate values of m, but for a given computer a practical limit is reached quite soon. Fortunately for larger m an indirect argument is possible. Consider in fact the accumulation value $c_{1a} \approx -1.401$ of the chain $k = 1.2^i$, $i \to \infty$ relatively to the total admissible parameter space $C : -2 \leqslant c \leqslant \frac{1}{4}$. Since for $c_{1a} < c < \frac{1}{4}$ there exist only a finite number of cycles, only one of which is stable for a fixed c, the dynamic behaviour of (1-21) is orderly, or regular. For conciseness let this part of C be designated by the mnemonic symbol rC(1,1), where the prefix r stands for regularity the first 1 inside the parentheses for the subscript 1 in $c_{1a}$, and the second 1 for the fact that the first (and in this particular case the only one) of the possible $N_k$ cycles of order $k = 1$ is involved. For the j-th of the $N_k$ cycles of order $k = m.2^i$, counted for example according to its parametric distance from $c_{1a}$, the analoguous symbol is rC(m, j). Contrarily to rC(1,1) inside the complementary part $-2 \leqslant c < c_{1a}$ of C there exists an infinity of cycles. This part of C is therefore irregular, or singular, and for this reason it is designated by sC(1,1). Hence, by definition, $C = C_1(1) = rC(1,1) + sC(1,1)$ for the cycle chain $k = 1.2^i$, with the separation point of $C_1(1)$ given by the accumulation value $c_{1a}$. A geometrical analysis of the positions of the curves $x_{n+k} = f_k(x_n, c)$, $k = 1, 2, \ldots$, relatively to the line $x_{n+k} = x_n$, confirms readily the properties (1-26), (1-27) and (1-28), and shows furthermore that the sum-representation $C_j(m) = rC(m, j) + sC(m, j)$ is valid for any cycle chain $k = m.2^i$, provided the $N_k$ "basic" cycles of the lowest order $k = m$ originate via the bifurcation (1-19). In fact considering the iterated recurrence $x_{n+k} = f_k(x_n, c)$, $k > 1$, it is seen [M 5] that the bifurcation structure in parameter space is exactly the same as in $x_{n+1} = f(x_n, c)$, because the cycles of order k of the latter are fixed points of the former, with $c_{1a}$ replaced by a set of $N < N_c(k)$ values of c, corresponding to the $N_k$ different cycles of order k. For example, two cycles independent of those already given in Tables 1 and 3 are $k = 5$, $c \approx -1.9854$ and $k = 10$, $c \approx -1.5362$. All intervals $C_j(m)$ of existence of the cycles of order $k = m.2^i$, $m > 2$, $i = 0,1, \ldots$ are located inside sC(1,1) and have the same structure

$$C_j(m) = rC(m, j) + sC(m, j), \; j = 1, 2, \ldots, N \leqslant N_c(m)$$

as the interval $C = C_1(1)$. Moreover, between the intervals $C_j(m-2)$ and $C_j(m-1)$ there exists always an interval $C_j(m)$. Hence, sC(1,1) contains also an enumerable set of intervals rC(m, j), inside of which the motion described by (1-21) is orderly. Except for $k = 3$, for which rC(3,1) is about 0.8 % of C, and about 3 % of sC(1,1), the parametric length of rC(m, j) is very small. It should be noted that all $C_j(m)$, $j = j_1, j_2, \ldots, j_r$, $m = m_1.m_2. \ldots m_s$, are located inside the interval $sCj_1(m_1)$, where all $j_i$ and $m_i$ as well as r, s and m are integers. Moreover, [M 5], $C_j(m) \subset C_p(r)$, where $j = j_1 \ldots j_s$, $m = m_1 \ldots m_s$, $p = j_1 \ldots j_{s-1}$, $r = m_1 \ldots m_{s-1}$. Figure 1-10 illustrates the relative position of the various intervals for the lowest values of j and m. An inspection of Fig. 1-10 suggests the name of box-within-a-box structure for the bifurcation set of (1-21) inside the singular part $sC_1(1)$ of C.

By definition one boundary point of the interval sC(m, j) is given by the accumulation value $c_{ma}$ of the cycle chain $k = m.2^i$, $i \to \infty$. When $m = 1$, the other

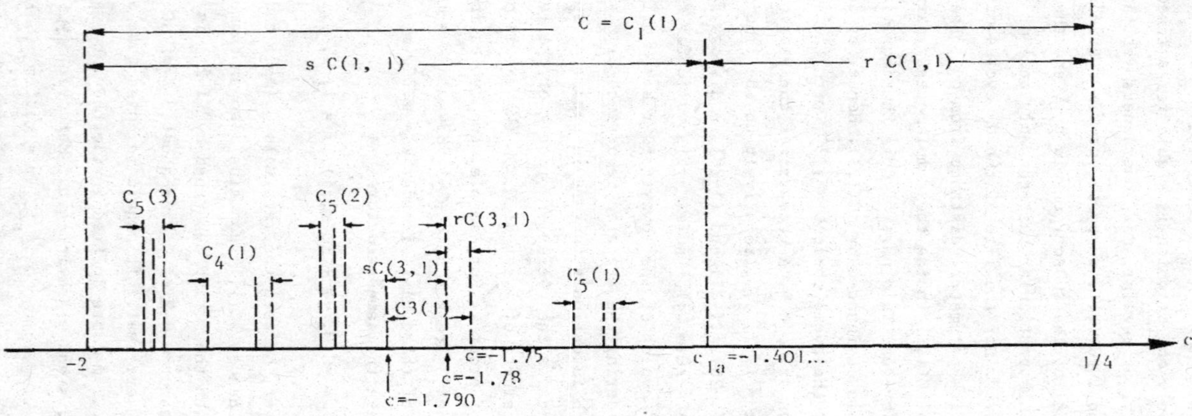


Fig. 1-10 : Relative positions of parametric existence-intervals of cycles of order k = 1, k = 3 and k = 5 of (1-21) (not to scale). The same relative positions hold when $C_1(1)$ is replaced by $C_j(m)$, $j = j_1, j_2, \ldots, j_r$ when $m = m_1, m_2, \ldots, m_s$; $j_i$, $m_i$, i, r ans s are integers.

limit ($c = -2$) results from the coalescence of a fixed point of (1-21) and a critical point of $f(x_n, c)$, i.e. of a consequent of a root $x_c$ of $f'(x_n, c) = 0$, giving rise to the invariant segment $X : -2 < x_n < 2$. An examination of the iterated recurrence (0-14) shows [M 5] that such a coalescence constitutes the other limit of all $sC(m, j)$. A cycle of order k, which originated via the bifurcation (1-20) coincides with a critical point when an invariant segment of order k comes into being. Let $c_k^*$ designate this double bifurcation : disappearance of a stable cycle and the appearance of a (cyclic) invariant segment of the same order. The set of the $c_k^*$ is of course enumerable ; some values are listed in Table 4.

| k | $c_k^*$ | $c_o$ (origin) | k | $c_k^*$ | $c_{ka}$ |
|---|---|---|---|---|---|
| $2^1$ | -1.543689013 | -1.25 | 1 | -2 | -1.401155189 |
| $2^2$ | -1.430357632 | -1.3107 | 3 | -1.790327493 | -1.780000107 |
| $2^3$ | -1.407405119 | -1.3816 | 4 | -1.942762011 | -1.940799806 |
| $2^4$ | -1.402492176 | -1.3969 | 5 | -1.633358704 | -1.625413716 |
| $2^5$ | -1.401441494 | -1.4002 | 5 | -1.862331091 | -1.860782522 |
| $2^6$ | -1.401216505 | -1.40096 | 5 | -1.985540378 | -1.98... |
| $2^7$ | -1.401168322 | -1.401138 | 9 = 3.3 | -1.786510545 | -1.786319890 |
| $2^8$ | -1.401158000 | -1.401146 | 15 = 3.5 | -1.783756262 | -1.78... |
| $2^9$ | -1.401155790 | -1.401146 | 27 = 3.3.3. | -1.786441527 | -1.786... |
| $2^{10}$ | -1.401155318 | -1.401146 | | | |
| $2^{11}$ | -1.401155220 | -1.401146 | | | |
| $2^{12}$ | -1.401155200 | -1.401... | | | |
| $2^i$, $i \to \infty$ | -1.401155189 | - | | | |

Table 4 - Some bifurcation values $c_k^*$ corresponding to invariant segments.

When $k = 1$, the invariant segment X of (1-21) coincides with its influence domain $X_i$ (c.f. Fig. 0-1 and Fig. 1-6a). When $k > 1$, then the cyclic segments form either a contiguous or a disjoint set of generally unequal individual segments (c.f. Fig. 1-11 and 1-12). This situation is entirely analoguous to the influence domains of individual points forming a cycle. A set of cyclical segments is similar to a set of points forming a cycle, except that the consequents wander all over each segment without any possibility of settling down, each cyclic segment admitting an infinity of unstable point singularities.

From Table 4, it can be seen that $c_k^* < c_{2k}^*$ for $k = 2^i$, $i = 1, 2, \ldots$ and $c_1^* = -2$, $c_\infty^* = c_{1a}$. Because all these $c_k^*$ are located inside the parametric interval $C(2^k) : c_2^* < c < c_{1a}$, and to each $c_k^*$ corresponds an even number of cyclic invariant segments, cycles of an odd order k cannot exist inside $C(2^k)$. The existence of an odd cycle would violate the geometric symmetry induced by the invariant segments, which is inconsistent with the continuous evolution of the singularity structure as a function of c. The singularity structure changes at each bifurcation, without, however, disrupting the global continuity of the phase space portrait.

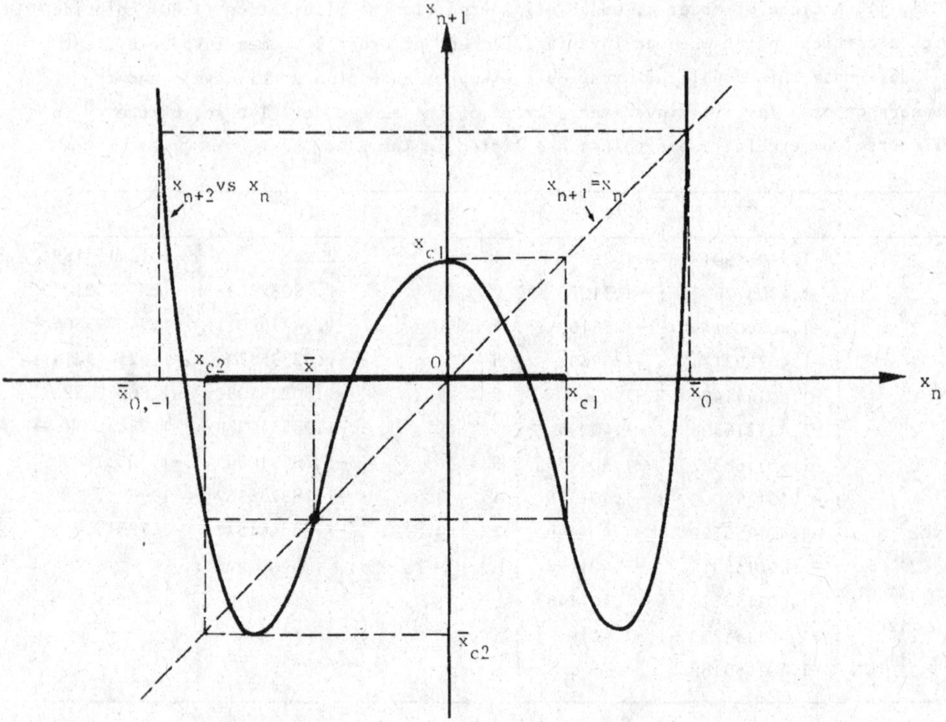


Fig. 1-11 - Contiguous cyclic invariant segments $X_1 : \bar{x} < x_n < x_{c1}$ and $X_2 : x_{c2} < x_n < \bar{x}$ of order k = 2 of (1-21), $x_{c1} = -1.5437$, $\bar{x} \simeq -0.839$. Influence domain $X_i : \bar{x}_{0,-1} < x_n < \bar{x}_o$ .

The value c = -2 in (1-21) is very special in the sense that a general solution in terms of elementary functions becomes possible. Let $x_n = 2 \cos y_n$, where $y_n$ is a new dependent variable. After insertion into (1-21), there results the recurrence $\cos y_{n+1} = \cos 2y_n$, which is modulo-$\pi$ equivalent to $y_{n+1} = 2y_n$. Since the solution of the last recurrence is $y_n = C.2^n$, C being an arbitrary constant, the general solution of (1-21), c = -2, is (see for example [L 9], [J 1], [P4], [H8],....)

$$x_n = 2 \cos y_n,\ y_n = 2^n \text{ arc cos } (x_o/2), \quad -2 < x_o < 2 , \tag{1-29}$$

where an appropriate branch of arc cos is chosen for each initial value $x_o$. No expression analoguous to (1-29) has been found so far other values of c, the $c_k^*$ included. The chaos generated by (1-21) for c = -2 amounts therefore to the selection of

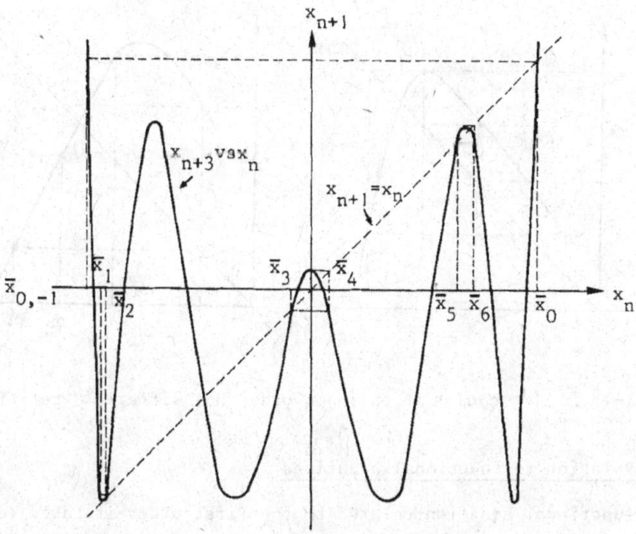


Fig. 1-12. Disjoint cyclic invariant segments
$X_1 : \bar{x}_1 < x_n < \bar{x}_2, X_2 : \bar{x}_3 < x_n < \bar{x}_4$ and $X_3 ; \bar{x}_5 < x_n < \bar{x}_6$ of order $k = 3$ of (1-21), $c_3^* \approx -1.790$. Influence domain $X_i : \bar{x}_{0,-1} < x_n < \bar{x}_o$.

ordinates of a periodic function at exponentially increasing discrete abscissae. The closeness of a given consequent half trajectory $\{x_n\}$, $n = 0, 1, 2, \ldots$, with respect to the neighbouring ones is described by the sensitivity coefficient

$$\partial x_n / \partial x_o = \left[-2^n / \sqrt{1-(x_o/2)^2}\right] \cdot \sin\left(2^n . \operatorname{arc} \cos(x_o/2)\right) .$$

This coefficient is a rapidly increasing oscillating function of n. Hence two initially close half trajectories diverge rapidly from each other as n increases. The same situation prevails for other recurrences involving a single extremum.

A remarkable feature of the complicated singularity structure inside $sC_1(1) = -2 < x_n < c_{1a}$ is the pronounced non uniqueness of cycles of an order higher than three (c.f. Table 2). These non unique cycles can be distinguished from each other by means of the following principal properties : 1) type of generating bifurcation and corresponding bifurcation value $c_o$, 2) current eigenvalue or parametric distance from either $c_o$ or $c_{1a}$, and in case of stability : type and extent of influence domain, 3) preceding bifurcation history if distinct from generating bifurcation, and finally 4) rotation sequence of the k points $x_1, \ldots, x_n, x_{n+1} = x_1$, constituting the cycle [M 5]. By the rotation sequence of a cycle is meant the algebraic order of its points on the $x_n$ axis. An example of two cycles of the same order but different rotation sequence in shown in Fig. 1-13.

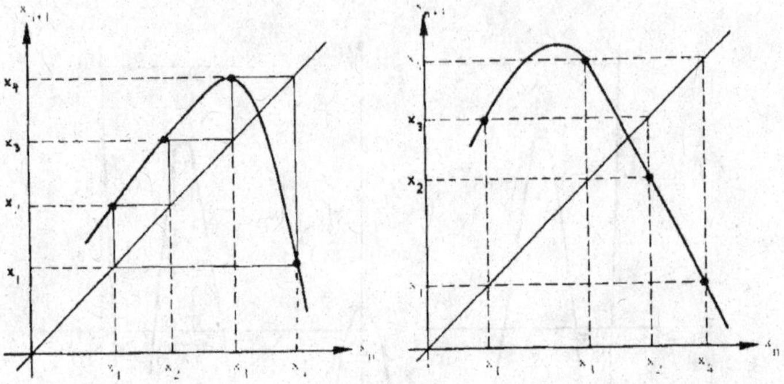

Fig. 1-13. Two cycles of the same order but different rotation sequence.

## 1.3 Relation to functional equations

The functional equation related to the first order iterated recurrence (0-14) in the simplest way is the equation of automorphic functions (c.f. 0-10)

$$w(f_k(x, c)) = w(x, c) \quad (1\text{-}30)$$

When f is a smooth function, and one looks for smooth solutions, then (1-30) is known to admit only the physically, biologically, etc. uninteresting solution $w(x, c) =$ const., which is unrelated to any specific f. For a given f, (1-30) admits however discrete (discontinuous) solutions in the ordinary sense, which correspond to fixed points and cycles of order k of (0-1), and it should admit continuous solutions in some generalized sense inside finite intervals of the x-axis, which correspond to cyclic invariant segments. For $k > 1$ equation (1-30) should admit in addition (induced) solutions consisting of excess antecedents of fixed points, cycles and invariant segments.

A slightly more complicated relation exists between (0-1) and the ordinary and iterated Schröder equations :

$$w(f(x, c)) = s\, w(x, c), \quad (1\text{-}31) \qquad w(f_k(x, c)) = s^k\, w(x, c),\ k > 1. \quad (1\text{-}31a)$$

Many local properties of (1-31) are known when f admits a convergent Taylor series in x. Let for example $\bar{x}$ be an asymptotically stable fixed point of (0-1) with the eigenvalue $s = f'(\bar{x}, c) = a_1$, and $R > 0$ the radius of convergence of

$$f(x, c) = \bar{x} + \sum_{i=1}^{\infty} a_i (x-\bar{x})^i,\quad a_i = a_i(c),\quad a_1 \neq 0,\ 1,\ -1\ . \quad (1\text{-}32)$$

The value $a_1 = 0$ has to be excluded because otherwise there is no Schröder equation, whereas for $a_1 = 1$ and $a = -1$ either (1-31) or (1-31a) degenerates into (1-30). The limit cases $a_1 \to 0$, $a_1 \to +1$ and $a_1 \to -1$ remain, however, meaningful. It is well known [K 5] that subject to (1-32), equation (1-31) possesses a convergent series solution

(1-33) $$w(x, c) = C \sum_{i=1}^{\infty} b_i (x-\bar{x})^i, \ b_1 = 1, \ b_i = b_i(c), \ i > 0, \ \max |x-\bar{x}| = R_1 \leq R,$$

where C is an arbitrary "integration" constant. When viewed through the properties of (0-1), the radius of convergence $R_1$ of (1-33) becomes related to that of (1-32) in an obvious manner. In fact, let $y_n = w(x_n, c)$ and for conciseness omit the explicit mention of the dependence on c, then

$$y_{n+1} = w(x_{n+1}) = w(f(x_n)) = s\, w(x_n) = s\, y_n,$$

i.e. a solution of the Schröder equation transforms formally the non linear recurrence (0-1) into a linear one :

(1-34) $$y_{n+1} = s\, y_n, \qquad s = s(\bar{x}, c).$$

The equivalence between a linear and a non linear recurrence being only possible in principle inside one phase cell, the radius of convergence $R_1$ of (1-33) is bounded by the distance between the fixed point $\bar{x}$ and the closest boundary point of the singly-connected phase cell inside of which $\bar{x}$ is located. But for an asymptotically stable fixed point the phase cell in question is simply the immediate influence domain $X_i$ ; hence the knowledge of $X_i$ defines $R_1$. The values of w(x, c) in the interval limited by $R_1$ and the farthest boundary point of $X_i$ can be obtained by analytical continuation

The meaning of the limiting cases $a_1 \to 0$ and $a_1 \to \pm 1$ becomes now also clear. From the stability point of view, nothing special happens at the limit $a_1 \to 0$, because if an immediate influence domain of $\bar{x}$ exists for $|a_1| < \varepsilon$, it exists also for $\varepsilon \to 0$ ; for $a_1 = \pm 1$, there arises a Liapunov critical case, and the influence domain of $\bar{x}$ may be of a finite size, or it may vanish. In the first case a solution of the Schröder equation is easily defined, although the series (1-33) will in general fail to converge. Consider as an example the recurrence

(1-35) $$x_{n+1} = f(x_n, c) = c\, x_n - x_n^3 + \frac{1}{4} x_n^5,$$

whose graph $x_{n+1}$ vs $x_n$ appears in in Fig. 1-14a, (c > 1) and Fig. 1-14b, (c = 1), respectively. When c > 1 the recurrence (1-35) admits five fixed points, two of which are asymptotically stable. As $c \to 1$, three fixed points merge, forming a composite asymptotically stable fixed point $\bar{x} = 0$ with the eigenvalue $\lambda = f'(0,1) = 1$ and the influence domain $-2 < x_n < 2$. The existence in (1-35) of the composite fixed point $\bar{x} = 0$ is the reason for the lack of convergence of the corresponding series (1-33), because otherwise w(x, c) would transform the composite fixed point into a simple fixed point of the linear recurrence (1-34). The solution of (1-31) must therefore be singular when c = 1 in (1-35). A similar example can be constructed when $a_1 = -1$,

(a) $\lambda > 1$

$\bar{x}_0,\ \bar{x}_2,\ -\bar{x}_2$ = unstable fixed points

$\bar{x}_1,\ -\bar{x}_1$ = stable fixed points

$-\bar{x}_2 < \bar{x}_n < \bar{x}_0$ : influence domain of $-\bar{x}_1$

$\bar{x}_0 < \bar{x}_n < \bar{x}_2$ : influence domain of $\bar{x}_1$

(b) $\lambda = 1$

$\bar{x}_0$ = asymptotically stable composite fixed point

$-\bar{x}_2 < x_n < \bar{x}_2$ : influence domain of $\bar{x}_0$

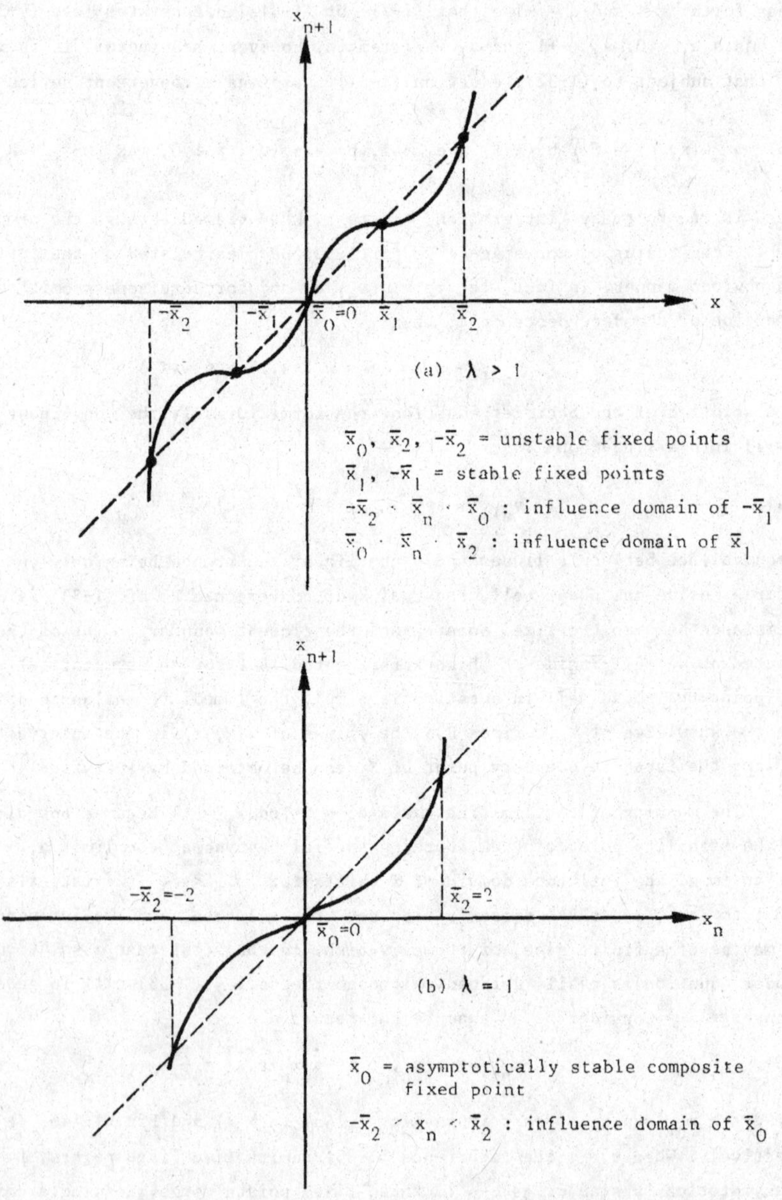

Fig. 1-14 : The appearance of an asymptotically stable fixed point with an eigenvalue $\lambda = +1$ and a finite influence domain. Fig. (b) results from Fig. (a) as $\lambda \to 1^+$.

except that the equation (1-31) is replaced by the iterated equation (1-31a) with $k = 2$. The limiting case $s = 1$ of the Schröder equation is simultaneously a limiting case of the equation of automorphic functions. The latter equation possesses therefore also piecewise continuous solutions characterized by the stable Liapunov critical case $\lambda = f'_k(x, c) = 1$ of the corresponding fixed point or cycle x. Such a critical case can occur either independently, or as a second iterate of the critical case $s = -1$.

An inspection of the special case of (0-1) illustrated in Fig. 1-4 leads to the rather obvious conclusion that the iterated Schröder equation (1-31a) admits solutions which are not simultaneously solutions of (1-31). In fact the f in question can be represented by a polynomial in $x_n$, with $\bar{x} = \bar{x}_2$ ; the series (1-32) is therefore convergent for all x. Assuming a uniform $x_n$-scale in Fig. 1-4, the radius of convergence of the series (1-33) is $R_1 \leq \bar{x}_2 - b$, $b = \min(a_1, x_1)$. The function defined by (1-33) can be continued analytically in the whole immediate influence domain $X_o : a_1 < x < \bar{x}_1$ of $\bar{x}_2$, but not into the disjoint part of the total influence domain. Since $X_o$ is a consequent of each of the segments $X_m : a_{2m+1} < x < a_{2m}$, $m = 1, 2, \ldots$, an analytical solution of (1-31a) exists inside each $X_m$ for $k > m$, and these solutions are distinct from that existing inside $X_o$. There is no straightforward way of relating a solution of (1-31) to the existence of an invariant segment X, because there appears to be no unique way of defining the constant s for the whole X (cf. Chapter VI).

Several variants are available for the determination of the coefficients $b_i$ in (1-33), the simplest of which consists in the insertion of (1-33) into (1-31), (1-32) and the identification of equal powers of $(x - \bar{x})^i$. The result is a sequence of algebraic expressions of rapidly increasing cumbrousness :

$$(1\text{-}36)\quad \begin{cases} b_2 = \dfrac{a_1}{s-s^2}\,,\ b_3 = \dfrac{1}{s-s^3}\left(a_3 + \dfrac{2sa_2^2}{s-s^2}\right), \\ b_4 = \dfrac{1}{s-s^4}\left[a_4 + \dfrac{a_1}{s-s^2}\left(a_2^2 + 2sa_3\right) + \dfrac{3a_2 s^2}{s-s^3}\left(a_3 + \dfrac{2sa_2^2}{s-s^2}\right)\right] , \\ \ldots\ldots \end{cases}$$

An alternate expression can be obtained by means of an intermediate small parameter expansion :

$$(1\text{-}37)\quad \begin{aligned} &f(x) = sx + \varepsilon g(x),\ s \neq 0, 1, \qquad 0 < \varepsilon \leq 1, \\ &w(x) = w_o(x) + \varepsilon w_1(x) + \varepsilon^2 w_2(x) + \ldots \quad , \end{aligned}$$

where, for simplicity, $\bar{x} = 0$ and the explicit dependence on c is omitted. Inserting (1-37) into (1-31) yields

$$w_o(f) + \varepsilon w_1(f) + \varepsilon^2 w_2(f) + \ldots = s(w_o(x) + \varepsilon w_1(x) + \varepsilon^2 w_2(x) + \ldots)$$

$$w_i(f) = w_i(sx) + \varepsilon g(x)\, w'_i(sx) + \frac{\varepsilon^2}{2!}\,[g(x)]^2 . w''_i(sx) + \ldots, w'_i(x) = \frac{d}{dx} w_i(x), \ldots$$

$$i = 0, 1, \ldots .$$

which is equivalent to the recursive system :

$$w_o(sx) - sw_o(x) = 0$$

$$w_1(sx) - sw_1(x) = h_1(x) = g(x)\, w'_o(sx)$$

(1-38)
$$w_2(sx) - sw_2(x) = h_2(x) = g(x)\, w'_1(sx) + \frac{1}{2!}\,[g(x)]^2 w''_o s(x)$$

$$\ldots\ldots\ldots\ldots\ldots\ldots\ldots$$

$$w_i(sx) - sw_i(x) = h_i(x) = \ldots$$

$$\ldots\ldots\ldots\ldots\ldots\ldots\ldots$$

where $w_o(x) = x$ and the $h_i(x)$, $i = 1, 2, \ldots$, are known functions. When an $h_i(x)$ admits a convergent Mc Laurin series $\sum_{m=0}^{\infty} a_m x^m$, then [G 8]

(1-39)
$$w_i(x) = \frac{a_o}{1-s} + a_1 \frac{x\log x}{s\log s} + \sum_{m=2}^{\infty} \frac{a_m}{s^m - s}\, x^m \quad .$$

In the special relationship between (1-32) and (1-33), it is implied that $a_1 = 0$ in (1-39) for each i, so that logarithmic terms do not appear as $\varepsilon \to 1$. The existence of a composite point singularity in (0-1) precludes the use of (1-37) for the determination of w(x) in the limit case $s \to 1$. A more general series development is required in that case. Let $f_n = f(f(\ldots(f(x)\ldots))$ be the n-th iterate of f(x), $n = 0, 1, 2$, with $f_o(x) = x$ and $f_1(x) = f(x)$. The particular solution sought is then given by the series (cf [G 8],[K 6])

(1-40)
$$w(x, c) = x + \sum_{n=1}^{\infty} s^{-n} f_n(x, c) \quad , \qquad s \to 1 .$$

Consider the recurrence (0-1) with an f admitting a composite stability domain $X_t$. An inspection of particular cases, like for example those shown in Fig. 1-7 suggests that the knowledge of existence of $X_t$ is only a first step towards the identification of the singularities existing inside it. $X_t$ is characterized by the property that for any initial point $x_o \in X_t$ it constitutes a region of no escape for the resulting consequent half-trajectory $\{x_n\}$, $n \geq 0$. This no-escape property gives rise to the following problem of mechanics [C 15] : if a fixed finite mass M is initially uniformly distributed inside $X_t$, i.e. if the initial density of M is $w_o(x) = M/d$, d = Euclidean length of $X_t$, and the motion of each point-particle $x_n$ of M is generated by (0-1), what happens to the mass density $w_m(x_n)$ after m iterations ? Does $w_m(x)$ converge in some mechanically meaningful sense to a limit function $w(x_n)$, $x_n \in X_t$ as $m \to \infty$ ?

Gedanken-experiments end invariably with positive conclusions, leaving only the particulars of the convergence to be identified. For each specific f in (0-1), this identification can be carried out by means of numerical experiments. For analytical f's without known stable point singularities, it was observed that the values of $w_m(x_n)$ oscillate considerably at almost all $x_n$, the oscillations following a regular but non-periodic pattern. It should be stressed that these oscillations are definitely not random or quasi random. The periodic evolution at the unstable cycle points $x_k \in X_t$, corresponding to unstable constant (k = 1) or periodic (k > 1) stationary states turns out to play a negligible role. For an analytic f the points $x_k$ are enumerable; collectively they are untypical. Several possibilities exist in principle for the characterization of the severity of oscillations at typical points, the simplest of which is the convergence of mean sums, i.e. the existence of the limit

$$(1\text{-}41) \qquad w(x_n) = w_\infty(x_n) = \lim_{m\to\infty} \frac{1}{m} \sum_{i=0}^{m-1} w_i(x_n), \qquad x_n \in X_t .$$

When the simple arithmetic-mean limit of the $w_i(x_n)$ exists, then $w(x_n)$ is independent of the initial mass distribution, provided the latter has a non-zero support. This property follows from the mechanically obvious (and essential) uniqueness of $w(x_n)$. Subject to (1-41), an equivalent definition of $w(x_n)$ consists in requiring that it be a particular solution of the Perron-Frobenius functional equation

$$(1\text{-}42) \qquad \frac{d}{dx}\int_{g(o)}^{g(x)} w(s)\,ds = w(x), \qquad \int_o^d w(s)\,ds = M,\ x \in X_t ,$$

where g is the (usually multivalued) inverse $f_{-1}$ of f. The subscript n in $x_n$ has been omitted. The integrals are to be understood in the sense of Lebesgue. If (1-41) is replaced by a less stringent mean limit, then (1-42) still holds in principle, except that the first integral has to be suitably modified to reflect the fact that w(s) is still a mechanical mass density. When $X_t$ contains only asymptotically stable point-singularities, then by definition w(x) is a sum

$$w(x) = \sum_i a_i\, \delta(x - \bar{x}_i), \qquad \sum_i a_i = M$$

of Dirac measures $\delta(x - \bar{x}_i)$, where $\bar{x}_i$, i = 1, 2, ... are the abscissae of the fixed points or cycles at which the fractions $a_i$ of the total mass M are finally concentrated. Orderly dynamics of (0-1) imply therefore singular solutions of the functional equation (1-42). On the contrary, when $X_t$ contains only invariant segments $X_k$, k = 1, 2, ..., then also by definition, w(x) is at least a piecewise continuous and piecewise differentiable function inside each $X_k$ ; outside of $\sum_k X_k$ and inside $X_t$ one has necessarily $w(x) \equiv 0$. The latter property follows from the stability of the $X_k$. Chaotic dynamics of (0-1) imply therefore rather regular solutions of (1-42).

Equation (1-42) has also been formulated in a much more abstract context (cf. [L 3]), called symbolic dynamics (for a recent review of this subject, see [M 8]). Although extremely fascinating in itself, symbolic dynamics provides only sufficient conditions for the validity of (1-42), i.e. less general conditions than ordinary mechanics. This decreased scope of validity is a consequence of the fact that necessary conditions depend on the concrete physical context of the dynamic problem, and cannot be supplied by purely mathematical considerations. Solutions of (1-42) play also a major role in the application of ergodic theory to the analysis of chaotic dynamics of discrete dynamic systems (see for example [L 7], [O 1]). The results obtained are substantial when the recurrence in question is piecewise linear (cf. [R 6], [C 9], [C 10], [P 1]), and rather limited otherwise. This is again due to the fact that when facing a specific recurrence of a physical, biological, etc. origin, very little specific information is supplied by an extremely general theory, in spite of considerable efforts expanded to show the contrary (for further details, see for example [F 2]). The main usefullness of a general theory consists in supplying a basic background against which a concrete problem can be situated, it does not supersede the study of this concrete problem and usually contributes very little to its solution.

Consider now the quadratic recurrence (1-21). Except for an enumerable set of values $c = c_k^*$ (cf. Table 4), corresponding to the existence of invariant segments of order k = 1, 2, ... of decreasing length as k increases, the dynamics of (1-21) are orderly, i.e. (1-21) for each $c \neq c_k^*$ equation (1-21) admits a single asymptotically stable fixed point or cycle whose influence domain is the total no-escape interval. The corresponding solution of (1-42) is composed entirely of Dirac measures. The support of these Dirac measures is given by the roots of the iterated recurrence (0-14). The fraction of the total mass found in the mean at each cycle point $\bar{x}_i$, i = 1, 2, ..., k, is proportional to the influence domain of the latter, as viewed via the iterated recurrence (1-24) at each $\bar{x}_i$. The contribution of (1-42) to the understanding of orderly dynamics of (1-21) is therefore somewhat marginal. This is however not so for the complex or chaotic dynamics, existing when c takes one of the $c_k^*$-values. The simplest possible case is $c_1^* = -2$ when (1-21) defines a single invariant segment of the type shown in Fig. (0-1a). Since the inverse to (1-21) admits only two distinct branches, equation (1-42) simplifies into the pure functional equation

(1-43) $$\frac{1}{2\sqrt{u}}\left[w(-\sqrt{u}) + w(\sqrt{u})\right] = \frac{1}{\sqrt{u}}\,w(\sqrt{u}) = w(x), \quad u = \frac{1}{2}\left(1 + \frac{1}{2}x\right),$$

which admits the relatively obvious general solution

(1-44) $$w(x) = A / \sqrt{1 - (x/2)^2}, \qquad A = \text{arbitrary constant.}$$

There exists an important relationship between the general solution of (1-42) and that of (1-21). Let the latter (cf. eq. (1-29)) be written in the form

(1-29a) $$x_n = g(y_n) = 2\cos y_n, \qquad y_n = 2^n \arccos(x_o/2) \quad ,$$

then

(1-45) $$w(x) = B\,\frac{d}{dx}\,g^{-1}(x), \qquad B = \text{arbitrary constant} \quad .$$

Equation (1-45) holds for any solution of (0-1), or (0-14), describing an invariant segment. As an example consider the piecewise linear recurrence (1-12), which defines an invariant segment of the type shown in Fig.(0-16), and for which (1-42) takes the form :

(1-46) $$\frac{1}{2}\left[w(-u) + w(u)\right] = w(u) = w(x),\ u = \frac{1}{2}(1+x) \ .$$

The only continuous solution of (1-46) is

(1-47) $$w(x) \equiv A = \text{arbitrary constant} \ .$$

The recurrence (1-12) admits a general solution which is periodic and linear inside one period :

(1-48)
$$x_n = 4\sum_{m=1}^{\infty} \frac{(-1)^{m+1}}{m} \sin(2\pi m y_n) = y_n \text{ for } -1 < y_n < +1 \ ,$$
$$y_n = C.2^n,\ C = \text{root of } x_o = 4\sum_{m=1}^{\infty} \frac{(-1)^{m+1}}{m} \sin 2\pi m C \ .$$

The relation (1-45) is clearly satisfied.

The recurrences (1-12) and (1-21) are both described by symmetrical functions with respect to the centre of their $x_n$-intervals of definition, and the corresponding solutions of (1-42) happen to conserve this symmetry. A conservation of symmetry is however a very exceptional property, as can be seen from the study of the symmetrical one parameter family of recurrences [G 15]

(1-49) $$x_{n+1} = 2^a |x_n|^b - 2,\ a = c/(1+c),\ b = (2+c)/(1+c),\ 0 \leq c < \infty \ ,$$

admitting for all c the same invariant segment X : $-2 < x_n < 2$. This family (Fig. 1-15) is bounded for c = 0 by the parabola (1-21) and for $c \to \infty$ by the "cut-hat" function (equivalent to (1-12))

(1-50) $$x_{n+1} = |x_n| - 2 \ .$$

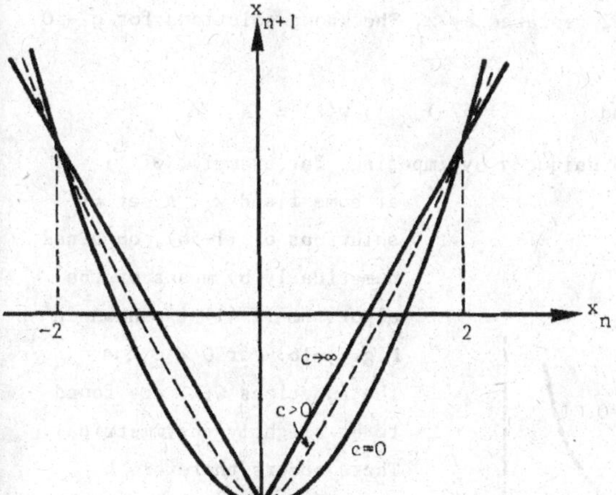


Fig. 1-15. A family of symmetric recurrences admitting the same invariant segment X : $-2 < x_n < 2$

The limiting recurrences (1-21) and (1-50) are both very special in the sense that their "microscopic" and "collective" solutions are known explicitly. The problem consists in determining the general solution of (1-49) and the corresponding solution of (1-42) when c > 0. The Ansatz

(1-51) $x_n = 2g(y_n)$,

(1-51a) $y_{n+1} = 2y_n$ ,

inspired by the symmetry of Fig. 1-15, converts (1-49) into the Picard functional equation :

(1-52) $$g(2y) = 2g^b(y) - 1 \ ,$$

where, for convenience, the subscript n has been omitted. Consider first the case

$0 \leqslant c \ll 1$, $b = 2 - c + ..$ . Since the positions of the point-singularities of (1-49) evolve smoothly near $c = 0$, a solution of (1-52) is at least asymptotically convergent in $c^m$, $m = 1, 2, ...$, i.e. it is permissible to write

$$g(y) = g_o(y) + c\, g_1(y) + ..., \qquad g_o(y) = \cos y \quad , \tag{1-53}$$

where the $g_m(y)$, $m > 0$ satisfy the recursive system

$$\begin{aligned} &g_1(2y) - 4g_o(y).g_1(y) = -2g_o^{\,2}(y).\log g_o(y) \\ &g_2(2y) - 4\, g_o(y).g_2(y) = ... \\ &..............................\,. \end{aligned} \tag{1-54}$$

It can be verified that a solution of the first equation in (1-54) is

$$\begin{aligned} &g_1(y) = y \sin y\, (A_1 + \sum_{n=0}^{\infty} \bar{f}(2^n y)) \quad , \\ &\bar{f}(z) = 2g_o^{\,2}(z)\, [\log g_o(z)]/(2z \sin 2z), \qquad z = 2^n y \quad , \end{aligned} \tag{1-55}$$

where $A_1$ is an arbitrary constant. The solutions are similar for $g_m(y)$, $m > 1$, except that the explicit expressions are more cumbersome to write out. The constants $A_m$ are defined without ambiguity by the initial condition $x_o = 2g(y_o)$.

The mass-density equation, i.e. the reduced form of (1-42) associated with (1-49) is

$$\frac{1}{2}\, u^{\alpha-1}\left[w(-u^\alpha) + w(u^\alpha)\right] = w(x), \qquad u = \frac{1}{2}\,(1+x) \quad , \tag{1-56}$$

where $\alpha = \frac{1}{2} + \frac{1}{4}\, c + ...$ near $c = 0$ and $\alpha = 1 - \frac{1}{c} + ...$ near $c \to \infty$. For simplicity the subscript n has been dropped and x/2 replaced by x. The known solutions for $c = 0$ and $c \to \infty$ are

$$w(x) = A/\sqrt{1-x^2} \quad \text{(1-57)} \qquad \text{and} \qquad w(x) \equiv A \quad , \quad \text{(1-57a)}$$

respectively. The constants A are fixed uniquely by imposing, for example, $w(x_o) = 1$ at some fixed $x_o$. A set of solutions of (1-56), obtained numerically by means of the algorithm (1-41) is shown in Fig. 1-16. For $0 < c < \infty$ the functions w(x) are found to be slightly unsymmetrical. There occurs therefore a deterministic parity violation. This intuitively unexpected property is easily confirmed by letting $r(x) = r(-x)$, $s(x) = -s(-x)$, $w(x) = r(x)+s(x)$. After insertion into (1-56), it is <u>not</u>

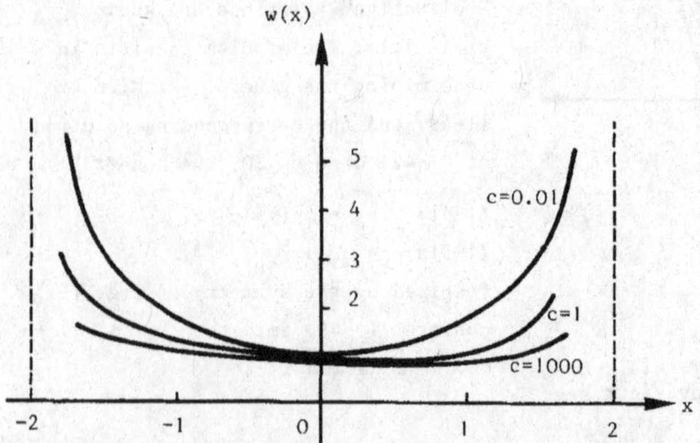


Fig. 1-16 : Numerically determined solutions of the mass-density equation (1-56).

possible to conclude that $s(x) \equiv 0$ ! The asymmetry of $w(x)$ leads necessarily to complications in a small-parameter expansion, because a qualitative difference between the generating and full solutions manifests itself via secular terms. Let in fact

$$(1\text{-}58) \qquad w(x) = w_o(x) + c\, w_1(x) + \dots \qquad w_o(x) = A_o / \sqrt{1-x^2} \quad ,$$

then for $0 \leqslant c \ll 1$

$$(1\text{-}59) \qquad \begin{aligned} &(2u)^{-1}.w_o(u) - w_o(x) = 0, \qquad u = \sqrt{\tfrac{1}{2}(1+x)} \quad , \\ &(2u)^{-1}.w_1(u) - w_1(x) = \tfrac{1}{2}\, w_o(x) + \tfrac{1}{2}\left[w_o(x) + A_o(1-x^2)^{-5/2}\right].\log u, \\ &\dots\dots \end{aligned}$$

The term $\frac{1}{2}\, w_o(x)$ in the second equation is clearly secular, because $w_o(x)$ is a solution of the homogeneous equation. Usual iteration methods (cf. [K 6]) of solving functional equations are therefore inadequate for (1-59), but there exists an at least asymptotically convergent series solution

$$(1\text{-}60) \qquad w_1(x) = w_o(x)\, [A_1 + v(x)] \quad , \qquad v(x) = \sum_{n=1}^{\infty} a_n\, x^n \; .$$

The constants $a_n$ are unknown explicitly and can only be determined numerically. It is however easy to see that the existence of (1-60) implies $a_1 \neq 0$, and thus $w(x) \neq w(-x)$. For $0 < \frac{1}{c} \ll 1$ the equations analoguous to (1-59) are

$$(1\text{-}61) \qquad \begin{aligned} &w_o\left(\tfrac{1+x}{2}\right) - w_o(x) = 0, \qquad w_o(x) = A \quad , \\ &w_1\left(\tfrac{1+x}{2}\right) - w_1(x) = A_o\left(1 + \log \tfrac{1+x}{2}\right) \quad , \\ &\dots\dots \end{aligned}$$

Since the second equation contains also a secular term, preventing the applicability of usual iterative methods, an alternate solution has to be sought. Such a solution exists also in the form of a series, which happens to contain both odd and even powers of x, and thus implies $w_1(x) \neq w_1(-x)$. The solutions of (1-56) are therefore unsymmetrical for all $0 < c < \infty$, in spite of the symmetry for $c = 0$ and $c \to \infty$.

The asymmetry of $w(x)$ arising in an otherwise symmetric problem, i.e. $w(x) \neq w(-x)$ when $f(x_n, c) = f(-x_n, c)$ in (0-1), limits severely the validity of the frequently used conjugate equivalence transformations. Two transformations $f : I \to I$ and $g : J \to J$ on x-intervals I and J, are said to be conjugate if there exists a single-valued uniquely inversible transformation $u : I \overset{\text{onto}}{\to} J$ such that $g(x) = u(f(u^{-1}(x)))$ [U 1], [H 2]). The temptation is great to use such transformations on two recurrences and on two associated mass densities by assuming a priori that the conjugate function $u(x)$ is sufficiently smooth (cf. [G 6], [R 6]). Unfortunately such a smoothness assumption is generally not justified because (1-42) contains a Lebesgue integral and in Lebesgue's integration theory u possesses three distinct parts

$$(1\text{-}62) \qquad u(x) = u_r(x) + u_{ws}(x) + u_{ss}(x) \quad ,$$

where the subscript stand for regular, weakly singular and strongly singular, respectively. Without advance knowledge it is not possible to assume that $u_{ws}(x) \equiv 0$

and $u_{ss}(x) \equiv 0$, because these singular functions are responsible for the non-conservation of symmetry between recurrences and mass-densities. An even more elementary example [G 3] than (1-49) and (1-58) is given by a symmetrical perturbation of the recurrence (1-12) at the ordinate $x_{n+1} = \frac{1}{2}$, so that the straight-line segments $0 < x_{n+1} < \frac{1}{2}$, $\frac{1}{2} < x_{n+1} < 1$ have the slopes $a > 1$ and $b > 1$ respectively (Fig. 1-17a). The corresponding mass densities are shown in Fig. 1-17b, and one of them is fundamentally unsymmetrical. Another example with a less pronounced asymmetry is given by the

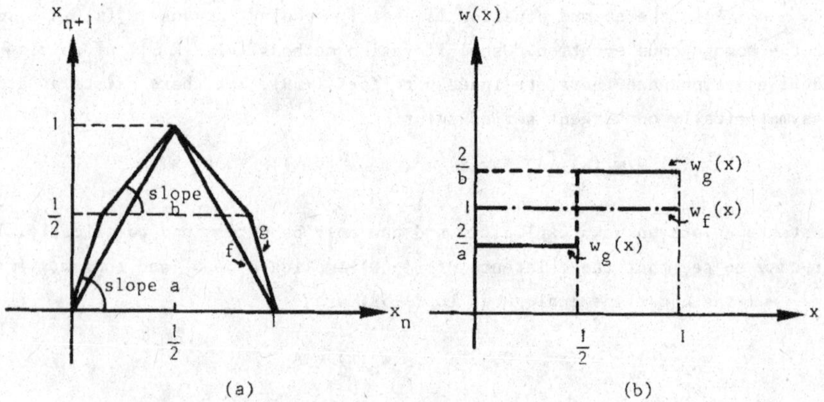


Fig. 1-17. Symmetrical cut-hat recurrences and the corresponding mass densities. Symmetry is not conserved in w(x).

quadratic recurrence (1-21) $c = c_2^* = -1.5437$ for which there exist two contiguous cyclic invariant segments (cf. Fig. 1-11). The corresponding numerically determined mass density is shown in Fig. 1-18 [C 12]. It possesses two segments $w(x) \equiv 0$ inside the influence domain $X_i$. Moreover, because of asymmetry, the non-zero part of w(x) disagrees with the shape obtained by means of smooth conjugate transformations. An analytical expression of w(x) is not yet available.

Mass densities are known for several other invariant segments of the quadratic recurrence (1-21), and in particular for [C 12]

$$c_3^* \equiv -1.7903, \qquad c_4^* \equiv -1.9428\ ,$$

$$c_5^* \equiv -1.6334, \qquad c_m^* \equiv -1.4304\ , \quad (m = 2^2)\ ,$$

and $c_m^* \equiv -1.4074$ $(m = 2^3)$. All these mass densities are unsymmetrical and possess segments $w(x) \equiv 0$ inside the influence domain. The non-zero parts of w(x) consist of a set of roughly parabolic sloped curves, similar to those shown in Fig. 1-18.

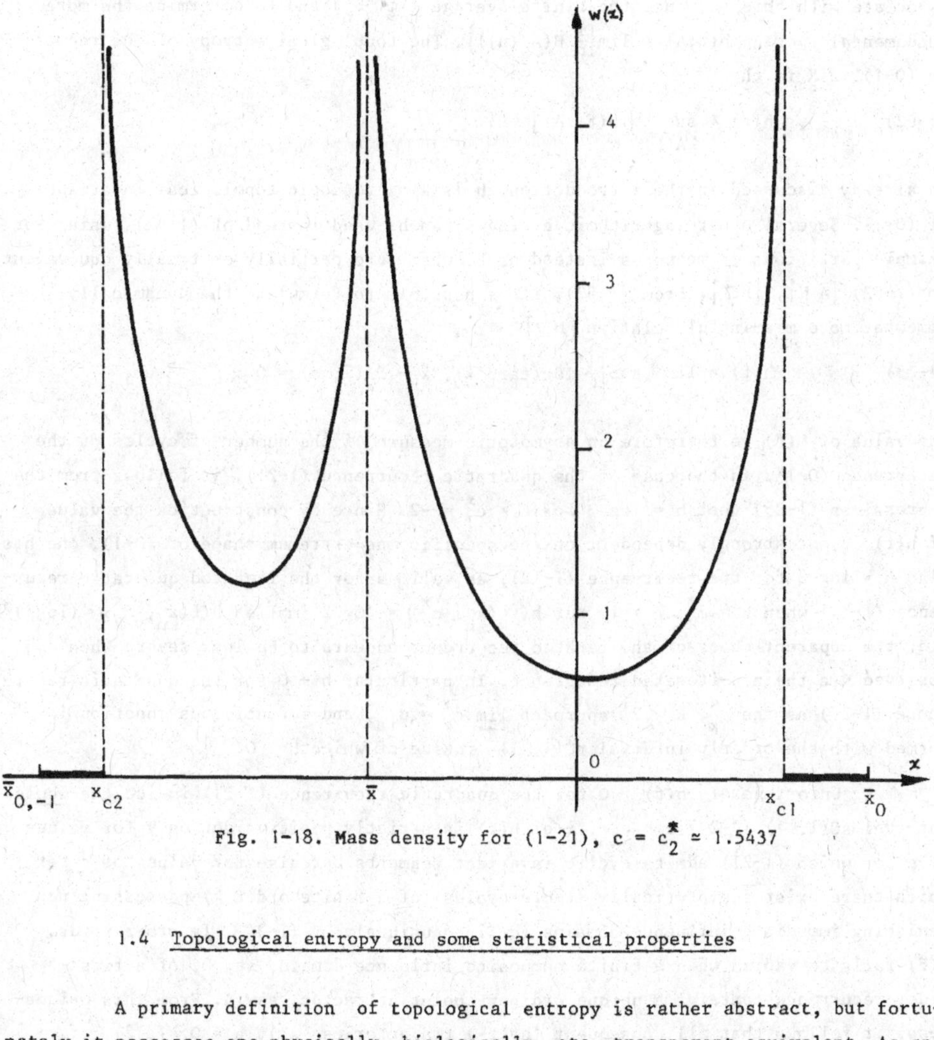


Fig. 1-18. Mass density for (1-21), $c = c_2^* \simeq 1.5437$

## 1.4 Topological entropy and some statistical properties

A primary definition of topological entropy is rather abstract, but fortunately it possesses one physically, biologically, etc. transparent equivalent. As an illustration, consider the recurrence (0-1) with $x_n$ restricted to a closed interval X. Let A(X) be the family of finite covers of X by open intervals and N(A) the smallest cardinal of a subcover from $A \in A(X)$. The positive number $H(A) \equiv \log N(A)$ is called an entropy of A. Let $A_k$, k = 0, 1, ..., n-1, be a finite family of elements of A(X). It is possible to construct a cover of X composed of n parts of X, each chosen inside a different $A_k$. This cover, ordered with respect to intersection, is denoted by the special symbol $\bigvee_{k=0}^{n-1} A_k$, [0 1]. Subject to f in (0-1) it is now possible to construct a set of iterated antecedent covers $f_{-1}(A)$, ..., $f_{-n}(A)$ as well as the intersection-ordered family of covers $A_a(n) = \bigvee_{k=0}^{n-1} f_{-k}(A)$, n = 1, 2, ..., which are also elements of A(X). The sequence of positive numbers $F_n = H(A_a(n))$ possesses the property of sub-additivity, i.e. $F_{n+m} \leqslant F_n + F_m$ for all integers n, m. Hence, it is possible to associate with this sequence the finite average $\lim_{n\to\infty}(\frac{1}{n}F_n)$ and to determine the more fundamental average $h(f,A) = \lim\left[\frac{1}{n} H(A_a(n))\right]$. The topological entropy of the recurrence (0-1) in X is then

(1-62) $$h(f) \equiv \sup_{A \in A(X)} \left[ h(f, A) \right] .$$

As already discussed in the introduction, h is a macroscopic topological invariant of (0-1). Several other logarithmic averages can be used instead of (1-62), using for example partitions or measures instead of X, which are partially or totally equivalent to (1-62) [A 1], [B 7]. From (1-62), it is possible to formulate the dynamically somewhat more meaningful relation [M 7]

(1-63) $$h(f) = \lim_{k\to\infty} \sup\left[\frac{1}{k} \log(\mathrm{card}\ \bar{x})\right],\ \bar{x} - f_k(\bar{x}, c) = 0 .$$

The value of h(f) is therefore an asymptotic measure of the number of cycles of the recurrence (0-1). In the case of the quadratic recurrence (1-21), it follows from the expressions (1-25) that $h = \log 2$ for $c = c_1^* = -2$. Since by construction the value of h(f) is not strongly dependent on the specific one-extremum shape of (0-1), one has also $h = \log 2$ for the recurrence (1-12), as well as for the iterated quadratic recurrence (1-24) when $c = c_k^*$, $k > 1$. But $h(f_k(x_n, c_k^*)) = \log 2$ implies $h(f(x_n, c_k^*)) = (\log 2)/k$, i.e. the apparent chaos of an iterated recurrence appears to be less severe when observed via the non-iterated recurrence. In particular $h \to 0$ for the quadratic recurrence (1-21) as the $c_k^*$, $k = 2^i$ approach $\lim_{i\to\infty} c_k^* = c_{1a}$, and a continuous junction is formed with the orderly interval rC(1, 1) inside of which $h \equiv 0$.

Unfortunately $h(f) > 0$ for the quadratic recurrence (1-21) inside the <u>whole</u> interval sC(1, 1) : $-2 \leqslant c < c_{1a}$, i.e. h(f) is strictly positive not only for values of c for which (1-21) admits cyclic invariant segments but also for values of c for which there exist asymptotically stable cycles (of a finite order k) possessing non vanishing immediate influence domains (defined uniquely by (1-24)). In other words, h(f) fails to vanish when a finite composite influence domain, say D, of a first-order recurrence contains a unique ordinary point attractor, say A. From this uniqueness, it follows that all consequent (half-) trajectories $\{x_n\}$, $n = 0, 1, 2, \ldots$, $x_0 \in D$ have no accumulation points besides those of A. In dynamical terms, this means that after a more or less prolonged transient a point $x_0$ of every consequent trajectory enters the (non-vanishing) immediate influence domain of A and then converges in an <u>orderly</u> fashion to the corresponding cycle points (the points $x_{n+km}$, $n \geqslant N \gg 1$, $m = 1, 2, \ldots$, N = fixed, converge to a fixed point of the iterated recurrence (1-24)). The topological entropy h(f) is therefore a dynamic qualifier of a transient to an attractor, and not of the attractor itself. Hence, the knowledge of the value of h(f) conveys vety little dynamic information. In most physical, biological, etc. situations, little, if anything, is gained by knowing the values of h(f), except when h(f) happens to vanish identically. In the latter case it is possible to conclude that the dynamics are orderly, in spite of possible appearances to the contrary.

A somewhat operationally more efficient logarithmic average is the variational entropy [C 11] $h_v(f)$ defined by

$$(1\text{-}64)\qquad \exp(h_v) = \lim_{n\to\infty} (|\lambda_2|.|\lambda_2|\ldots|\lambda_n|)^{1/n} \quad ,$$

where the $\lambda_i$, $i = 1, 2, \ldots, n$ are successive pseudo-eigenvalues $\lambda_i = f'(x_i, c)$, $' = \frac{\partial}{\partial x_i}$, of a non-degenerate (half-) trajectory of consequents $\{x_i\}$ of the recurrence (O-1). Non-degenerary means that $\{x_i\}$ neither coincides with a cycle nor attains a cycle after a finite number of iterations. For numerical computations a more convenient form of (1-64) is

$$(1\text{-}65)\qquad h_v(f) = \lim_{n\to\infty} h_n(f),\ h_{n+1} = (\log|\lambda_{n+1}| + nh_n)/(n+1),\ n = 1, 2, \ldots,\quad h_o = \log|\lambda_o| .$$

When $\{x_i\}$ approaches an attractive cycle, then $\exp(h_v)$ converges to the eigenvalue of the latter. $h_v$ is therefore non-positive in a dynamically orderly situation. In principle it can be also used to discover the existence of asymptotically stable cycles, but practically the computations based on (1-65) are not very efficient because when the order of the cycle is unknown, it is difficult to ascertain when the points $x_i$ of $\{x_i\}$ have reached the immediate influence domain of a cycle point. Moreover, the accuracy of the numerical computations is very difficult to control when the number of iterations increases inordinately (the maximum allowable number of iterations depends mainly on the number of digits carried and on the sensitivity coefficients of the algorithm used).

In order to determine the disorderliness of specific recurrences, a certain number of its consequent half-trajectories $\{x_i\}$, $i = 1, 2, \ldots$ were subjected to conventional pseudo-randomness tests (cf. [K 3], [K 9]). It turned out that compared to, say Lehmer's recurrence (O-16), the number sequences $\{x_i\}$, $i = 1, 2, \ldots$ generated by (1-21) are quite regular [G 7]. For example, when $-1.791 \leqslant c \leqslant -1.780$ in (1-21), strong regularities are disclosed by the $\chi^2$ test, and strong peaks are present in the spectra of the discrete autocorrelation functions $C_i$ of $\{x_i\}$, for $1000 \leqslant \max i \leqslant 50000$. The main periodicities of the $C_i$ are $k = 2$ and $k = 3$, eventhough the cycles of the same order are all unstable. A similar behaviour is observed when $-2 \lessapprox c \leqslant -1.990$. The existence of accumulations of cycles in a recurrence is therefore not a strong source of stochasticity, at least in the sense of (practical pseudo-) randomness. A plausible criterion of (pseudo-) randomness in a deterministic dynamic system like (O-1) is the amount of loss of correlation in the iterates $x_n$ with increasing n, as expressed by a sequence of autocorrelation coefficients of increasing order. In the case of finite sequence $\{x_i\}$ generated by (1-21) the correlations between the $x_{n+m}$, $m = 1, 2, \ldots$ decrease first with decreasing values of c, but a relatively constant saturation level is reached soon. Further decreases of c lead only to minor fluctuations around this level. The lack of pronounced higher order periodicities in the spectra of the autocorrelation coefficients can be explained as follows : the rather strong periodicity k = 2 has a double source. On one hand, there is a memory effect of the cycle of order k = 2, due to the analyticity of $f(x_n, c)$ with respect to c, and on the other hand, all cycle families have the structure $k = m.2^i$. In other words, during bifurcation each cycle point $\bar{x}_i$ splits into two points $\bar{x}_{i1}$, $\bar{x}_{i2}$ which remain close to $\bar{x}_i$. Due to the analyticity of $f(x_n, c)$ with respect to $x_n$, the positions of the bifurcated points change more slowly than the associated eigenvalues, responsible for the bifurcations. Higher order families of cycles ($m > 3$ in $k = m.2^i$, i = 0, 1, ...) cluster near the unstable cycles of order k = 2 and k = 3, and their periodicities have thus only a weak collective (or macroscopic) effect.

The statistical properties of the quadratic recurrence (1-21), $-2 \leq c \leq -1.99$ are not markedly different from those of the piecewise linear recurrence (1-12), or the exponential one (1-2). A quantitative illustration of the results of the $\chi^2$ test is shown in Table 5, with c = -1.99 in (1-21) and c = 2.925 in (1-2). The lack of (pseudo-) randomness is quite apparent and rather similar. When c is closer to -2 in (1-21), only the last digit in the $\chi^2$ and $\sigma$ columns are slightly affected. The lack of sensitivity of the statistical tests to the precise value of c suggests that the more chaotic invariant segments are "macroscopically stable", with respect to a finite number of iterations, i.e. their presence manifests itself on the collective level not only for $c = c_k^*$ but also (via prolonged transients) in a certain interval around $c = c_k^*$. The situation is similar to that of non linear resonances, which also exist strictly only for an enumarable set of frequency ratios. The analogy consists in the fact that the synchronization and phase stability of the latter are replaced by a sort of transient inertia of the former, reflecting essentially the smoothness of $f(x_n, c)$ in (0-1).

| Recurrence (1-21), c=-1.99 (a) | | | Recurrence (1-12) (b) | | | Recurrence (1-2), c=2.925 (c) | | |
|---|---|---|---|---|---|---|---|---|
| k | $\chi^2/2500$ | $\sigma/100$ | k | $\chi^2/2500$ | $\sigma/100$ | k | $\chi^2/2500$ | $\sigma/100$ |
| 2 | 248 | 87.4 | 2 | 208 | 73 | 2 | 407 | 134 |
| 5 | 59.2 | 20.6 | 5 | 30 | 10 | 5 | 106 | 37.1 |
| 10 | 19.4 | 6.49 | 10 | 2.4 | 0.48 | 10 | 43.8 | 15.1 |
| 20 | 17.8 | 5.92 | 20 | 2.2 | 0.42 | 20 | 55.7 | 19.3 |
| 50 | 17.8 | 5.94 | 50 | 2.2 | 0.43 | | | |

Table 5, $x_{n+k}$ vs $x_n$. $\chi^2$ and standard deviation on a relative scale $|\Delta x_n/x_n| = 0.02$. Two $x_n$ are considered to be the same if they differ by less than 2 %. A uniformly distributed sequence $\{x_n\}$ would yield unity in the $\chi^2$ and $\sigma$ columns.

# CHAPTER II

# SOME PROPERTIES OF SECOND ORDER RECURRENCES

## 2.0 Introduction

Direct descriptions of evolution processes by means of second order recurrences arise more frequently in physics than those by first order recurrences. An elementary example is the study of charged particles constrained to move inside toroïdal surfaces, like those existing in contemporary accelerators and storage rings. When the transverse and longitudinal motions are not strongly coupled, then the non-dimensional equations of motion can be brought to the standard form

(2-1) $$x_{n+1} = y_n + F(x_n), \quad y_{n+1} = -x_n + F(x_{n+1}), \quad F(0) = 0, \quad F(1) = 1 \quad ,$$

where F is a known smooth function. For the longitudinal motion in an alternating gradient accelerator [G 12]

(2-2) $$F(x) = x - \frac{1-\mu}{\cos\varphi_s}\sin(bx+\varphi_s) - \sin\varphi_s \; , \; b = \pi - \varphi_s, \; 0 \leqslant \varphi_s < \frac{\pi}{2} \; ,$$
$$-1 < \mu < 1 \; ,$$

where $\varphi_s$ is the so-called equilibrium phase and $\mu$ a parameter depending on the type of accelerated particle, the amplitude of the sinusoïdal accelerating voltage and on the focussing properties of the ring. In a microtron, only one of the parameters is free, because of the oparating constraint $\varphi_s = \pi/2 - \text{arc tg}\,\frac{1-\mu}{\pi}$. For transverse motion in one plane in the presence of thin sextupoles or octupoles [G 11] the function F(x) in (2-2) is replaced by

(2-3) $$F(x) = \mu x + (1-\mu)x^2 \quad \text{and (2-4)} \quad F(x) = \mu x + (1-\mu)x^3 \quad ,$$

respectively. The recurrence (2-1) has the special property that the Jacobian determinant of its right-hand sides is identically equal to unity. For this reason it is called conservative, or sometimes Hamiltonian, to reflect the fact that in an equivalent description by means of differential equations, the latter would derive from a Hamiltonian function.

A second example from physics is the analysis of open resonators by means of laws of geometrical optics. In the case of a two-dimensional resonator formed by a plane and a curved quartic mirror one encounters the non-conservative recurrence [M 3]

(2-4) $$u_{n+1} = u_n - 2\,v_n + P_3(u_n, v_n), \quad v_{n+1} = 2\,s\,u_n + (1-4s)v_n + Q_3(u_n,v_n),$$
$$P_3(u, v) = \sum_{m=0}^{3} a_m\, u^m\, v^{3-m}, \quad Q_3(u, v) = \sum_{m=0}^{3} b_m\, u^m . v^{3-m} \quad ,$$

where $a_m$, $b_m$ and s are constants, and $u_n$, $v_n$ represent the tangent of the reflection angle and the normalized distance between the mirrors, respectively.

In population dynamics an extension of the Nickolson-Bailey host-parasitoid relation gives rise to the non-conservative recurrence (see [B2-3] and references therein)

(2-5) $$x_{n+1} = x_n . f_1(x_n) . f_2(x_n, y_n), \quad y_{n+1} = y_n . f_3(x_n, y_n) \quad ,$$

where $x_n$ and $y_n$ are host and parasitoid densities. The $f_1(x_n)$ describes the per capita rate of increase of the host as a function of its own density, $f_2(x_n, y_n)$ the proportional survival of $x_n$ hosts confronted by $y_n$ parasitoids, and $f_3(x_n, y_n)$ the per capita rate of increase of the parasitoid as a function of its own, and its hosts, densities. A basic form of (2-5), when both host and parasitoid have discrete synchronised generations is [B2]

(2-6) $$\begin{aligned} x_{n+1} &= x_n \exp\left[c(1 - x_n/K)\right] . \exp(-a\, y_n) \\ y_{n+1} &= b\, x_n \left[1 - \exp(-a\, y_n)\right] \quad , \end{aligned}$$

where a, b, c, and K are positive constants. c represents the intrinsic growth rate of the host and K the carrying capacity of the environment, a and b being coupling constants.

A recurrence from economics describes the Samuelson accelerator-multiplier model [S2] for the national income $x_n$ :

(2-7) $$x_n = k\, x_{n-1} + c(x_{n-1} - x_{n-2}) + A_n + y_n \quad ,$$

where the first term represents the consumption, the second acceleration investment, the third autonomous investment and the fourth net government outlay. When the government tries to follow a contracyclical policy then $y_n$ is made to depend on $x_n$ or on its antecedents. The simplest four cases are

(2-8a) $y_n = a\,(x_{n-1} - x_{n-2})$ = compensation of income trends ,
(2-8b) $y_n = b\,(E - x_n)$ = adjustment to a specified value E via present income,
(2-8c) $y_n = b(E - x_{n-1})$
(2-8d) $y_n = b(E - x_{n-2})$ } adjustment to a specified value E via past income .

E represents usually a desired income level (corresponding, for instance, to full employment), a is a real constant and b a positive one. An implicit assumption behind (2-7) and (2-8) is that the state of the economy is very close to a state of stable equilibrium, so that non linear dependences do not need to be taken into account. This assumption is very convenient, but hardly realistic, at least in the contemporary economy.

The use of indirect descriptions of evolution processes by means of second order autonomous recurrences is motivated by the following two properties :
a) compared to smooth continuous formulations (for example, in terms of differential

equations, with continuous coefficients), the number of variables is reduced by at least one unit, and

b) for piecewise smooth formulations (for example, in terms of differential equations with piecewise continuous coefficients) there is a gain of smoothness. As an illustration of the first property, consider the rather well known case of a linear differential equation with smooth periodic coefficients :

(2-9) $$\ddot{x}(t) + q(t)\,\dot{x}(t) + p(t)\,x(t) = 0 \quad ,$$

where p, q, are at least continuously differentiable and $p(t+T) = p(t)$, $q(t+T) = q(t)$, $T > 0$. Let $x_1(t)$ and $x_o(t)$ be two solutions of (2-9) verifying the initial conditions

(2-10) $$x_1(0) = 1,\ \dot{x}_1(0) = 0, \qquad x_2(0) = 0,\ \dot{x}_2(0) = 1 \ .$$

Because of linearity, the following relations hold

(2-11) $$x_1(t+T) = a_1x_1(t) + a_2x_2(t), \qquad x_2(t+T) = b_1x_1(t) + b_2x_2(t) \ ,$$

where, because of (2-10), the coefficients a, b, are given by

$$a_1 = x_1(T),\ a_2 = \dot{x}_1(T), \qquad b_1 = x_2(T),\ b_2 = \dot{x}_2(T) \ .$$

Except for the non standard notation, (2-11) is an autonomous recurrence of order two, whose characteristic equation at the fixed point ($x = 0$, $x = 0$) is

(2-12) $$\begin{vmatrix} a_1 - \lambda & b_1 \\ a_2 & b_2 - \lambda \end{vmatrix} = \lambda^2 - A\lambda + B = 0 \quad ,$$

where $A = x_1(T) + \dot{x}_2(T)$, $\quad B = x_1(T)\,\dot{x}_2(T) - x_2(T)\,\dot{x}_1(T)$ .

The parallelism between (2-9) and (2-11) is in principle easily extendable to non linear differential equations, except that the functions f, g, in the corresponding recurrence $x_{n+1} = g(x_n, y_n)$, $y_{n+1} = f(x_n, y_n)$ are more difficult to express explicitly. In general, f and g are non-classical franscendental functions and a recourse to series expansions or to numerical computations becomes unavoidable. Two apparently inoccuous illustrative examples of this kind are

(2-13) $$\ddot{x}(t) + 2a\,\dot{x}(t) + \omega_o^2(1-h\cos\omega t)x(t) + cx^2(t) = 0$$

and

(2-14) $$\ddot{x}(t) + 2a\,\dot{x}(t) + \omega_o^2\,x(t) + c\,x^3(t) = B\cos\omega t \quad ,$$

where a, h, c, B, are real constants and $\omega_o$, $\omega$ positive ones (all different from zero). Equation (2-13) possesses an unusual (parametric) subharmonic resonance [G 14] and (2-14) a large numbers of (external) harmonic and subharmonic ones [H 5],[H 6].

An illustration of the second property is furnished by the evolution equations of piecewise linear control systems with hysterisis (see for example chapter 8 of [A 5 ]). One of the more elementary cases is

(2-15) $$\ddot{x}(t) + \dot{x}(t) = -u(z),\ z(x) = x(t) + b\dot{x}(t),\ u(z) = -u(-z),\ b \neq 0,$$

where the dependence between u and z for $u > 0$, $z > 0$ is shown in Fig. 2-1.

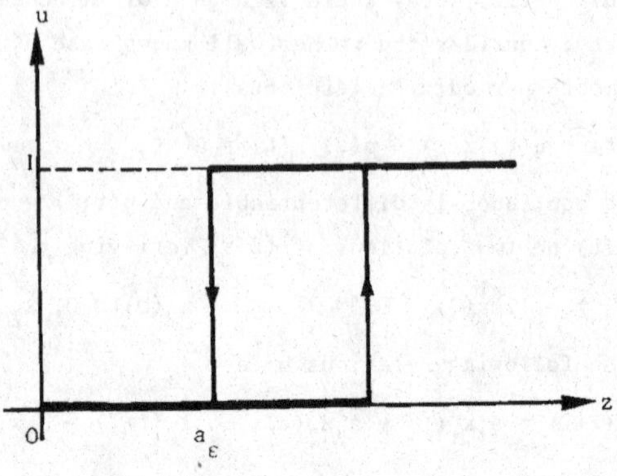


Fig. 2-1. Hysteresis loop of a piecewise linear control system. $0 < a < 1$, $\varepsilon > 0$ .

The phase space of the continuous dynamic system (2-15) consists of three partially overlapping surfaces which are each single-valued in the coordinates x, $\dot{x} = y$ and z. In the formulation of (2-15), it is specified that all trajectories x(t), y(t), z(t) are continuous at the edges of these surfaces. Any one of these edges can thus be chosen as a "surface of section without contact" to the trajectories. On every surface the trajectories are uniquely defined, and can even be expressed explicitly, because (2-15) is a linear equation with constant coefficients. The determination of the second order recurrence equivalent to (2-15) reduces therefore to the continuity-matching of three linear component recurrences, followed by the elimination of z. The final result is cumbersome to write out, but quite simple in structure (see pp. 604-621, [A 5] for the various possible forms depending on certain inequalities between the parameters a, b and ε). One unexpected feature of the phase portrait of (2-15) is the coexistence of an asymptotically stable limit cycle with an asymptotically stable (finite) segment of fixed points. This occurs when

$$\frac{1+a}{1-b} < \frac{1}{\varepsilon} \left(1 - \frac{e^{-b}}{1-b} \; e^{-2\varepsilon}\right) \quad .$$

The boundary of the respective influence domains is not an unstable limit cycle, but a pair of continuous piecewise smooth spirals. The apparent violation of "structural continuity" in phase space (absence of an unstable limit cycle) is a consequence of the partial overlapping of component phase surfaces describing the presence of hysterisis.

An illustration of a reduction of the number of variables and of a simultaneous gain of smoothness is given by the differential equation with piecewise constant periodic coefficients

$$\text{(2-16)}\qquad \begin{aligned} &\ddot{x}(t) + 2a\,\dot{x}(t) + \left[b + c\,u(t)\right] x(t) = d \quad , \\ &u(t+2\pi) = u(t), \quad u(t) \equiv 1, \quad 0 < t < \pi; \quad \equiv 0, \quad \pi < t < 2\pi , \end{aligned}$$

where $x(t)$ and $\dot{x}(t)$ are continuous for all $t > 0$, and $a$, $b$, $c$, $d$ are positive constants. Equation (2-16) arises in the study of gear vibrations. Assume for simplicity that $a^2 > b$ and $a^2 > b+c$, then the solution of (2-16), written in terms of two smooth components, is :

$$\text{(2-17)}\qquad x(t) = \begin{cases} x_1(t) = A\,d^{s_1 t} + B\,e^{s_2 t} + d/(b+c), & s_{1,2} = -a\pm\sqrt{a^2-b-c},\ 0 \leqslant t \leqslant \pi, \\ x_2(t) = C\,e^{s_3 t} + D\,e^{s_4 t} + d/b, & s_{3,4} = -a\pm\sqrt{a^2-b}, \quad \pi \leqslant t \leqslant 2\pi , \end{cases}$$

where A, B, C, D, are integration constants. Two of these constants are fixed by the continuity conditions $x_1(\pi) = x_2(\pi)$, $\dot{x}_1(\pi) = \dot{x}_2(\pi)$, and the other two by the initial conditions $x_1(0) = x_0$ , $\dot{x}_1(0) = \dot{x}_0$. Letting $t_1 = 2\pi$, $x(t_1) = x_1$, $\dot{x}(t_1) = \dot{x}_1$, the recurrence equivalent to (2-16) becomes

$$\text{(2-18)}\qquad x_1 = g(x_0, \dot{x}_0), \qquad y_1 = f(x_0, \dot{x}_0) \quad ,$$

where $g(x_0, \dot{x}_0) = x(2\pi)$ and $f(x_0, \dot{x}_0) = \dot{x}(2\pi)$. The gears described by (2-16) will run silently under the loading d if the corresponding solution of (2-16) is periodic, or equivalently, if the recurrence (2-18) admits a fixed point $x_1 = x_0$, $\dot{x}_1 = \dot{x}_0$. The existence of such a fixed point is equivalent to the existence of a real root A, B, C, D, of the four linear equations :

$$\begin{aligned} &A + B + d/(b+c) = C\,e^{2\pi s_3} + D\,e^{2\pi s_4} + d/b \\ &A\,e^{s_1\pi} + B\,e^{s_2\pi} + d/(b+c) = C\,e^{s_3\pi} + D\,e^{s_4\pi} + d/b \\ &s_1 A + s_2 B = s_3 C\,e^{2\pi s_3} + s_4 D\,e^{2\pi s_4} \\ &s_1 A\,e^{s_1\pi} + s_2 B\,e^{s_2\pi} = s_3 C\,e^{s_3\pi} + s_4 D\,e^{s_4\pi} \quad . \end{aligned}$$

## 2.1 Elementary properties of linear recurrences

The simplest autonomous recurrence of second order is a linear one

$$\text{(2-19)}\qquad x_{n+1} = a\,x_n + b\,y_n, \quad y_{n+1} = c\,x_n + d\,y_n, \quad n = 0, \pm1, \pm2, \ldots ,$$

where a, b, c, d are real constants. This recurrence is priviledged in the sense that it admits an explicit general solution (see eq. (0-23)), expressible in terms of elementary functions. It should be stressed that (2-19) is not to be confused with a nominally identical finite difference equation, because otherwise the notion "general solution" involves some semantic misunderstandings (cf. Introduction). The availability of an explicit general solution permits to answer all questions pertaining to problems like stability, and existence or uniqueness of invariant curves, in a deductive manner, without any need of intermidiate simplifying assumptions.

Let $\Delta_o = ad - bc$ designate the Jacobian determinant of the right-hand sides of (2-19). The recurrence (2-19) admits a single fixed point $x_n = y_n = 0$ in the finite part of the phase plane, which is characterized by its eigenvalues $\lambda_{1,2}$. The $\lambda$ are real or complex roots

(2-20) $$\lambda_{1,2} = \frac{1}{2}(a+d) \pm \left[(a+d)^2/4 - \Delta_o\right]^{\frac{1}{2}}$$

of the characteristic equation

(2-21) $$\Delta(\lambda) = \begin{vmatrix} a-\lambda & b \\ c & d-\lambda \end{vmatrix} = \lambda^2 - (a+d) + \Delta_o = 0 \ .$$

Consider first $\Delta_o \neq 0$, which implies that $\lambda_1 . \lambda_2 \neq 0$.

When $\lambda_1$ and $\lambda_2$ are real and distinct, the general solution of (2-19) is

(2-22) $$\begin{cases} x_n = A\,\lambda_1^n + B\,\lambda_2^n, \quad y_n = \frac{1}{b}(\lambda_1 - a)A\,\lambda_1^n + \frac{1}{b}(\lambda_2 - a)B\,\lambda_2^n, & b \neq 0, \\ x_n = A\,a^n, \quad y_n = B\,d^n + (a-d)^{-1}\,A\,a^n, \quad \lambda_1 = a, \quad \lambda_2 = d, & b = 0, \end{cases}$$

where, (contrary to a finite difference equation) A and B are arbitrary real constants (and not arbitrary real-valued periodic functions of period T = 1). The values of A and B are unambiguously defined, for example, by the initial conditions $x_n = x_o$, $y_n = y_o$ when n = 0. The inequality $\lambda_1 \neq \lambda_2$ implies $a \neq d$, so that the second set of equations in (2-22) is never indeterminate. The first set of equations in (2-22) holds also when $\lambda_1$ and $\lambda_2$ are complex, with the proviso that the constants A and B are also complex, ans so chosen that the dependence between $x_n$, $y_n$ and $x_o$, $y_o$ is purely real for all n. This is always possible, because when $\lambda_1$, $\lambda_2$ are complex, they are necessarily complex conjugate (cf. eq. (2-20)).

When $\lambda_1 = \lambda_2 = \lambda$ is a double root of (2-21), the general solution of (2-19) becomes

(2-23) $$\begin{cases} x_n = A\,\lambda^n + Bn\lambda^n, \quad y_n = \frac{A}{b}(\lambda - a)\lambda^n + \frac{B}{b}\left|(n+1)\lambda - n\right|\lambda^n, & b \neq 0, \\ x_n = A\,a^n, \quad y_n = Ba^n + Ana^n, \quad \lambda = a = d, & b = 0, \end{cases}$$

where again A and B are real constants.

Equations (2-22) and (2-23) are degenerate when either one or both $\lambda$ take the values 0, +1 or -1. The case $\lambda = 0$ is excluded by the inequality $\Delta_o \neq 0$. When $\lambda = +1$ or $\lambda = -1$, it is possible to transform the recurrence (2-19) (by means of a non-singular transformations) into a form where at least one equation becomes an identity. This identity gives rise to the impression that the fixed point $x_n = y_n = 0$ is either non unique ($\lambda = +1$), or that the recurrence (2-19) possesses cycles of order two ($\lambda = -1$).

When $\Delta_o = 0$, the right-hand sides of (2-19) are not linearly independent i.e. they contain a common factor, say $a\,x_n + b\,y_n$. It is therefore possible to write (2-19) in the form

$$y_{n+1} = r\, x_{n+1}, \qquad n = 0,\ 1,\ \ldots\ ,$$

where r is a known constant. Since this relation permits to express $x_n$ by $y_n$, and vice versa, the recurrence (2-19) separates into two identical first order recurrences with the eigenvalue $\lambda = a+d$. A further degeneracy occurs when $a+d = 0$, because then $\lambda_1 = \lambda_2 = 0$, and the recurrence (2-19) becomes equivalent to $x_n \equiv 0$, $y_n \equiv 0$. It is therefore convenient to assume $\lambda \neq 0, \pm 1$ in what follows. If a linear recurrence with one of the critical value $\lambda = 0, +1, -1$, arises in the study of a non linear recurrence, then the properties of the former are generally unrelated to the properties of the latter, and no real loss occurs by omitting the linear cases $\lambda = 0, +1$ and $-1$.

A stability analysis of the fixed point $x_n = y_n = 0$ can be carried out directly on the basis of the solutions (2-22) and (2-23). By inspection it is obvious that $x_n = y_n = 0$ is asymptotically stable when $|\lambda_1| < 1$, $|\lambda_2| < 1$, and unstable when either $|\lambda_1| > 1$ or $|\lambda_2| > 1$ (or both). When $\lambda_{1,2}$ are complex conjugate and $|\lambda_1| = |\lambda_2| = 1$, i.e. $\lambda_{1,2} = e^{\pm i\varphi}$, $0 < \varphi < \pi$, then $x_n = y_n = 0$ is stable without being stable asymototically. In mechanics, the constant $\varphi$ is usually called a rotation angle (sometimes a phase advance).

Compared to the information content of (2-22) or (2-23), the knowledge of stability or instability of $x_n = y_n = 0$ is relatively meager, i.e. it describes only an asymptotic property of the half-trajectories $\{x_i, y_i\}$, $i = 0, 1, 2, \ldots$ or $i = 0, -1, -2, \ldots$ of the recurrence (2-19). In fact, from the general solutions (2-22) and (2-23), it is possible to deduce the existence of phase plane curves on which all points of any given half-trajectory (or full trajectory) of (2-19) are located. Two or more half-trajectories are obviously dynamically equivalent if their points are located on the same phase plane curve. Such curves are therefore dynamically more relevant than any discrete set of half-trajectories. This property was already noticed by Poincaré, who called these priviledge curves "analytical" invariant curves. The reason for the qualifier analytical will become apparent later. From a physical, biological, etc. point of view, analytical invariant curves are conceptually equivalent to phase plane trajectories of a continuous dynamic system, described by differential equations.

It was already found by Birkhoff [B 4] that if an analytical invariant curve $\mathcal{C}$ is described by a function $F(x_n, y_n) = C$ = constant, then $F(x_n, y_n)$ is a solution of the functional equation of automorphic functions [F 3]

(2-24) $$w(x_{n+1}, y_{n+1}) = w(x_n, y_n) .$$

More particularly, if the half-trajectories $\{x_i, y_i\}$, $i = 0, 1, 2, \ldots$ whose points lie on $\mathcal{C}$ are generated by the recurrence

(0-22a) $$x_{n+1} = g(x_n, y_n),\ y_{n+1} = f(x_n, y_n)\ ,$$

where f, g, are reasonably smooth single-valued functions, then (2-24) can be written in the more explicit form

(2-25) $$w(g(x,y),\ f(x,y)) = w(x,\ y)\ ,$$

and the insertion of $w = F(x, y)$ into (2-25) must lead to an identity, say in y for all admissible x (or vice versa). The functional equations (2-24) and (2-25) are coordinate-dependent, because x and y have to be the same as in their generating recurrence (0-22a). Moreover, the interval X of admissible x, i.e. the interval inside of which (2-25) should be an identity, is not necessarily the same for every analytical invariant curve $\mathcal{C}$, and in particular the boundaries of X are unknown, unless the corresponding $\mathcal{C}$ is indirectly specified. The inverse problem of determining $\mathcal{C}$ via a solution $w(x, y)$ of (2-25) is theregore generally extremely difficult, because one does not even know before-hand inside of what interval X of x (or Y of y) one should search for such a solution. The determination of $w(x, y)$ and X (or Y) appear thus as two inseparable parts of the same problem. The dynamically important study of the functional equation (2-25) is further complicated by the almost complete absence of global existence, uniqueness and continuity theorem, as well as by the absence of any "general" numerical algorithms. The situation is therefore entirely different from that of continuous dynamic systems, where there exist usually several alternate means of determining a trajectory passing through a given phase-space point.

Some light can be thrown on the above-mentioned inverse problem by examining particular forms of the recurrence (2-19). Let for example

(2-26) $$x_{n+1} = a\,x_n + y_n,\quad y_{n+1} = (a^2-1)x_n + a\,y_n\ ,$$

where the constant a is different from ±1, but otherwise arbitrary. By substitution it can be verified that

$$w(x,\ y) = y^2 + (1-a^2)x^2$$

is a solution of

$$w(ax + y,\ (a^2-1)x + ay) = w(x,\ y)\ ,$$

and so is

$$w(x,\ y) = F(z),\qquad z = y^2 + (1-a^2)x^2\ ,$$

where F is an arbitrary continuous function. The functional equations (2-24), (2-25) possess therefore more similarity to partial differential equations than to ordinary ones. For this reason particular solutions of (2-25) cannot generally be found by means of successive iterations, unless the admissible shapes of F are suitably narrowed down, till uniqueness is achieved. In some cases this can be done by requiring that F be sufficiently smooth, but in others the non-uniqueness will persist even if F is analytic in x and y, as can be seen from the preceding example. If $F(z)$ is such that the maximally admissible intervals X and Y of a particular solution $w(x, y) = z$ are not diminished, then the non-uniqueness of F has in principle no influence on the uniqueness of the corresponding analytical invariant curves, because regardless of the form of F, the function $w(x, y) = F(z) = C$ describes the same locus of points as $w(x, y) = z = C_1$, provided the constants C and $C_1$ are suitably chosen, for example so that the corresponding curves pass through the same "initial" point $(x_o, y_o)$.

The example (2-26) shows that $w(x, y)$ need not be single valued in y inside its x-interval of existence X. For this reason it is generally more convenient to separate $\mathscr{C}$ into branches $\mathscr{C}_\theta$ and replace (2-25) by a certain number of equations of the form $y_{n+1} = \theta(x_{n+1})$, which subject to (0-22a) can be written

(2-27) $$f(x, \theta(x)) = \theta(g(x, \theta(x))), \quad y = \theta(x), \quad y_o = \theta(x_o) \quad ,$$

where $\theta(x)$ is a single-valued function, and each solution $\mathscr{C}_\theta : y = \theta(x)$ describes only one branch of $\mathscr{C}$. Like in the case of (2-25), the x-intervals of existence of the various $\theta(x)$ are not known a priori. In contrast to (2-25), the solutions of (2-27) are generally unique, i.e. they do not depend on any arbitrary function. This property results not only from the non linearity of (2-27), but also from the fact that for a given $\theta$, $y_{n+1} = \theta(x_{n+1})$ implies $y_n = \theta(x_n)$ for all $x_n$ for which $\theta$ is single-valued. In the particular case of the recurrence (2-26), there exist two single valued branches $\mathscr{C}_\theta$, described by

$$y = + C\sqrt{-(1-a^2)x^2} \quad \text{and} \quad y = - C\sqrt{-(1-a^2)x^2} \quad ,$$

respectively. Except for the fixed point (0,0), there passes only a single invariant curve $\mathscr{C}$ through every point $(x_o, y_o)$ of the phase plane. Two invariant curves pass through (0,0) when $|a| > 1$, and none when $|a| < 1$. At the fixed point, there is thus a violation of the uniqueness of invariant curves. It is for this reason that a fixed point is called a singular point of the recurrence, and the corresponding curves, if they exist, are called singular invariant curves (at the fixed point). The statement in parenthesis is often omitted.

From the non-uniqueness of the solutions of (2-25) it might be concluded that invariant curves can also be constructed by the following procedure, which amounts essentially to interpreting the equations (0-22a) as a system of finite difference equations instead of as a recurrence :

1°) Choose as an initial point the point $(x_n, y_n)$ on any freely chosen consequent half-trajectory $\{x_i, y_i\}$, $i = 0, 1, 2, \ldots$ Assume for example that $x_{n+1} > x_n$.

2°) Join the points $(x_n, y_n)$ and $(x_{n+1}, y_{n+1})$ by a continuous curve segment $\mathscr{C}_n$, described by a single-valued function $y = \varphi_n(x)$, $x_n \leqslant x \leqslant x_{n+1}$, where $\varphi_n$ is freely chosen. If the functions f, g in the recurrence (0-22a) are single-valued (as it is usually assumed), then $\mathcal{C}_n$ has an unambiguously defined unique consequent $\mathcal{C}_{n+1}$, described parametrically by $x = g(\bar{x}, \varphi_n(\bar{x}))$, $y = f(\bar{x}, \varphi_n(\bar{x}))$, $x_n \leqslant \bar{x} \leqslant x_{n+1}$. Assume for convenience that $x_{n+2} > x_{n+1}$ and that the curve segment $\mathcal{C}_{n+1}$ is such that it can be described by $y = \varphi_{n+1}(x)$, $x_{n+1} \leqslant x \leqslant x_{n+2}$, where $\varphi_{n+1}$ is also single-valued. The successive consequents $\mathcal{C}_{n+k}$, $k > 1$, of $\mathcal{C}_n$ can be continued indefinitly (f, g, are single valued) without any inherent limitations, except that at some stage $\mathcal{C}_{n+k}$ may no longer be describable by a single-valued function $\varphi_{n+k}(x)$. The union $\mathcal{C}_\varphi$ of all $\mathcal{C}_{n+k}$ appears thus as a continuous invariant curve of the equations (0-22a), in spite of the fact that by construction it depends on an arbitrary function. If all the intervals $x_{n+k} < x < x_{n+k+1}$ are disjoint, then the functional equations (2-24) and (2-25) are obviously satisfied by the $\varphi_{n+k}(x)$, $k = 1, 2, \ldots$ The curve $\mathcal{C}_\varphi$ is, however, not an invariant curve of the recurrence (0-22a) in the sense of the stricter functional equation (2-27), i.e. it is not "analytical" in the sense of Poincaré, because it violates the simultaneous validity of $y_{n+1} = \theta(x_{n+1})$ and $y_n = \theta(x_n)$ on the union of $\mathcal{C}_n$ and $\mathcal{C}_{n+1}$. In fact, $\mathcal{C}_n$ and $\mathcal{C}_{n+1}$ being on the same single-valued invariant curve branch, one should have $\varphi_n(x) = \varphi_{n+1}(x)$, $x_n \leqslant x \leqslant x_{n+2}$, which is impossible unless $\varphi_n(x) = \theta(x)$ and $\theta(x)$ satisfies (2-27). The requirement that $\varphi_n(x)$ and $\varphi_{n+1}(x)$ be parts of the same solution $\theta(x)$ of (2-27) eliminates the discontinuities of slope, curvature, etc. at the points $(x_{n+k}, y_{n+k})$, $k = 1, 2, \ldots$ of $\mathcal{C}_\varphi$, which arise otherwise by construction. If $\mathcal{C}_\varphi$ is to be an analytical invariant curve of the recurrence (0-22a), the curve segment $\mathcal{C}_n$ must therefore be a maximally smooth one, subject to a variation of $\varphi_n$ (in the sense of the calculus of variations). If for some k the intervals $x_{n+k} < x < x_{n+k+1}$ and $x_{n+k+1} < x < x_{n+k+2}$ are not disjoined, then $\varphi_{n+k}(x)$ and $\varphi_{n+k+1}(x)$ do not satisfy the functional equation (2-25) for all x, and again $\mathcal{C}_\varphi$ is not an analytical invariant curve. For this reason no further mention will be made of invariant curves like $\mathcal{C}_\varphi$. The qualifying adjective analytical will be omitted henceforth, unless the context should dictate otherwise.

Invariant curves $\mathcal{C}$ of the linear recurrence (2-19) are of course described parametrically by the general solution (2-22) or (2-23), after the constants A and B have been expressed in terms of the initial values $x_o$ and $y_0$. In order to arrive at the form $w(x, y) = C$ or $y = \theta(x)$ it is sufficient to eliminate from (2-22) or (2-23) the parameter n. Although n takes only discrete values in (2-19), the right-hand sides of (2-22) and (2-23) possess obviously an unambiguous extension to continuous values of n, except possibly for the limits $n \to \pm\infty$. In other words the general solution (2-22) or (2-23) solves also the problem of continuous (as well as fractional) iterates associated with the recurrence (2-19). The property that there exist continuous sets of initial conditions $(x_o, y_0)$ such that the corresponding discrete trajectories $\{x_n, y_n\}$, n = integers, lie all on the same (invariant) curve, reflects the possibility of replacing certain smooth variations of $x_o$ and $y_0$ by smooth variations of n, i.e. it reflects the possibility of imbedding discrete n's into continuous ones. An elimination of n from (2-22) and (2-23) implies the existence of such an imbedding.

Expressed in another way, this means that the problem of continuous iterates of the recurrence (0-22a) can be solved locally (i.e. on curves) by means of particular solutions of the functional equations (2-25) and (2-27). If the corresponding invariant curves are non singular at all points $(x_o, y_o)$ of a phase plane region G (i.e. if they do not traverse any fixed points or points of cycles), then this problem can be solved globally in G. Whether G can be extended to the whole phase plane depends on the nature of singular points, and singular loci of points (i.e. it depends on the nature of the point set, where invariant curves fail to be unique).

The general solutions (2-23) and (2-24) of the recurrence (2-19) are too compact to be easily used in a discussion of geometrical shapes of the corresponding invariant curves. For this reason it is worthwhile to examine (2-19) in an alternate way. Assume first that the eigenvalues of (0,0) are real and distinct. By means of the linear transformation

$$x = \alpha u + \beta v, \qquad y = p_1 \alpha u + p_2 \beta v \quad ,$$

where $\alpha$, $\beta$ are constants and

(2-28) $$p_{1,2} = (\lambda_{1,2} - a)/b, \; b \neq 0 \; ; \quad p_{1,2} = c/(\lambda_{1,2} - d), \; b = 0 \quad ,$$

the recurrence (2-19) can be brought into the (canonical) diagonal form

(2-29) $$u_{n+1} = \lambda_1 u_n, \qquad v_{n+1} = \lambda_2 v_n \quad .$$

The rectangular axes $u_n$, $v_n$ are called the principal axes of (2-19) and $p_{1,2}$ the eigendirections of (0,0). When (2-31) is transformed back to the coordinate frame $x_n$, $y_n$, the $u_n$ and $v_n$-axes have the slopes $p_1$ and $p_2$, respectively. The general solution of the recurrence (2-31) is obviously

(2-30) $$u_n = \lambda_1^n \cdot u_o, \quad v_n = \lambda_2^n \cdot v_o, \qquad n = 0, \pm 1, \ldots \; ,$$

and no obstacle appears in its extension to continuous n. When $\lambda_1$, $\lambda_2$ are both positive, n is easily eliminated, and one obtains an explicit representation of the invariant curves $v = \theta(u)$ covering the phase plane

(2-31) $$v = v_o \cdot |u/u_o|^{\nu} \cdot \operatorname{sgn}(\frac{u}{u_o}), \qquad \nu = (\log \lambda_2)/(\log \lambda_1) \quad .$$

When $\lambda_1$ or $\lambda_2$ is negative, it is sufficient to iterate (2-30) once before the elimination of n. The shapes of invariant curves are unaffected, but the points of a discrete trajectory $\{x_i, y_i\}$, $i = 0, \pm 1, \ldots$, lie now alternately on two distinct invariant curve segments. When $0 < |\lambda_1|, |\lambda_2| < 1$ or $|\lambda_1|, |\lambda_2| > 1$ the invariant curves $v = \theta(u)$ resemble parabolas. They are tangent to $v \equiv 0$ at (0,0), and symmetric with respect to $u \equiv 0$. By analogy with the trajectories of differential equations the point (0,0) is called a (non-critical or a simple) node. In order to convey the information about the signs of $\lambda_1$ and $\lambda_2$, a node is said to be of type 1 when $\lambda_1 > 0$, $\lambda_2 > 0$, of type 2 when $\lambda_1 \cdot \lambda_2 < 0$, and of type 3 when $\lambda_1 < 0$, $\lambda_2 < 0$.

The corresponding locations of two points $(x_i, y_i)$ and $(x_{i+1}, y_{i+1})$ of $\{x_i, y_i\}$, $i = 1, 2, \ldots$ are shown in Fig. 2.2. A node is either asymptotically stable ($|\lambda_1|, |\lambda_2| < 1$) or unstable ($|\lambda_1|, |\lambda_2| > 1$).

When $0 < |\lambda_1| < 1$, $|\lambda_2| > 1$ (or vice versa), the invariant curves of the recurrence (2-19) have a hyperbolic shape, except for the two asymptotes $u \equiv 0$ and $v \equiv 0$ (Fig. 2.3). By analogy with differential equations the fixed point (0,0) is called a saddle (of type 1, 2 or 3, depending on the signs of $\lambda_1$, $\lambda_2$). A saddle is considered to be unstable in spite of the fact that there exist two invariant curve branches on which the points $(x_i, y_i)$ of discrete half-trajectories approach (0,0) asymptotically as $i \to \infty$ or $i \to -\infty$.

Suppose now that the $\lambda_{1,2}$ are complex. It is often convenient to express them in a polar form

(2-32) $$\lambda_{1,2} = \alpha \pm i\beta = \sigma . e^{\pm i\varphi}, \qquad 0 < \varphi < \pi \quad ,$$

where $\sigma$ is called the growth factor and $\varphi$ the rotation angle. The complex valued linear transformation

$$\xi = u + iv, \qquad \eta = -i(u-v)$$

converts the diagonal complex-valued recurrence (2-29) to the real-valued canonical form

(2-33) $$\xi_{n+1} = \sigma(\xi_n \cos\varphi - \eta_n \sin\varphi), \qquad \eta_{n+1} = \sigma(\xi_n \sin\varphi + \eta_n \cos\varphi) \quad .$$

When $\sigma = 1$ the general solution of (2-33) is

(2-34) $$\xi_n = a \cos(n\varphi + \varphi_0), \; \eta_n = a \sin(n\varphi + \varphi_0), \; \xi_o = a \cos\varphi_0, \; \eta_o = a \sin\varphi_0,$$

or after the elimination of n,

(2-34a) $$\xi^2 + \eta^2 = a^2 = \xi_o^2 + \eta_o^2$$

The invariant curves of (2-33) are therefore concentric circles, and those of (2-19) concentric elliples (Fig. 2-4a). The fixed point (0,0) is called a centre. It is stable without being stable asymptotically. Consider now a discrete trajectory $\{x_i, y_i\}$, $i = 0, \pm 1, \ldots$ When the rotation angle is commensurable with $2\pi$, i.e. when

(2-35) $$\varphi = 2\pi\, r/k, \qquad r, k = \text{positive integers} \quad ,$$

then $\{x_i, y_i\}$ consists only of k distinct points, although the corresponding invariant curves (cf. eq. 2-34a) represent continua. Such a degenerate case is called (loosely) a resonance, or (more precisely) an exceptional case in the sense of Cigala [C 4]. When $\varphi$ is incommensurable with $2\pi$, the points $(x_i, y_i)$ fill the whole invariant curve as $i \to +\infty$ or $i \to -\infty$. It should be noted that the invariant curves (2-34a) depend neither on the rotation angle (i.e. they are the same whether one has an exceptional case or not) nor on any auxiliary ordering of points on a discrete

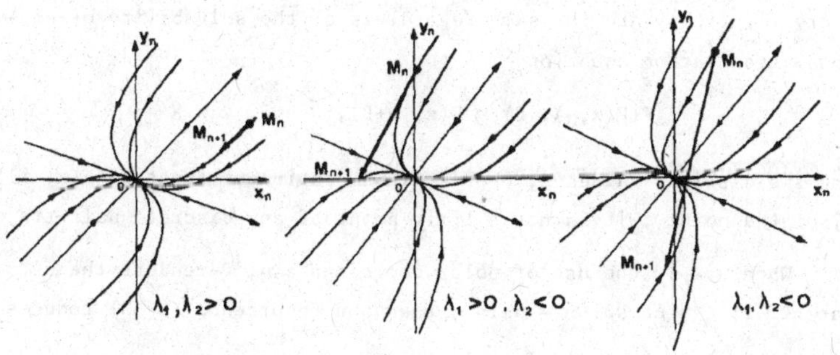


Fig. 2-2. Nodes of type 1, 2 and 3.

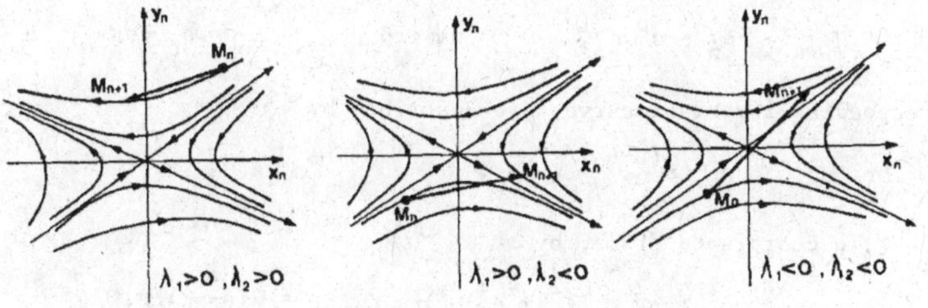


Fig. 2-3. Saddles of type 1, 2 and 3.

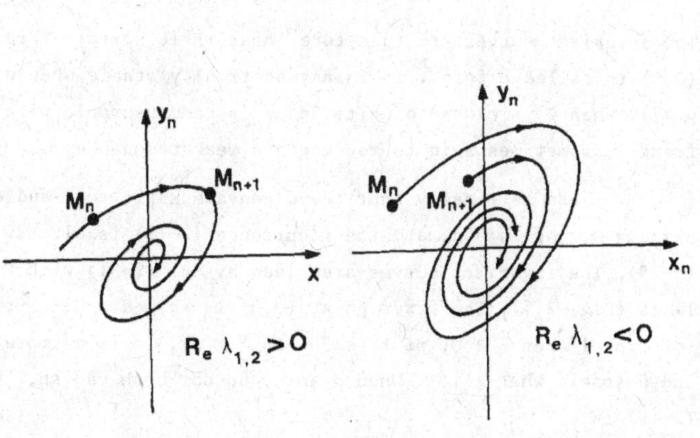


Fig. 2-4a. Two stable foci.

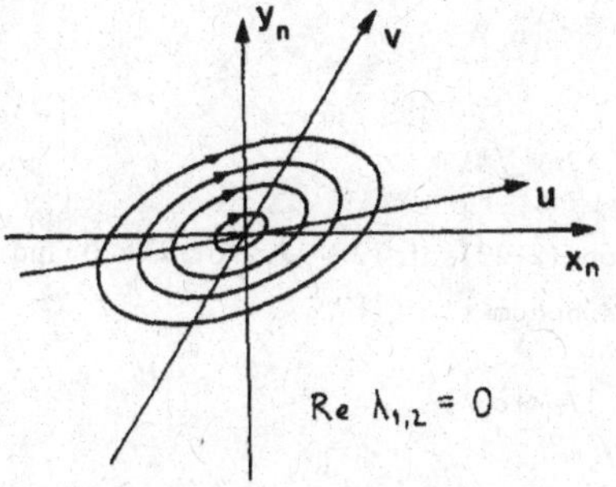


Fig. 2-4a. A centre

trajectory (i.e. they are the same regardless of the solvability of an associated functional translation equation

$$F(F(x,\alpha), \beta) = F(x, \alpha+\beta), \qquad x \in X \quad ,$$

relating two distinct values $\alpha$, $\beta$ of $i$ via a continuous function $F(x, i)$). A centre is an isolated point ; it is not a limit point of any discrete half-trajectory.

When $\sigma \neq 1$, the use of polar coordinates $\rho$, $\theta$ renders the situation more transparent. If $\xi = \rho\cos\theta$, $\eta = \rho\sin\theta$, then the recurrence (2-33) reduces to

(2-33a) $$\rho_{n+1} = \sigma\rho_n, \qquad \theta_{n+1} = \theta_n + \varphi \quad ,$$

whose solution is

(2-36) $$\rho_n = \sigma^n.\rho_o, \qquad \theta_n = n\varphi + \theta_o \quad , \qquad \xi_o = \rho_o \cos\theta_o, \quad \eta_o = \rho_o \sin\theta_o \; .$$

The corresponding invariant curves are described by

(2-37) $$\rho = \sigma^{(\theta-\theta_o)/\varphi} . \rho_o,$$

or in the $\xi$, $\eta$ coordinate frame, by

(2-37a) $$\begin{cases} \xi^2 + \eta^2 = (\xi_o^2 + \eta_o^2)\sigma^\alpha, & \alpha = 2(\theta-\theta_o)/\varphi, \\ \theta = \text{arc tg } (\xi/\eta), & \theta_o = \text{arc tg } (\xi_o/\eta_o) \; . \end{cases}$$

The invariant curves are therefore logarithmic spirals (Fig. 2.4b). The fixed point (0,0) is called a focus. It is asymptotically stable when $\sigma < 1$, and unstable when $\sigma > 1$. When $\sigma$ is close to unity and $\varphi$ is commensurable with $2\pi$, the corresponding focus is sometimes said to represent a weak resonance.

Assume finally that the eigenvalues are real and equal $\lambda_1 = \lambda_2 = \lambda$. If $b = c = 0$, then $a = d$, and the recurrence (2-19) is already in the canonical form (2-29). The invariant curves are given by eq. (2-31) with $\nu = 1$, and they are straight lines (Fig. 2.5). The fixed point (0,0) is called a bi-critical node or a star-node (of type 1 when $\lambda > 0$, of type 3 when $\lambda < 0$). It is asymptotically stable when $|\lambda|<1$, and unstable when $|\lambda|>1$. When b and c do not both vanish, the canonical form of (2-19) is

(2-38) $$u_{n+1} = \lambda u_n - v_n, \qquad v_{n+1} = \lambda v_n \quad .$$

Its general solution

(2-39) $$v_n = \lambda^n.v_o, \qquad u_n = \lambda^n(u_o - nv_o/\lambda)$$

is only slightly simpler than the general solution (2-23) of (2-19). For $\lambda > 0$ and $\frac{v}{v_o} > 0$, the explicit equation of invariant curves becomes

(2-40) $$u = v.(u_o/v_o) - v.\left[\log(v/v_o)\right]/(\lambda\log\lambda) \quad .$$

A mirror-symmetric expression holds for $v/v_o < 0$. These invariant curves have a parabolic shape and in x, y-coordinates they are tangent at (0,0) to the line y = px, where $p = (\lambda-a)/b$ if $b \neq 0$ and $p = c/(\lambda-a)$ if $b = 0$. The line y = px is also an asymptote of $y = \theta(x)$ for large $|x|$ (Fig. 2.6). The fixed point (0,0) is called a critical node. It is asymptotically stable when $|\lambda| < 1$, and unstable when $|\lambda| > 1$.

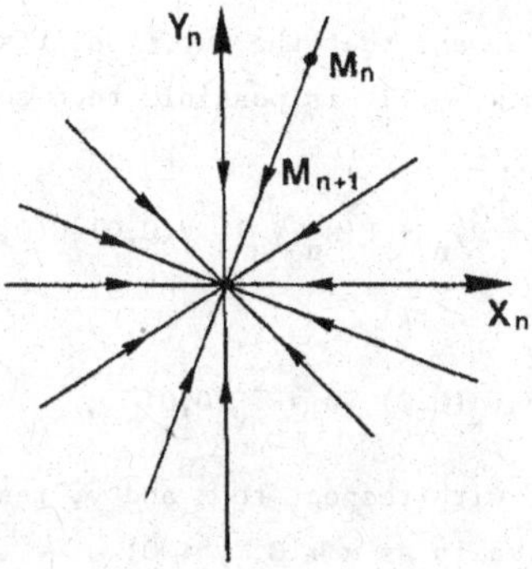


Fig. 2-5. A star-node or bi-critical node

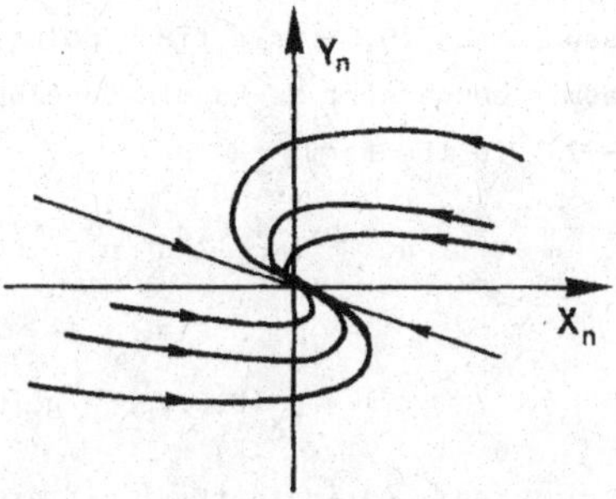


Fig. 2-6. A critical node

The fixed point of the recurrence (2-46) is either a centre ($|a| < 1$) or a saddle ($|a| > 1$). The values $|a| = 1$ have been excluded because they imply $|\lambda| = 1$ at (0,0).

Since the constants a, b, c, d, in the recurrence (2-19) are necessarily imprecise when it represents a physical, biological or other natural evolution process, two general situations have to be distinguished : i) a small change of a, b, c, d, leaves the phase portrait of (2-19) qualitatively invariant, and the invariant curves are modified only quantitatively, ii) a small change of a, b, c, d, changes the phase portrait qualitatively. In the first case the recurrence (2-19) is said to be inert or grossier (or more loosely, structurally stable), and in the second case non-inert. Non-inertness arises obviously when (0,0) is a centre, a critical or bi-critical node, and when at least one $\lambda$ is equal to 0, -1 or +1. In a physical context it may happen that the admissible changes of a, b, c, d are not arbitrary, for example that they have to satisfy the constraint $\Delta_o = 1$ (conservation of phase plane area), then a non-inert recurrence becomes conditionally inert (subject to the constraint in question). If by definition $\Delta_o = 1$, then a centre of the recurrence (2-19) is inert.

## 2.2 The effect of non linear perturbations

The determination of the phase plane portrait of the non linear recurrence (0-22a) would procede along the same lines as that of the linear recurrence (2-19), if the general solution of the former could be expressed in terms of elementary or known transcendental functions. Unfortunately the functions involved are unknown, and a recourse to approximate methods, both analytical and numerical becomes unavoidable. Assume that (0,0) is a fixed point of (0-22a) and that the functions f(x, y), g(x, y) admit convergent Mc Laurin developments in x, y. It is possible then to write (0-22a) in the form

(2-41) $$x_{n+1} = ax_n + by_n + G(x_n,y_n),\ y_{n+1} = cx_n + dy_n + F(x_n,y_n),\ F(0,0)=G(0,0) = 0,$$

where

(2-42) $$a = g_x(0,0),\ b = g_y(0,0),\ c = f_x(0,0),\ d = f_y(0,0)\quad ,$$

the subscripts representing partial differentiation with respect to x and y, respectively. The function F, G decrease at least quadratically as $x \to 0$, $y \to 0$.

By a continuity argument it is to be expected that when the phase portrait of the linear approximation of (2-42) is a node, a focus or a saddle, then this phase portrait remains at least locally the same in the presence of F and G, i.e. it is inert with respect to the structural perturbations F and G. This conjecture is confirmed by an analytical argument, which is simple in principle but rather cumbersome in practice (see [L 4] for details). In the case of a centre the phase portrait is generally not inert, because F and G may involve sources of damping or anti-damping, which affect the invariant curves at an arbitrary small distance from (0,0). Even if the Jacobian determinant

(2-43) $$J(x,y) = (a+G_x)(d+Fy) -(b+G_y)(c+F_x)$$

of (2-41) is identically equal to unity, the non linear terms F and G are usually dominant. Similarly to second order differential equations, the survival of a centre of the recurrence (2-41) requires some rather special circumstances (cf. section 2.5).

When (0,0) is a saddle, a (simple) node or a focus for $F = G \equiv 0$, then the known solutions (2-31), (2-37) of the functional equations (2-25) and (2-27) can be used to define small segments of the invariant curves of (2-41) with any prescribed accuracy, provided the initial point $(x_o, y_o)$ is sufficiently close to (0,0) and these segments are sufficiently short. A continuation by means of numerically computed consequents or antecedents is possible, provided (2-41) and its inverse (or inverses) are iterated N times. The maximally permissible value of N is rather small, because the initial error grows usually rather rapidly with N. Invariant curve segments of appreciable length require therefore highly accurate initial segments. For conciseness the latter are called germs of the corresponding invariant curves.

When the eigenvalues of (0,0), evaluated subject to (2-42), are real and distinct, two germs can be determined in the form of a Mc Laurin development :

(2-44) $$y = \theta(x) = \beta_1 x + \beta_2 x^2 + \dots \quad , \qquad \theta(0) = 0 \quad ,$$

where the $\beta_i$, $i = 1, 2, \dots$ are real constants. Inserting (2-44) into the functional equation (2-27) yields a recursive system of linear algebraic equations for the coefficients $\beta_i$. The equation defining $\beta_1$ coincides with that of the eigendirections $p_1$ and $p_2$ at (0,0). Hence, one may choose either $\beta_1 = p_1$ or $\beta_1 = p_2$ (cf. eq. (2-28)). With $\beta_1$ fixed, the determination of $\beta_i$, $i > 1$, is straightforward and unambiguous :

(2-45) $$\beta_2 = \beta_2(\beta_1) = \frac{1}{2}\left[(G_{xx}+\beta_1^2 G_{yy}+2\beta_1 G_{xy})\beta_1 - (F_{xx}+\beta_1^2 F_{yy}+2\beta_1 F_{xy})\right] / \left[c-\beta_1 b-(a+\beta_1 b)^2\right],$$

$$\beta_3 = \beta_3(\beta_1, \beta_2) = \dots \quad ,$$

$$\beta_4 = \beta_4(\beta_1, \beta_2, \beta_3) = \dots \quad ,$$

where all derivatives of F and G are evaluated at (0,0). The explicit expressions of $\beta_i$ increase very rapidly in notational bulk as i increases, and for this reason they are indicated only symbolically. The convergence of the series (2-44) has been proven analytically for sufficiently small $|x|$ by constructing a majorating series [L 4]. Since the method of proof is non constructive, the radius of convergence is unknown. Carefully controlled high-accuracy computations have shown that this radius, although sometimes quite small, is sufficiently large to render the development (2-44) highly useful. An iterative method of determining the coefficients $\beta_i$, $i > 1$, is also known ([H 1], [P 2], see also the discussion in [M 10]). Since the accuracy of the $\beta_i$ so determined increases with the number of iterations, practically useful results can only be obtained by means of a computer [G 4].

The same reasoning applies when (0,0) is not a fixed point but one point of a cycle of order k. It is sufficient to replace the recurrence (0-22a) by its k-th iterate, and the functional equation (2-27) by

(2-27a) $$f_k(x, \theta(x)) = \theta(g_k(x, \theta(x))) \quad .$$

The determination of the iterated functions $f_k$, $g_k$ is usually quite tedious without the use of a computer.

One germ of the form (2-44) can also be obtained when (0,0) is a critical node. Because the proof of convergence is also of the majorizing type [L 5], [D 1], the radius of convergence is unknown. Like in the case of a saddle or a simple node, it is small, but usually not unusable.

The singular invariant curves generated by the preceding analytical germs are simply the axes u = 0 and v = 0 (for the critical node, only u = 0) of the linear canonical forms, distorted by the non linear terms F and G in (2-41). Distortions of other invariant curves of the linear approximation cannot be determined in the same manner, because their logarithmic dependence on $\lambda$ and v precludes the existence of a Mc Laurin series $\theta(x)$ at $x = 0$. Corrections of the linear approximation, independent of the existence of a Mc Laurin series at $x = 0$, can be obtained by means of a small parameter development. As an illustration consider (2-41) and assume that $(x_o, y_o)$ is sufficiently close to (0,0). If $\lambda_1 \neq \lambda_2$, and $\lambda_1$, $\lambda_2$ are real, then (2-41) can be transformed to the canonical form

$$(2\text{-}41a) \qquad u_{n+1} = \lambda_1 u_n + \varepsilon G(u_n, v_n), \qquad v_{n+1} = \lambda_2 v_n + \varepsilon F(u_n, v_n) ,$$

where $\varepsilon > 0$ is a suitably chosen parameter characterizing the smallness of F and G with respect to linear terms. Since the local phase plane portrait of (2-41a) is the same as that of (2-29), a solution of the functional equation (2-27), with the initial condition $(u_o, v_o)$, can be sought in the form

$$(2\text{-}46) \qquad v = \theta(u) = \theta_o(u) + \varepsilon\theta_1(u) + O(\varepsilon^2), \qquad v_o = \theta(u_o) \quad ,$$

where $O(\varepsilon^2)$ represents the error of the approximation of $\theta(u)$ by $\theta_o$ and $\theta_1$. Inserting (2-46) into (2-41a) leads to a system of two linear functional equations :

$$(2\text{-}47) \qquad \begin{aligned} &\lambda_2\theta_o(u) - \theta_o(\lambda_1 u) = 0, && \theta_o(u_o) = v_o \quad , \\ &\lambda_2\theta_1(u) - \theta_1(\lambda_1 u) = \theta'_o(\lambda_1 u).G(u,\theta_o(u))-F(u,\theta_o(u)), && \theta_1(u_o) = 0 , \end{aligned}$$

where $\theta'(z) = \frac{d}{dz}\theta(z)$. The solution of the first equation is known. The second equation is the same as the first, except that it is non homogeneous, and has a different initial condition. In order to determine $\theta_1(u)$ it is necessary to find a maximally (smooth) particular solution $\theta_1$ in the presence of the right-hand side. This cannot be done in general because, contrarily to differential equations, no constructive algorithm is available to express $\theta_1$ in terms of the homogeneous solution $\theta_o$. For this reason, $\theta_1$ can be found explicitly only for particular F and G. Even if $\theta_1$ is found, a similar difficulty will usually arise in the determination of a higher-order approximation

$$(2\text{-}46a) \qquad v = \theta(u) \simeq \theta.(u) + \varepsilon\theta_1(u) + \ldots + \varepsilon^m\theta_m(u), \qquad m \geqslant 2 \quad .$$

When $\lambda_1$, $\lambda_2$ are complex, (2-41) can be brought into the form :

$$(2\text{-}41b) \qquad \begin{aligned} \xi_{n+1} &= \sigma(\xi_n \cos\varphi - \eta_n \sin\varphi) + \varepsilon G(\xi_n, \eta_n) \\ \eta_{n+1} &= \sigma(\xi_n \sin\varphi + \eta_n \cos\varphi) + \varepsilon F(\xi_n, \eta_n) \quad , \end{aligned}$$

where $\varepsilon$ plays the same role as in (2-41a). A first order approximation of an invariant curve passing through the point $(\xi_o, \eta_o)$

$$\eta = \theta(\xi) = \theta_o(\xi) + \varepsilon\theta_1(\xi) + O(\varepsilon^2), \qquad \eta_o = \theta(\xi_o) \quad ,$$

is given by a solution of the following two functional equations :

$$(2\text{-}47a)\begin{cases} \sigma(\xi\sin\varphi + \theta_o(\xi)\cos\varphi) - \theta_o(\zeta) = 0 , & \theta_o(\xi_o) = \eta_o , \\ \sigma(\cos\varphi + \sin\varphi.\theta_o'(\zeta)).\theta_1(\xi) - \theta_1(\zeta) = \theta_o'(\zeta).G(\xi,\theta_o(\xi)) - F(\xi,\theta_o(\xi)), & \theta_1(\xi_o) = 0 , \\ \zeta = \sigma(\xi\cos\varphi - \dot\theta_o(\xi)\sin\varphi) . \end{cases}$$

Contrarily to (2-47), the first equation is non linear. But since its solution $\theta_o(\xi)$ is known (cf. eq. (2-37) and (2-37a)), this non linearity is not a source of additional difficulties. The second equation is still a linear one, although with non-constant coefficients. Similarly to (2-47), an explicit solution $\theta_1(\xi)$ can only be found for particular forms of F and G. For some F and G it is more expedient to transform (2-41b), (2-46a) and (2-47a) to polar coordinates. The practical success of a development like (2-46) and (2-46a) depends thus on the explicit solvability of a non-homogeneous linear functional equation with known variable coefficients

$$(2\text{-}48) \qquad a(x)\,w(x) - w(b(x)) = c(x) \quad , \quad w(x_o) = 0 \quad ,$$

where a(x), b(x) depend on the linear part of the recurrence (2-41), and c(x) on its non linear perturbations F and G. When the initial point $(x_o, y_o)$ is neither a fixed point nor sufficiently close to a fixed point (or a point of a cycle) of (0-22a), then the study of the corresponding "regular" invariant curves cannot be based on the particular form of the recurrence (2-41). The properties of such regular invariant curves are still quantitatively unknown. Their qualitative properties can, however, be often inferred from the singular invariant curves, determined by means of Mc Laurin series like (2-44), and the almost-singular invariant curves, determined by means of (truncated) small-parameter series like (2-46a). In fact, some separatrices are determined by the former, and the contents of the resulting phase plane cells by the latter.

When (0,0) is an asymptotically stable node of the recurrence (0-22a), with distinct positive eigenvalues $\lambda_1$, $\lambda_2$, then it is possible in principle to determine two invariant curves initialized at (0,0) by means of analytical solutions of the two Schröder equations

$$(2\text{-}49) \quad \begin{aligned} w_1(g(x,y),\, f(x,y)) &= \lambda_1\, w_1(x,y), & w_1(0,0) &= 0 \quad , \\ w_2(g(x,y),\, f(x,y)) &= \lambda_2\, w_2(x,y), & w_2(0,0) &= 0 \quad . \end{aligned}$$

If the f, g possess convergent Mc Laurin series in x and y, then so do the $w_1$ and $w_2$, although the radius of convergence of the latter is quite small and not known in advance (see for example pp. 55-61, [L 4]). It is obvious by inspection that $w(x,y)=0$ implies $w(g(x,y), f(x,y)) = 0$, i.e. the curves described by $w_1(x, y) = 0$ and $w_2(x, y) = 0$ are invariant curves of the recurrence (0-22a). These two curves turn out to be the same as those described by the series (2-44) with $\beta_1 = p_1$ and $\beta_1 = p_2$. The use of the Schröder equation (2-49) is therefore ineffective in opening up new possibilities. Equations (2-49) are also valid when (0,0) is a stable focus, except that $w_1$, $w_2$ are complex-valued.

When (0,0) is an asymptotically stable critical node, then the independent Schröder equations (2-49) are replaced by the coupled set

$$\text{(2-50)} \qquad \begin{aligned} w_1(g(x,y),\, f(x,y)) &= \lambda w_1(x,y) - w_2(x,y), & w_1(0,0) = 0 \;, \\ w_2(g(x,y),\, f(x,y)) &= \lambda w_2(x,y) \;, & w_2(0,0) = 0 \;, \end{aligned}$$

and there exist again analytical solutions with a small (unknown) radius of convergence [L 5]. One of these solutions is equivalent to (2-44) with $\beta_1 = p$.

When the fixed point (0,0) of the recurrence (2-41) is a star-node ($\lambda_1 = \lambda_2 = \lambda$), the Schröder equations (2-49) degenerate into a single equation

$$\text{(2-51)} \qquad w(\lambda x + G(x,y),\, \lambda y + F(x,y)) = \lambda w(x,y), \qquad w(0,0) = 0 \;,$$

where x, y has been written instead of u, v. For $G \equiv 0$, $F \equiv 0$, the invariant curves of a star-node are straight lines with an arbitrary slope ; hence the solutions of the corresponding Schröder equation must be of the form

$$w(x,y) = Ax + By \;,$$

where A, B are arbitrary constants. This property differentiates essentially the star-node from the simple and critical nodes. If the perturbations F and G admit convergent Mc Laurin developments, then a natural extension of the linearized solution is

$$\text{(2-52)} \qquad \begin{aligned} &w(x,y) = A\varphi(x,\, y) + B\psi(x,\, y) \;, \\ &\varphi(x,y) = x + \sum_{i+j=2}^{\infty} \alpha_{ij} x^i y^j, \quad \psi(x,\, y) = y + \sum_{i+j=2}^{\infty} \beta_{ij} x^i y^j \;, \end{aligned}$$

where A, B are also arbitrary constants. The coefficients $\alpha_{ij}$ and $\beta_{ij}$ can be determined recursively by inserting (2-52) into (2-51). For exemple, when

$$\text{(2-53)} \qquad G(x,y) = a_1x^2 + a_2xy + a_3y^2, \; F(x,y) = b_1x^2 + b_2xy + b_3y^2 \;,$$

then

$$\text{(2-54)} \qquad \begin{aligned} \varphi(x,y) = x &+ \frac{1}{\lambda(1-\lambda)}(a_1x^2+a_2xy+a_3y^2) + \frac{1}{\lambda(1-\lambda)(1-\lambda^2)}\Big[(2a_1^2 + a_2b_1)x^3 + \\ &+ (3a_1a_2+a_2b_2+2a_3b_1)x^2y + (a_1a_3+a_2^2+a_2b_3+2a_3b_2)xy^2 + (a_2a_3+2a_3b_3)y^3\Big] + . \\ \psi(x,y) = y &+ \frac{1}{\lambda(1-\lambda)}(b_1x^2+b_2xy+b_3y^2) + \frac{1}{\lambda(1-\lambda)(1-\lambda^2)}\Big[(2a_1b_1+b_1b_2)x^3 + \\ &(2a_2b_1+a_1b_2+b_2^2+2b_1b_3)x^2y + (a_3b_1+a_2b_2+3b_2b_3)xy^2 + (a_3b_2+2b_3)y^3\Big] + \ldots \end{aligned}$$

Since the general form of the coefficients $\alpha_{ij}$, $\beta_{ij}$ of $w = A\varphi + B\psi$ is the same as that of Mc Laurin series solutions of (2-49), the same type of majorizing method can be used to show that the Mc Laurin series of $\varphi(x,y)$ and $\psi(x,y)$ have a non-zero radius of convergence. Like in the case of the solutions of (2-49) and (2-50, no a priori estimate of this radius is available.

In contrast to (2-44) and (2-46), no solutions of the Schröder equations (2-49) are known when (0,0) is a saddle. The reason for this somewhat peculiar absence will become apparent in section 2.3.

Compared to second order differential equations the most striking feature of the (non-linear) recurrence (0-22) or (0-22a) is the absence of constructive algorithms permitting the determination of regular invariant curves. The algorithms available for singular and weakly singular invariant curves are generally ill conditioned when an extension of their germ is attempted. A numerical extension requires computations with many significant digits. Reliable numerical results can rarely be obtained without the inclusion of some analytically formulated redundancy tests, because at present there are no explicit global estimates of x- and y- errors incurred after, say, N iterations of the recurrence (0-22a). Such tests have been used in the numerical determination of all invariant curves presented in this monograph. The detailed nature of these tests is unfortunately not algorithm-independent, and for this reason their discussion must await another occasion.

## 2.3 Influence domain of a stable point singularity

Consider a stable focus or node $\bar{M} = (\bar{x},\bar{y})$ of the recurrence (0-22a). Because the invariant curves near $\bar{M}$ are qualitatively the same as in the case of a linearized recurrence, $\bar{M}$ is an isolated singular point, i.e. it is not infinitesimally close to another singular point of (0-22a). Let M = (x, y) be a (non singular) point close to $\bar{M}$, so that the Euclidean distance

$$s(M, \bar{M}) = \left[(x-x)^2 + (y-y)^2\right]^{\frac{1}{2}} < \varepsilon \qquad , \tag{2-55}$$

where $\varepsilon > 0$ is a sufficiently small number. Designate the k-th consequent of M by $M_k$, k = 1, 2, .... The asymptotic stability of $\bar{M}$ implies the inequality

$$s(M_k, M) < \varepsilon_k(\varepsilon) \qquad , \tag{2-56}$$

where $\varepsilon_k \to 0$ as $k \to \infty$, and max $\varepsilon_k \to 0$ as $\varepsilon \to 0$. (2-56) restates simply the existence of a small (possibly infinitesimal) neighbourhood $D_\varepsilon$ of $\bar{M}$, such that M and $M_k$ are also points of $D_\varepsilon$, and the $M_k$ approach $\bar{M}$ asymptotically (which is guaranteed by the existence of invariant curves "terminating" at $\bar{M}$). A physically, biologically, etc. more relevant problem consists, however, in the determination of the maximal phase plane region D containing $\bar{M}$, such that all M and $M_k$ belong to D and $\lim_{k\to\infty} M_k = \bar{M}$. D is called the total influence domain of $\bar{M}$. Like in the case of first order recurrences it need not be singly connected. The phase plane region $D_\varepsilon$ defined by $s(M,\bar{M}) < \max \varepsilon$ is a first (and generally a rather modest) estimate of D.

In the case of a stable focus and a stable node (simple, critical or bi-critical), the value of max $\varepsilon$ can be estimated from a (generalized) Liapunov V-function (see for example [N 1]). Let $V(x, y)$ be a continuous positive-definite function, vanishing only at $M = (x, y)$. If the inequality

(2-57) $$\Delta V = V(g(x,y), f(x, y)) - V(x, y) < 0$$

holds inside the region defined by (2-55), then the region $D_\varepsilon$ : $V(x, y) < \varepsilon$ is located inside the influence domain D of M. In fact, the inequality (2-57) is a consequence of $s(M_{k+1}, \bar{M}) < s(M_k, \bar{M})$ and $\lim_{k\to\infty} M_k = \bar{M}$.

The inverse problem of deducing the asymptotic stability of $\bar{M}$ from the inequalities $V > 0$ and $\Delta V < 0$ is much more complicated, because these inequalities are local in character and convey no information on invariant curves passing close to $\bar{M}$, but originating outside of the region $D_\varepsilon$. Such a situation arises frequently when the recurrence (0-22a) possesses stochastic properties. Hence, a stability argument based on Liapunov functions can be used only if it is known beforehand that an $\varepsilon$-neighbourhood of a given fixed point is free of "foreign" invariant curves. The content of this restriction will become clearer after the study of some properties of non conservative heteroclinic points (Chapter IV).

A much better estimate, and often even the exact extent of the influence domain D of a stable fixed point can be deduced from the knowledge of singular invariant curves. Consider as an example the recurrence [G 4]

(2-58) $$x_{n+1} = y_n, \qquad y_{n+1} = \frac{4}{5} x_n + x_n^2 + y_n^2 .$$

By means of a computation of numerous discrete half-trajectories it was established that the dynamics of (2-58) are orderly, i.e. there are no stochastic effects. The fixed point $O = (0,0)$ of (2-58) is a stable star-node of type 2 ($\lambda_1 = 2/\sqrt{5}$, $\lambda_2 = -2/\sqrt{5}$). If $V = x^2 + \frac{3}{2} y^2$, then

$$\Delta V = -\frac{1}{25} x^2 - \frac{1}{2} y^2 + \frac{12}{5} x(x^2+y^2) + \frac{3}{2}(x^4 + y^4) ,$$

and $\Delta V < 0$, provided $|x| < 0.1$, $|y| < 0.1$. A part of the influence domain D of O is therefore the interior $D_\varepsilon$ of the phase plane ellipse $V(x_n, y_n) = x_n^2 + \frac{3}{2} y_n^2 = 0.02$. The recurrence (2-58) admits, however, a second fixed point $P = (0.1, 0.1)$, which is a saddle of type 2 ($\lambda_{1,2} = -0.1 \pm \sqrt{1.01}$, $p_{1,2} = -0.1 \mp \sqrt{1.01}$). The inverse recurrence of (2-58) is doublevalued

(2-58a) $$x_n = -0.4 \pm (0.16 + y_{n+1} - x_{n+1}^2)^{1/2} , \qquad y_n = x_{n+1} .$$

The fixed points O and P have thus the excess antecedents $O_{-1} = (-0.8, 0)$ and $P_{-1} = (-0.9, 0.1)$. Let the evolution on an invariant curve of a discrete consequent half trajectory $\{x_i, y_i\}$, $i = 0, 1, 2, \ldots$ be indicated by arrows, pointing in the direction of increasing i. Consider the four (singular) invariant curve branches crossing a saddle. A branch on which the arrows point away from a saddle is called a consequent branch, and one on which the arrows point towards the saddle, an antecedent branch. Since one of the consequent branches crossing P goes to infinity and the other to the star-node O, the saddle P lies on the boundary $\mathcal{C}$ of the influence domain D of O, and so does its excess antecedent $P_{-1}$. Other points of $\mathcal{C}$ lie on the antecedent invariant curve branches crossing P, and on the corresponding excess antecedent curves crossing $P_{-1}$. The excess antecedent curve of the consequent segment $P \to O$ goes of course from $P_{-1}$ to $O_{-1}$. The complete phase portrait of (2-58) is shown in Fig. 2-7. The antecedent, and excess antecedent invariant curve branches of P form a closed contour $\mathcal{C}$, which turns out to be by construction the boundary of the immediate influence domain D of the star-node O. $\mathcal{C}$ consists of four distinct segments : $\widehat{PC}$, $\widehat{PBC_{-1}}$, $\widehat{P_{-1}C}$ and $\widehat{P_{-1}C_{-1}}$ (see Fig. 2-7), where $C_{-1}$ is the antecedent of C. D is thus much larger than the $D_\varepsilon$ deduced from $V > 0$ and $\Delta V \leqslant 0$.

The points C and $C_{-1}$ are singular in the sense that they are boundary points of two distinct contiguous antecedent curve segments with different arrow directions. In order to convey the information about the arrow reversal concisely, points like C and $C_{-1}$ are called points of alternance. The point B is also a boundary point of two distinct contiguous antecedents curve segments, but it is not a point of alternance. Consider the Jacobian determinant J(x, y), eq. (2-43), of the recurrence (0-22a), or (0-22), and the curve $J_0$ defined by $J(x, y) = 0$. Let $J_n$, n = 1, 2, ... designate the successive consequents of the curve $J_0$. The curves $J_n$ are called critical curves (of rank n) of the recurrence (0-22a) or (0-22). Critical curves are natural generalizations of critical points of first order recurrences. Points of alternance are always located on critical curves of a finite rank. In the case of the recurrence (2-58), $C_{-1}$ is located on $J_0 : x_n = -2/5$, and C on $J_1 : y_n = x_n^2 - 4/25$ (see Fig. 2-7). The influence domain $D_\varepsilon$ of O, obtained by means of a Liapunov V-function, is not only much smaller than the total influence domain D, but also smaller than the "analytical" influence domain $D_a$ of O, bounded by the invariant curve segment $\widehat{CPB}$ and the critical curve segment $\widehat{CB}$ (dashed in Fig. 2-7).

Consider now the Schröder equation (2-51) associated with the fixed point O of the recurrence (2-58). Since a solution of a Schröder equation transforms a non linear recurrence ("centred" on O) into a linear recurrence (also centred on O), the radius of convergence of the series solution w(x, y), eq. (2-52) cannot exceed $R = \sqrt{2}/10$, i.e. it cannot exceed the distance between the fixed points O and P. An analytical continuation of w(x, y) is obviously possible to the whole of $D_a$, but not to the whole of D, except if the Schröder equation is suitably generalized to involve excess antecedents. The situation is therefore similar to that of first order recurrences and that of one-variable Schröder equations (cf. eq. (1-33) - (1-40)).

A more complicated boundary of the influence domain of a stable fixed point exists in the case of the recurrence [G 4], [G 10],

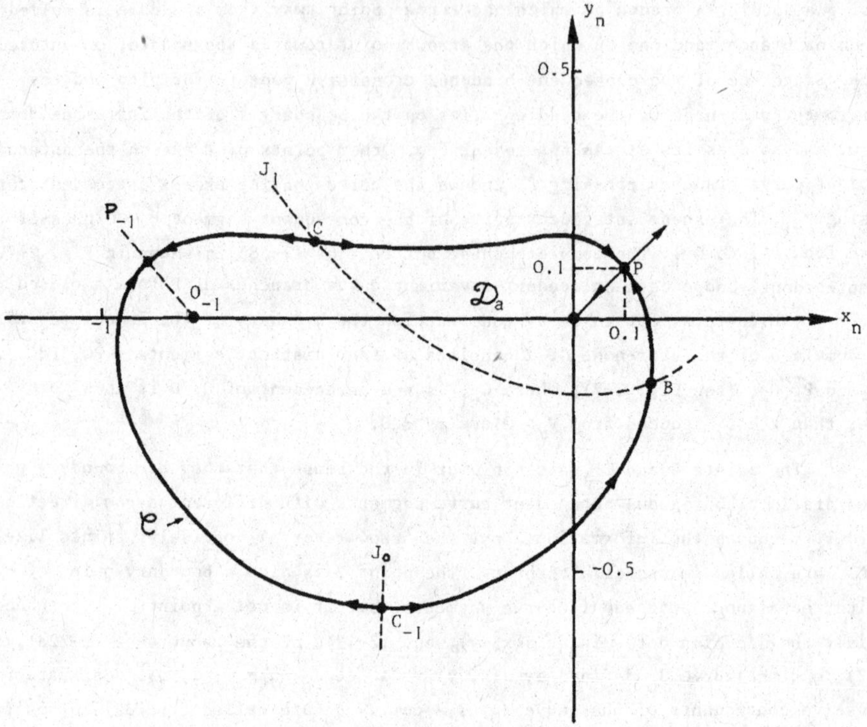


Fig. 2-7. Influence domain of the star-node O of the recurrence (2-58).
$\mathcal{D}_a$ : analytical influence domain.

(2-59) $$x_{n+1} = y_n, \qquad y_{n+1} = \frac{1}{2}x_n + \frac{1}{10}x_n^2 + c\,x_n y_n ,$$

where c is a parameter. At least for $0.09 \leqslant c \leqslant 0.11$, the phase plane portrait of (2-59) is essentially stochaticity-free. The fixed point $O = (0,0)$ is a stable star-node with the eigenvalues $\lambda_{1,2} = \pm\sqrt{2}/2$, independently of the value of c. The inverse of (2-59) is double-valued

(2-59a) $$y_n = x_{n+1}, \qquad x_n = -5\left(\frac{1}{2} + cy_n\right) \pm 5\left[\left(\frac{1}{2} + cy_n\right)^2 + \frac{2}{5}y_{n+1}\right]^{\frac{1}{2}} .$$

O possesses one excess antecedent $O_{-1} = (-5,0)$. For $c = 0.09$, another fixed point of (2-59) is the unstable node $F = (\bar{x}, \bar{y})$ of type 2 ($\bar{x} \simeq 2.613$, $\bar{y} \simeq 2.613$, $\lambda_1 \simeq 1.249$, $\lambda_2 = -1.012$, $p_1 \simeq 1.25$, $p_2 \simeq -1.01$). Near F there exists a cycle D, E of order two, which is a saddle of type 1 with the eigenvalues $\lambda_1 \simeq 1.500$, $\lambda_2 \simeq 0.950$. The corresponding coordinates and eigenslopes are : $D = (5,0)$, $p_1 \simeq 0.00$, $p_2 \simeq -1.22$ ; $E = (0,5)$,

$p_1 \simeq 99.0$, $p_2 \simeq -0.78$. The three singular points D, E, F possess the excess antecedents $D_{-1} = (-10, 0)$, $E_{-1} \simeq (-9.500, 5)$, $F_{-1} \simeq (-10, 2.613)$, whose antecedents are $D_{-2} \simeq (-5.155, -9.500)$, $D'_{-2} \simeq (9.065, -9.500)$, $E_{-2} = (0, -10)$, $E'_{-2} = (4, -10)$, $F_{-2} \simeq (-3.506, -10)$ and $F'_{-2} \simeq (7.506, -10)$. Tracing the invariant curve branches which cross D and E, and taking into account the successive excess antecedents, yields the phase plane portrait shown in Fig. 2-8.



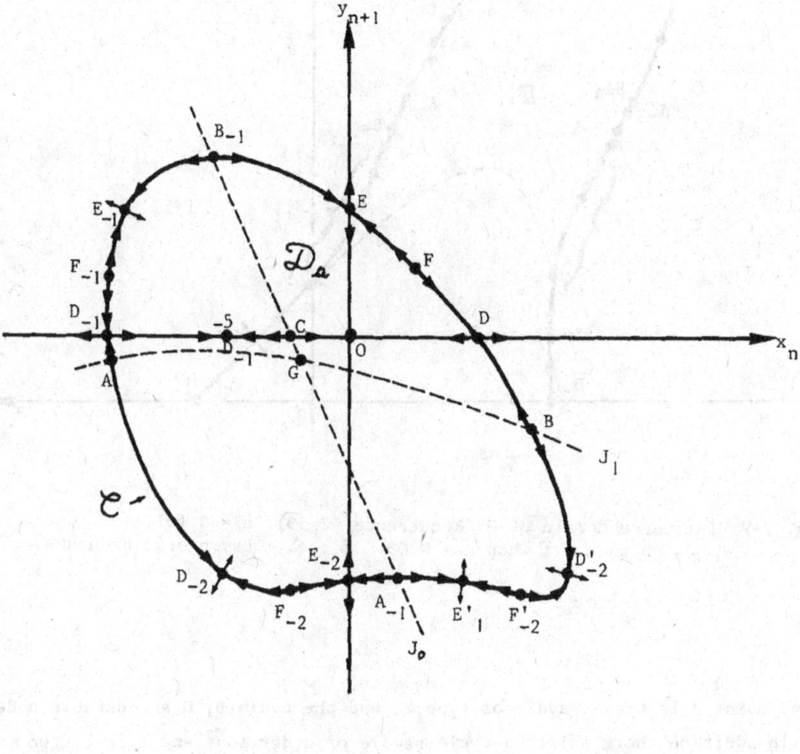

Fig. 2-8. Influence domain of the star-node O of the recurrence (2-59) for c = 0.09. $\mathcal{D}_a$ = analytical influence domain.

There exist four points of alternance A, B, $A_{-1}$, $B_{-1}$ on the closed contour $\mathcal{C}$, defining the boundary of the immediate influence domain $\mathcal{D}$ of O. The analytical influence domain $\mathcal{D}_a$ of O consists only of the area enclosed by the curve segments $\widehat{BDFEB}_{-1}$, $\widehat{B_{-1}CG}$ and $\widehat{GB}$ (see Fig. 2-8). Like in the preceding example, an analytical solution of the associated Schröder equation (2-51) exists inside the whole of $\mathcal{D}_a$.

The recurrence (2-59) possesses the peculiar property that the size of $\mathcal{D}$ varies rapidly with c. For c = 1.01, a part of the boundary $\mathcal{C}$ of $\mathcal{D}$ is shown in Fig. 2-9.

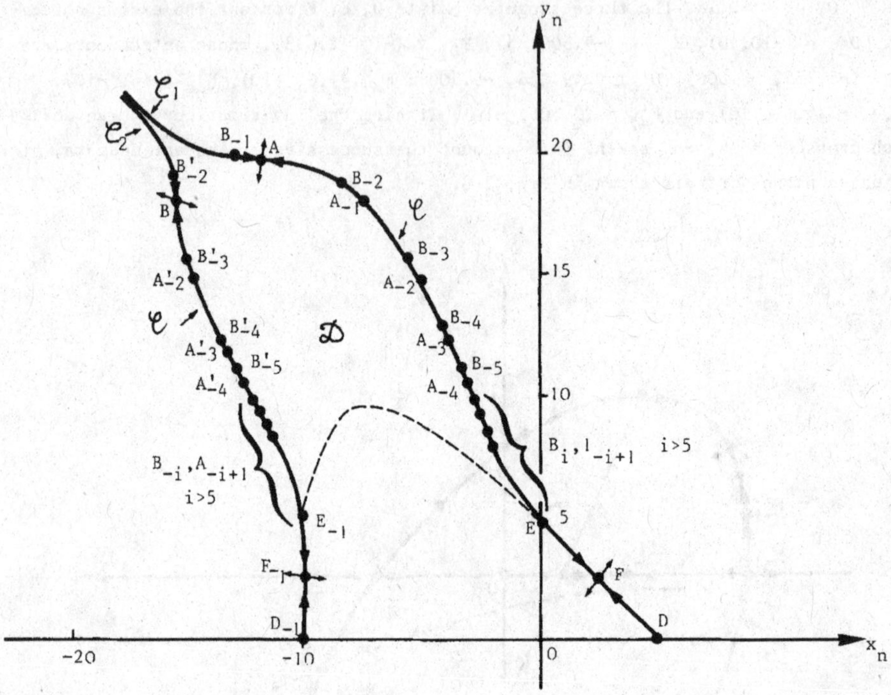


Fig. 2-9. Influence domain of O. Recurrence (2-59), $c = 0.101$.
--- Boundary $\mathcal{C}$ when $c = 0.09$. $\mathcal{C}_1$, $\mathcal{C}_2$ : asymptotic branches.

The fixed point F is now a saddle of type 2, and the cycle D, E an unstable node of type 1. In addition there exists a cycle saddle of order four and type 2, two points of which are denoted by A and B. The boundary $\mathcal{C}$ of $\mathcal{D}$ contains an infinity of antecedents of A, B (denoted in Fig. 2-9 by $A_i$, $B_i$, $A'_i$, $B'_i$, $i = -1, -2, \ldots$) possessing two accumulation points : the node E and its excess antecedents $E_{-1}$. $\mathcal{C}$ has also two unbounded asymptotic branches $\mathcal{C}_1$ and $\mathcal{C}_2$. Compared to the case $c = 0.09$ (dashed in Fig. 2-9), the area of $\mathcal{D}$ has increased substantially. This increase does not occur gradually as c increases from 0.09 to 1.01, but suddenly at $c = 0.1$. When c increases further, the size of $\mathcal{D}$ changes slowly but the boundary curve $\mathcal{C}$ becomes more and more jagged (example : $c = 0.11$, Fig. 2-10). The complicated shape of $\mathcal{C}$ is due to the existence on $\mathcal{C}$ of a very large number of cycles, whose sequence of generating bifurcations is similar to that of a first-order recurrence. For $c = 0.11$ several cycle sets of the type $k = m.2^i$, $m > 4$, $i = 1, 2, \ldots$, have been identified [G 4].

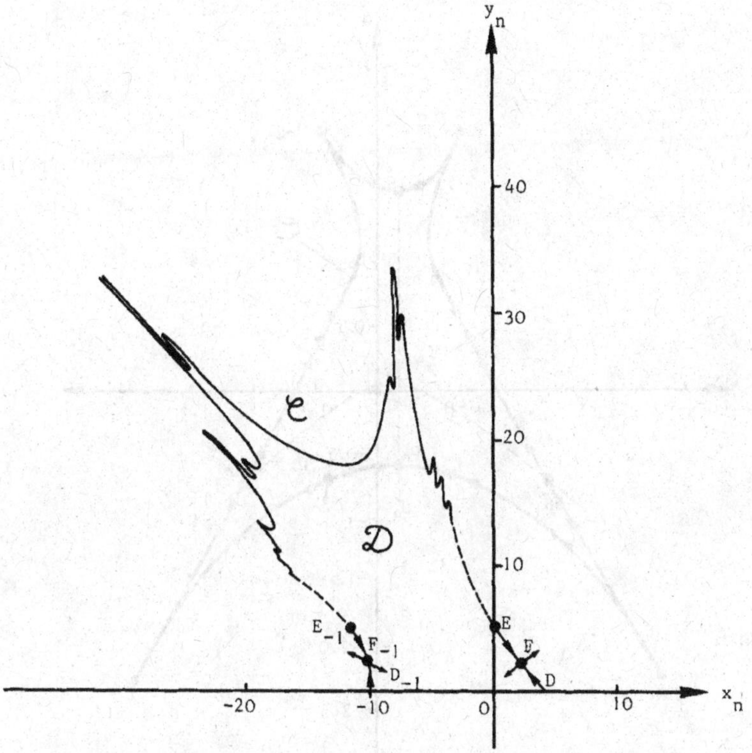


Fig. 2-10. Influence domain of O. Recurrence (2-59), c = 0.11.

The influence domain of the fixed point O = (0,0) of the recurrences (2-58) and (2-59) is always singly-connected, i.e. the total influence domain coincides with the immediate influence domain (which is larger than the analytical influence domain). Similarly to what was observed in the case of first-order recurrences (cf. Fig. 1-4), an influence domain may consist of several disjoint parts. Consider in fact the recurrence [G4], [G 10]

(2-60) $$x_{n+1} = y_n, \qquad y_{n+1} = 2c\,x_n - y_n - x_n^2 \quad ,$$

where c is a parameter. Its inverse is

(2-60a) $$y_n = x_{n+1}, \qquad x_n = c \pm \left[c^2 - y_n - y_{n+1}\right]^{\frac{1}{2}}$$

When c = -0.15, the fixed point O = (0,0) is a stable node of type 1, with distinct eigenvalues ($\lambda_1 \simeq 0.947$, $\lambda_2 \simeq 0.0528$). The influence domain D of O is shown in Fig.2-11

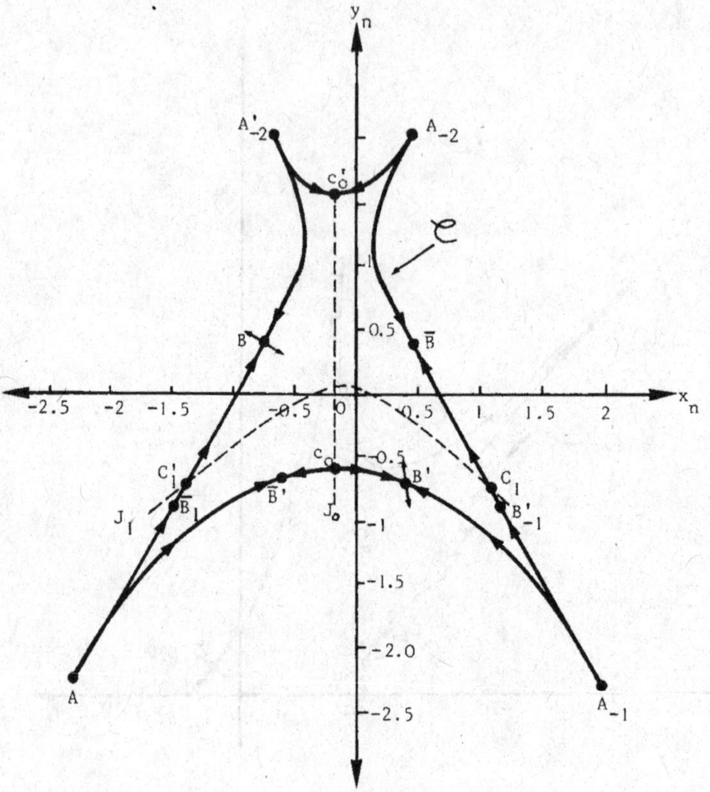


Fig. 2-11. Influence domain of the node O. Recurrence (2-60), c = -0.15.
A : unstable node, B, B' : cycle saddle of type 3.

The boundary $\mathcal{C}$ of D is given by two consequent invariant curves starting at the unstable node A = $(\bar{x}, \bar{y})$ of type 2 ($\bar{x} = -2 + 2c$, $\bar{y} = -2 + 2c$, $\lambda_1 \simeq -2.488$, $\lambda_2 \simeq 1.488$), and the antecedent invariant curves crossing the cycle saddle B, B' of order two. The coordinates of the points B, B' are given by

$$x = c \pm \sqrt{c^2 - 2c} \quad , \qquad y = -(1 - 2c)x - x^2 .$$

For c = -0.15, one has B ≃ (-0.7179, 0.4179), B' ≃ (0.4179, -0.7179), $\lambda_1 \simeq -2.604$ and $\lambda_2 \simeq -0.769$, i.e. the saddles are of type 3. The fixed point A has an excess antecedent $A_{-1}$ = (2, -2.3), which in turn has two distinct antecedents $A_{-2}$ ≃ (0.4179, ,2), $A'_{-2}$ ≃ (-0.7179, 2). The excess antecedents of B, B' are $\bar{B}$ ≃ (0.4179, 0.4179), $\bar{B}'$ ≃ (-0.7179, -0.7179), $B_{-1}$ ≃ (1.0576, -0.7179) and $B'_{-1}$ ≃ (-1.3576, -0.7179). The contour $\mathcal{C}$ contains four points of alternance $C_o$, $C'_o$, $C_1$, $C'_1$, located on the critical curves $J_o : x_n = c$ and $J_2 : y_n + \frac{1}{2}x_n^2 - cx_n + \frac{1}{2}c^2 = 0$, respectively. The critical curve $J_1 : y_n + x_n - c^2 = 0$, which is also described by the vanishing discriminant of the inverse recurrence (2-60a), divides the phase plane into two regions, one admitting two antecedents and the other none.

As c increases, the unstable node A and its antecedents are little affected, whereas the cycle B, B' undergoes a qualitative change. When c = -0.05, the points and the eigenvalues of the latter become B ≃ (-0.3702, 0.2702), B' ≃ (0.2702, -0.3702) and $\lambda_1 \simeq -1.178$, $\lambda_2 = 0.179$, respectively. The saddles are now of type 2. The antecedent invariant curves crossing B and B' have undergone a bifurcation. After coming closer at the waist of $\mathcal{D}$ (see Fig. 2-11), they become tangent to the critical curve $J_1$, and then they separate. In this process, the boundary $\mathcal{C}$ of the influence domain $\mathcal{D}$ of O becomes doubly connected, and $\mathcal{D}$ separated into two disjoint parts (Fig. 2-12).

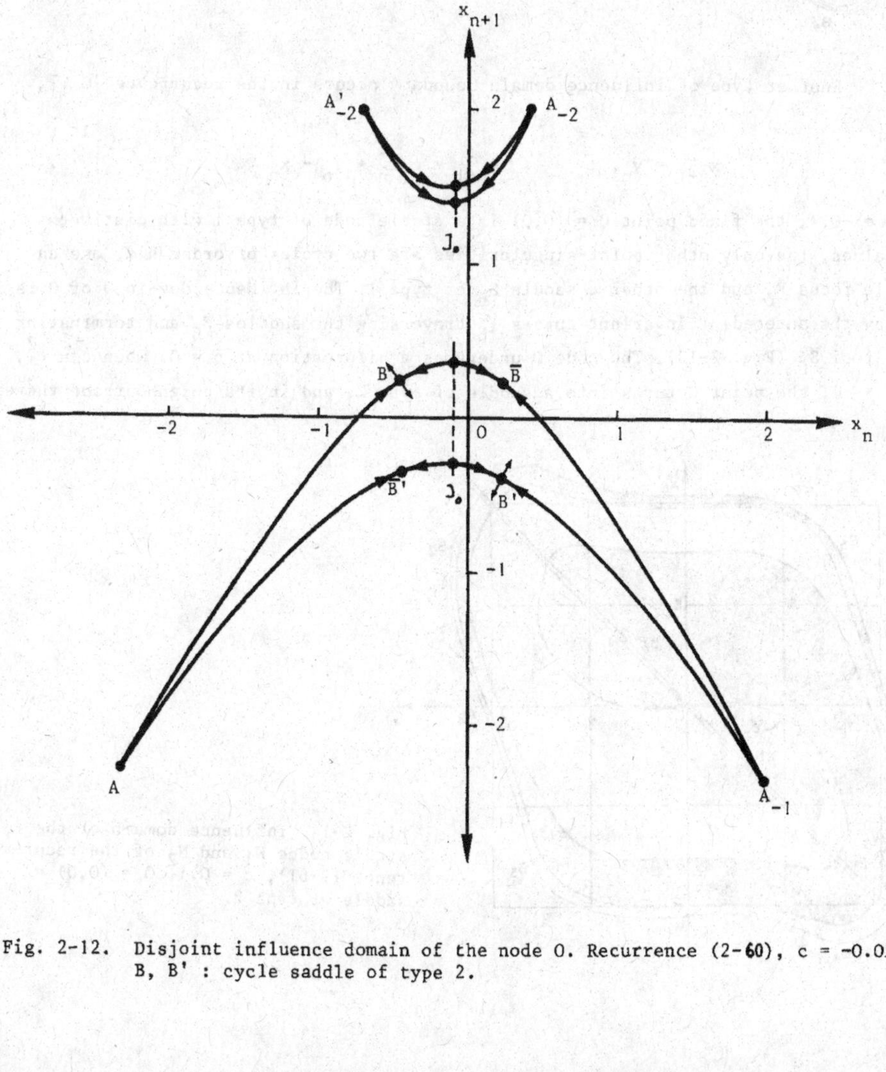


Fig. 2-12. Disjoint influence domain of the node O. Recurrence (2-60), c = -0.05. B, B' : cycle saddle of type 2.

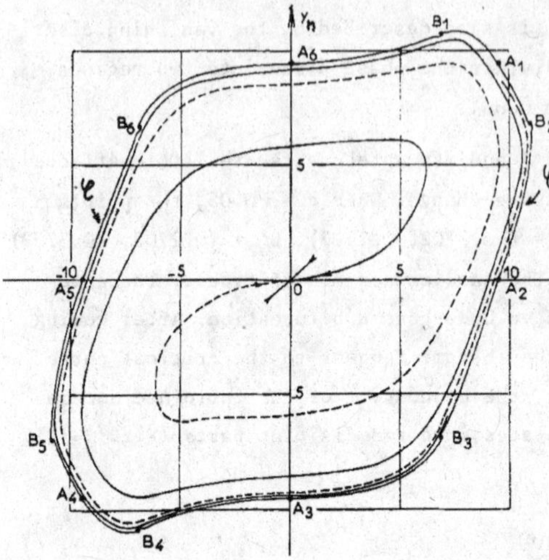


Fig. 2-13. Stability boundary defined by a pair of cycles. $A_i$ = saddles, $B_i$ = unstable foci. Recurrence (2-61), $c = -0.1$, $0 = (0,0)$ = stable node.

Another type of influence domain boundary occurs in the recurrence [G 4], [G10]

(2-61) $$x_{n+1} = y_n, \qquad y_{n+1} = c\,x_n + y_n - x_n^3 \quad .$$

When $c = -0.1$, the fixed point $0 = (0,0)$ is a stable node of type 1 with distinct eigenvalues. The only other point-singularities are two cycles of order six, one an unstable focus $F_6$ and the other a saddle $S_6$ of type 1. The influence domain D of 0 is given by the antecedent invariant curves $\mathcal{C}_6$ traversing the saddles $F_6$ and terminating at the foci $S_6$ (Fig. 2-13). The node 0 undergoes a bifurcation at $c = 0$. When $c = \varepsilon^2$, $0 < \varepsilon^2 \ll 1$, the point 0 turns into a saddle of type 2, and in its neighbourhood there

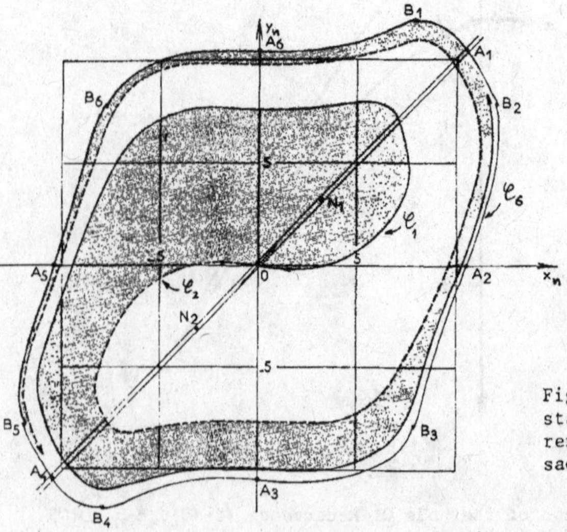


Fig. 2-14. Influence domain of the stable nodes $N_1$ and $N_2$ of the recurrence (2-61), $c = 0.1$, $0 = (0,0)$ = saddle of type 2.

appear two stable nodes $N_1$ and $N_2$, whose abscissae are $x = \pm\varepsilon$. The cycle pair $F_6$, $S_6$ is qualitatively unchanged. The influence domains $D_1$, $D_2$ of $N_1$, $N_2$ are defined on one hand by the invariant curves $\mathcal{C}_6$, and on the other hand by the antecedent invariant curves $\mathcal{C}_1$ and $\mathcal{C}_2$ crossing the saddle O. The curves $\mathcal{C}_1$ and $\mathcal{C}_2$ spiral towards the curve $\mathcal{C}_6$. An illustration ($\varepsilon = 0.1$) of the shape of these curves is given in Fig. 2-14. A similar situation occurs also in the case of the recurrence [G 4], [G 10] :

$$(2\text{-}62) \qquad x_{n+1} = y_n,\ y_{n+1} = \frac{1}{10}(2y_n - 5x_n)(x_n^2 + y_n^2 - \frac{6}{5}x_n - c\,y_n + \frac{1}{2}),\quad \frac{11}{10} \leqslant c \leqslant \frac{13}{10}\ ,$$

except that the invariant curves $\mathcal{C}_1$, $\mathcal{C}_2$ crossing O pass through additional point-singularities, and the fixed points $N_1$, $N_2$ are replaced by cycles. The analysis of D is as straightforward as in the preceding examples, but it involves too many tedious details to be described here (cf. [G 10]).

In the examples (2-58) to (2-62), the key to the determination of the influence domain D of a stable fixed point was the existence of point-singularities which happened to lie on the boundary $\mathcal{C}$ of D. It is, however, also possible to encounter a situation where the boundary $\mathcal{C}$ of D does not traverse any point-singularity. Consider as an example the predator-prey recurrence (2-6) with positive parameters a, b, c, K. (2-6) admits three fixed points O = (0,0), A = $(\bar{x}, \bar{y})$ and B(K, 0) where the coordinates $\bar{x} > 0$, $\bar{y} > 0$ are roots of two algebraic equations which cannot be solved explicitly (cf. Chapter V). Assume that the parameters a, b, c, K are so chosen that A is a stable focus, while the fixed points O and B are unstable. When a, b, K are kept fixed and c is made to increase, the focus A becomes unstable beyond a certain "critical" value $c = c_0$. When $0 < c-c_0 \ll 1$, the unstable focus A is surrounded by what appears to be an asymptotically stable regular invariant curve $\mathcal{L}$. The size of $\mathcal{L}$ is found to contract to A as $c-c_0 \to 0^+$. The shapes of some $\mathcal{L}$ are shown in Fig. 2-15. When this situation is viewed via one of the inverse branches of (2-6), the focus A is stable and the curve $\mathcal{L}$ constitutes the boundary of its influence domain. Is should be noted that the closed "invariant" curves $\mathcal{L}$ have not been determined by solving the functional equation (2-25) or (2-27). The entities designated by $\mathcal{L}$ in Fig. 2-15 consist merely of a few thousand individually unresolved points of a discrete consequent half-trajectory $\{x_i, y_i\}$, i = 1, 2, ..., where the first few hundred points have been omitted. The omitted points of $\{x_i, y_i\}$ are deemed to represent a transient approaching asymptotically the dynamic stationary state described by $\mathcal{L}$.

## 2.4 Approximate determination of an isolated closed invariant curve

By analogy with the theory of differential equations, it is possible to study the bifurcation of a centre (in linear approximation), when the latter forms a Liapunov critical case in the transition stable focus → unstable focus. Assume that the coefficients a, b, c, d, in the recurrence (2-41) depend on an absolutely small parameter $\varepsilon$, and their values are such that the fixed point O = (0,0) of the linear recurrence (2-19) is a stable focus when $\varepsilon < 0$, en centre when $\varepsilon = 0$ and an unstable focus when $\varepsilon > 0$. By means of linear transformation, the recurrence (2-41) can be transformed to the canonical form

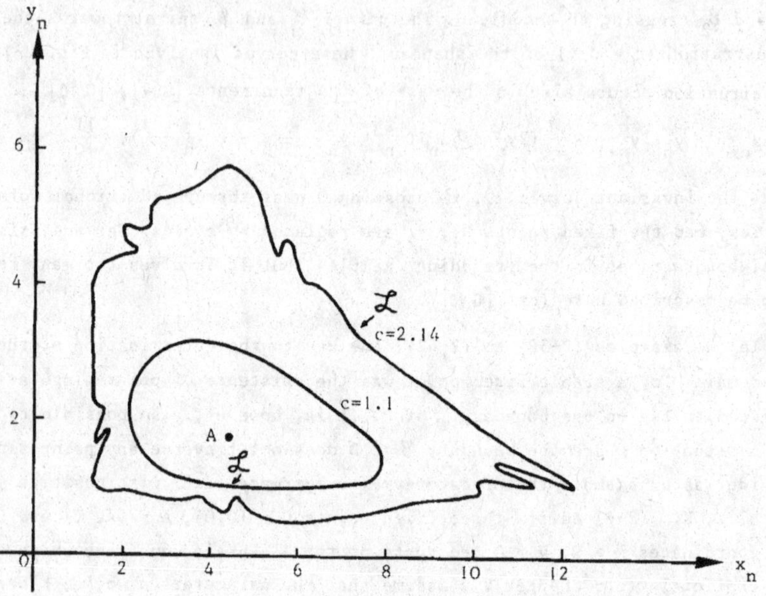


Fig. 2-15. A stability boundary $\mathcal{L}$ without point-singularities. One inverse branch of the recurrence (2-6). A : fixed point. Position of A depends on c, a = b = 1, K = 10.

$$\text{(2-63)}\quad \begin{aligned} r_{n+1} &= (1+\alpha)(r_n \cos\varphi - s_n \sin\varphi) + R(r_n, s_n, \varepsilon), \quad R(0,0,\varepsilon) = 0,\ \alpha = \alpha(\varepsilon), \\ s_{n+1} &= (1+\alpha)(r_n \sin\varphi + s_n \cos\varphi) + S(r_n, s_n, \varepsilon), \quad S(0,0,\varepsilon) = 0,\ \alpha(0) = 0, \end{aligned}$$

where the functions R, S, admit convergent Mc Laurin developments, which start with at least quadratic terms. It is well known that there exist non linear almost-identity transformations, with the help of which the recurrence (2-63) can be brought into the partially decoupled polar form (cf. [C 4], [M 4])

$$\text{(2-64)}\quad \begin{aligned} \rho_{n+1} &= (1+\alpha)\rho_n + G_1 \rho_n^3 + \rho_n^5 . p(\theta_n, \varepsilon) + \dots , \quad p(\theta_n + 2\pi, \varepsilon) = p(\theta_n, \varepsilon) , \\ \theta_{n+1} &= \theta_n + \varphi + H_1 \rho_n^2 + \rho_n^4 . q(\theta_n, \varepsilon) + \dots , \quad q(\theta_n + 2\pi, \varepsilon) = q(\theta_n, \varepsilon) , \end{aligned}$$

where $G_1$, $H_1$ are constants and p, q known trigonometric polynomials in $\theta_n$. The omitted terms are of higher order in $\rho_n$. The (so called Cigala) constants $G_1$, $H_1$ depend of course on $\varphi$ (and sometimes also on $\varepsilon$). Assume for simplicity that the functions R, S are not too special, so that $G_1 \neq 0$. If $G_1 = 0$, then (2-64) can be replaced by another partially decoupled recurrence. If $G_1 \neq 0$, then in general $O(G_1) = O(\varepsilon^\circ)$, $O(H_1) = O(\varepsilon^\circ)$, where $O(a) < O(b)$ is an abreviation for the inequality $|a| < k|b|$, k = positive constant independent of b.

It is plausible (but unfortunately not yet absolutely certain) that the solution of (2-63), and thus of (2-64), is always a smooth function of $\varepsilon$. If it is, a particular family of solutions passing through a given initial point $(\bar{\rho}, \bar{\theta})$, can be sought in the form

$$\text{(2-65)} \qquad \rho_n = \rho_o + \nu(\varepsilon)\rho_o^3 u_n + \dots \quad , \quad \theta_n = n\varphi + \theta_o + \mu(\varepsilon)\rho_o^2 v_n + \dots, \; \rho_o = \rho_o(\varepsilon) ,$$

where $\theta_o$ is an arbitrary constant, $\rho_o$, $\nu$, $\mu$ undetermined functions of $\varepsilon$, and $u_n$, $v_n$ undetermined functions of n. Moreover, $u_n$ is assumed to be a bounded periodic function of $(n\varphi + \theta_o)$ of period $2\pi$. Inserting (2-65) into (2-64) and collecting terms of the lowest possible order in $\varepsilon$, yields

$$\alpha\rho_o + G_1\rho_o^3 = 0 .$$

If $G_1 > 0$, then the only (real) solution of this algebraic equation is $\rho_o = 0$. If $G_1 < 0$, then

$$\text{(2-66)} \qquad \rho_o = +\sqrt{-\alpha/G_1} \qquad \text{and} \qquad O(\rho_o^2) = O(\alpha) \quad ,$$

the dependence of $\alpha$ on $\varepsilon$ being known by definition. Analyzing the order of magnitude of the remaining terms, one finds $O(\nu) = O(\rho_o^2)$. Several possibilities exist for the orders of $\mu$ and $p(\theta_n, \varepsilon)$. In the least degenerate case of the recurrence (2-63), one has $O(\mu) = O(H_1)$ and $O(p(\theta_n,0)) = O(\varepsilon^o)$, and thence

$$\text{(2-67)} \qquad u_{n+1} = u_n + p(n\varphi + \theta_o, 0), \qquad v_{n+1} = v_n + H_1$$

The solution (2-65) describes parametrically an isolated closed curve when $u_n$ is a periodic function of $(n\varphi + \theta_o)$, as above. The constant $\theta_o$, as well as the arbitrary constant contained in $v_n$, are defined uniquely by the initial condition $(\rho_n, \theta_n)_{n=0} = (\bar{\rho}, \bar{\theta})$. Summarizing the argument it can be said that an approximate description of a closed invariant curve, analoguous to a limit cycle of a differential equation, is given by

$$\text{(2-68)} \qquad \begin{aligned} \rho_n &= \rho_o + \rho_o^5 u_n + \dots \quad , \qquad \rho_o = \sqrt{-\alpha/G_1} \; , \\ \theta_n &= n(\varphi + H_1\rho_o^2) + \theta_o + \dots \quad , \\ u_n &= \text{periodic solution of } u_{n+1} = u_n + p(n\varphi + \theta_o, 0) \quad . \end{aligned}$$

In an alternate method, (2-64) is inserted into the functional equation of invariant curves $\rho_{n+1} = w(\theta_{n+1})$ :

$$\text{(2-69)} \qquad (1+\alpha)\rho + G_1\rho^3 + \rho^5.p(\theta, \varepsilon) + \dots = w(\theta + \varphi + H_1\rho^2 + \rho^4.q(\theta,\varepsilon) + ..), \; w(\bar{\theta}) = \bar{\rho} ,$$

and a solution of the form

$$\rho = w(\theta) = \rho_o + \mu(\varepsilon)\, \rho_o^3 .\, w_3(\theta) + \dots \quad , \qquad w_3(\theta+2\pi) = w_3(\theta) \quad ,$$

is sought. After carrying out an order-of-magnitude analysis, one finds a result entirely analogous to (2-68) :

(2-70)
$$\rho = w(\theta) = \rho_o + \rho_o^5 w_3(\theta) + \ldots \quad , \quad \rho_o = \sqrt{-\alpha/G_1} \quad ,$$
$$w_3 = \text{periodic solution of } w_3(\theta+\varphi) - w(\theta) = p(\theta, 0)$$

It is noteworthy that (2-70) does not depend on $H_1$ and q. Like in the case of the closed invariant curves described by

(2-34a)
$$w(\rho_n, \theta_n) = \rho_n^2 = \xi_n^2 + \eta_n^2 = \xi_o^2 + \eta_o^2 \quad ,$$

and

(2-26)
$$w(x_n, y_n) = y_n^2 + (1-a^2)x_n^2 = \rho_n^2 - a^2\rho_n^2\cos^2\theta_n \quad , \quad |a| < 1 \quad ,$$

this reflects merely the independence of $\rho_n = w(\theta_n)$ of any auxiliary ordering of "successive" $\rho_n$, $\rho_m$ by means of a solution $W(\rho, n)$ of the translation equation

$$W(W(\rho, n), m) = W(\rho, n+m)$$

Once an invariant curve $\bar{\mathscr{C}}: y = \bar{\theta}(x)$ is known, its local stability can be determined in principle in a completely straightforward manner. Consider, in fact, the recurrence (0-22a), the functional equation (2-27) and let $\varphi(x) = \theta(x) - \bar{\theta}(x)$ be a solution of the "perturbed" functional equation

(2-71)
$$G(x, \varphi(x)+\bar{\theta}(x)) = (F(x, \varphi(x)+\bar{\theta}(x))) + \bar{\theta}(F(x, \varphi(x)+\bar{\theta}(x))), \quad \varphi(\bar{x}_o) = \bar{y}_o \quad ,$$

where $(\bar{x}_o, \bar{y}_o)$ is an initial point sufficiently close to $\bar{\mathscr{C}}$ (but not on $\bar{\mathscr{C}}$, i.e. $\bar{y}_o \neq \bar{\theta}(\bar{x}_o)$). Let $\mathscr{C}$ be the invariant curve described by $y = \varphi(x)$, and $(x_o, y_o)$ an arbitrary point on $\mathscr{C}$. Consider a small neighbourhood $\mathscr{D}(x,y)$ of the curve $\bar{\mathscr{C}}$, and the parametric family of consequent half-trajectories $\{x_i, y_i\}$, $i = 0, 1, 2, \ldots$, initialized at $(x_o, y_o)$. If for every $(\bar{x}_o, \bar{y}_o) \in \mathscr{D}$, sufficiently close to $\bar{\mathscr{C}}$, the points of all $\{x_i, y_i\}$ approach $\bar{\mathscr{C}}$ asymptotically as $i \to \infty$, then $\bar{\mathscr{C}}$ is asymptotically stable. If, on the contrary, the points of $\{x_i, y_i\}$ leave $\mathscr{D}$ for at least one $\mathscr{C}$, no matter how close to $\bar{\mathscr{C}}$ the initial point $(\bar{x}_o, \bar{y}_o)$ of $\mathscr{C}$ is chosen, then $\bar{\mathscr{C}}$ is unstable. If $\bar{\mathscr{C}}$ is neither asymptotically stable nor unstable, then it is simply but not necessarily Liapunov-stable. In many cases local asymptotic stability (or instability) of $\bar{\mathscr{C}}$ can be deduced from the linear approximation of (2-71), i.e. from the variational equation of $\bar{\mathscr{C}}$:

(2-72)
$$\left[G_y(x,\bar{\theta}(x)) - F_y(x,\bar{\theta}(x)).\bar{\theta}'(F(x,\bar{\theta}(x)))\right]\varphi(x) = \varphi(F(x, \bar{\theta}(x))), \quad \varphi(\bar{x}_o) = \bar{y}_o .$$

Generalized eigenvalues, or more precisely, (Liapunov) characteristic numbers of $\bar{\mathscr{C}}$ can be defined in terms of the logarithmic rate of convergence of $\mathscr{C}$ to $\bar{\mathscr{C}}$ (or divergence of $\mathscr{C}$ from $\bar{\mathscr{C}}$). A necessary condition for a bifurcation of $\bar{\mathscr{C}}$ is then, as expected, the occurence of a degenerate characteristic number. Like in the case of other functional equations encountered so far, no general methods of solving (2-71) or (2-72) are presently available.

In order to illustrate the intermediate steps between (2-63) and (2-66), consider the particular recurrence

$$\text{(2-73)}\qquad \begin{aligned} r_{n+1} &= (1+\varepsilon)(r_n\cos\varphi - s_n\sin\varphi) + a(r_n^2 - s_n^2) \quad , \quad |\varepsilon| \ll 1, \quad 0 < \varphi < \pi \; , \\ s_{n+1} &= (1+\varepsilon)(r_n\sin\varphi + s_n\cos\varphi) - 2ar_n s_n \quad , \quad a = \text{constant} \; , \end{aligned}$$

which does not contain any cubic terms. In the initial linear step, $2r = x+y$, $2s = i(x-y)$, two complex-conjugate recurrences are produced :

$$\text{(2-73a)}\qquad x_{n+1} = (1+\varepsilon)e^{i\varphi}.x_n + ay_n^2, \qquad y_{n+1} = (1+\varepsilon)e^{-i\varphi}.y_n + ax_n^2 .$$

The next step consists of an almost-identity transformation

$$\xi = x + \alpha\eta^2, \qquad \eta = y + \alpha^*\xi^2 ,$$

where $\alpha$ is an undetermined complex constant, and the asterisk * represents a complex conjugate. The recurrence (2-73a) becomes

$$\text{(2-73b)}\qquad \begin{aligned} \xi_{n+1} &= (1+\varepsilon)e^{i\varphi}.\,\xi_n + \gamma\xi_n^2\eta_n + \dots \; , \\ \eta_{n+1} &= (1+\varepsilon)e^{-i\varphi}.\,\eta_n + \gamma^*\xi_n\eta_n^2 + \dots \; , \end{aligned}$$

where the omitted terms are of fourth order in $\xi_n$, $\eta_n$. The constant $\alpha$ and $\gamma$ are chosen so that

$$\text{(2-73c)}\qquad \begin{aligned} (1+\varepsilon)\left[(1+\varepsilon)e^{-i2\varphi} - e^{i\varphi}\right]\alpha &= -a \; , \\ e^{i2\varphi}\left[(1+\varepsilon)e^{-i3\varphi} - 1\right]\gamma &= -2a^2 \; . \end{aligned}$$

This particular choice makes (2-73b) free of quadratic terms. The final step, leading to a form like (2-66), requires the introduction of polar coordinates $\xi = \rho.e^{i\theta}$, $\eta = \rho e^{-i\theta}$ :

$$\text{(2-73d)}\qquad \begin{aligned} &\rho_{n+1} = (1+\varepsilon)\rho_n + G_1\rho_n^3 + \dots, \qquad \theta_{n+1} = \theta_n + \varphi + H_1\rho_n^2 + \dots \; , \\ &G_1 = \tfrac{1}{2}(\gamma e^{-i\varphi} + \gamma^* e^{i\varphi}), \qquad H_1 = \tfrac{i}{2}(\gamma e^{-i\varphi} - \gamma^* e^{i\varphi}) \; . \end{aligned}$$

The quantity of intermediate algebraic manipulations increases very rapidly with the number of terms in (2-73d), because additional almost-identity transformations become necessary, and for this reason the determination of $p(\theta, \varepsilon)$ and $q(\theta, \varepsilon)$ has been omitted. The sign of $G_1$ in (2-73d) depends on the value of $\varphi$ and the sign of the constant a in the recurrence (2-73).

It should be stressed that the approximation (2-68) and (2-70) apply only in a sufficiently small neighbourhood of the unstable focus (0,0), i.e. for $\rho_o \ll 1$. There exists, however, considerable numerical evidence that closed invariant curves continue to exist at considerable parametric distances $|\varepsilon|$ from their generating bifurcation $\varepsilon = 0$ (stable focus → unstable focus), and that they undergo several types of bifurcations before degenerating (for example, becoming non rectifiable) or disappearing entirely (for example by breaking up into disjoint segments, or by developing a contact with a singular point). The corresponding implications for the solvability of the functional equations (2-25) and (2-27) are not yet entirely clear (cf. Chapter V). Moreover, the stable focus → unstable focus bifurcation is apparently not the only one which gives rise to the appearance of a closed regular invariant curve.

## 2.5 Some relationships between recurrences, continuous iterates and differential equations

Consider the set of first-order recurrences

(2-74) $$x_{n+k} = f_k(x_n), \quad f_1(x) = f(x), \qquad k = \pm 1, \pm 2, \ldots, \; n = 0, \pm 1, \ldots,$$

$f(x)$ = single valued inside the admissible interval X of x, their explicit general solution

(2-75) $$x_{n+k} = F(n+k, x_o)$$

and the set of functional iterates

(2-76) $$f_k(x) = f_{k-1}(f(x)), \quad f_1(x) = f(x), \; f_o(x) = x \quad ,$$

where negative k and n represent antecedents (or in an equivalent vocabulary, inverse recurrences and inverse iterates). Although antecedents are generally not unique, they are also represented by (2-75), provided the function $F(n, x_o)$ is unambiguously defined for $n = -1, -2, \ldots$ and $x_o \in X$. If the function $F(n, x_o)$ happens to be also unambiguously defined for fractional and continuous values of n, then (2-75) is also a solution of the problem of fractional and continuous iterates associated with (2-76).

The solution of the functional equations (2-76), expressed in this monograph in terms of the theory of first-order recurrences $x_{n+1} = f(x_n)$, is more general than that considered in the contemporary abstract theory of fractional and continuous iterates. As an illustration of the increased scope of validity so attained, consider the Myrberg recurrence (0-11), whose k-th iterates, k = 1, 2, 4, are written for convenience in a slightly modified form :

(2-77) $$x_{n+\frac{1}{2}} = x_n^2 + c$$

(2-77a) $$x_{n+1} = x_{n+\frac{1}{2}}^2 + c = (x_n^2 + c)^2 + c$$

(2-77b) $$x_{n+2} = x_{n+1}^2 + c = (x_{n+1}^2 + c)^2 + c = \{\left[(x_n^2+c)^2+c\right]^2 + c\}^2 + c \quad .$$

When $-1.25 < c < -1.3816$ (see Table 3, Chapter I), the recurrence (0-11) possesses two fixed points, two cycles of order two, and two cycles of order four, and no cycles of a higher order. The recurrence (2-77a) possesses therefore four fixed points and two cycles of order two (fixed points of (2-77b) which are not fixed points of (2-77a)). Hence, according to a well known contemporary abstract theorem (see for example [I 1], [K 7]), (2-77a) does not admit any iterative square root (or half-iterate), in spite of the fact that it was constructed from the latter. The existence theorems in question exclude iterative square roots like (2-77), because their proofs require implicitly that cycles be simultaneously fixed points, which is of course impossible by definition. In other words, the points of the cycles of order two of (2-77a) possess no iterative square roots of the postulated type. It should be noted however, that the points of the cycles of order two of (2-77a) are the only points where (2-77) fails to be a "regular" half-iterate of (2-77a). The contemporary abstract theory of fractional and continuous iterates is therefore technically correct, but conceptually unnecessarily restrictive. Like some other contemporary abstract theories (cf. Introduction, pp. 18-21 and Chapter I, pp. 57-59), it is highly interesting in itself, but rather irrelevant in a wider context (and especially in one of an applied origin). Since the properties of recurrences are based on the existence of singularities (both with respect to the independent variable x and the parameter n), all meaningful solutions are intrinsically "regular" (in the sense of contemporary abstract theory) only almost everywhere and not everywhere, the "irregular" elements coinciding of course with the singularities. More complicated types of irregularities will be discussed in Chapter VI.

It is well known that the problem of fractional and continuous iterates (2-75) can also be solved by means of functions satisfying the Abel and the Schröder functional equations

(2-78) $$v(f(x)) = v(x) + 1 \quad ,$$

(2-78a) $$w(f(x)) = cw(x) \quad ,$$

when the value of $c \neq 0, \pm 1$ is properly chosen. After k = 1, 2, ... iterations of (2-78) and (2-79), there results

(2-79) $$v(f_k(x)) = v(x) + k \quad ,$$

(2-79a) $$w(f_k(x)) = c^k w(x) \quad .$$

When the solutions $y = v(x)$, $y = w(x)$ possess at least one inverse $x = v^{-1}(y)$, $x = w^{-1}(y)$, then a particular solution of the functional equations (2-76) is given by

(2-80) $$f_k(x) = v^{-1}(v(x) + k) \quad ,$$

(2-80a) $$f_k(x) = w^{-1}(c^k w(x)) \quad .$$

The solutions (2-80), (2-80a) coincide of course for k = 0, 1, 2, ... with the solution (2-75) of the recurrence (2-74). If the right-hand sides of (2-80), (2-80a) conserve their meaning for fractional and continuous (positive) values of k, then they represent also fractional and continuous iterates of f(x), except when a specific x is a singular element of the recurrence (2-74), or $k \to +\infty$. When $k < 0$, (2-80) and (2-80a) conserve their validity, provided $v^{-1}$, $w^{-1}$ designate all possible inverses.

Since in general v(x) and w(x) are complicated transcendental functions, as a rule, the study of $v(x)$, $w(x)$ is carried out locally. For example, a preference is given to a stable fixed point, say $x = 0$ of $x_{n+1} = f(x_n)$, and after choosing $c = f'(0)$, the solution of (2-78a) is determined in the form of a Mc Laurin series. Although the radius of convergence of this series is small, $w(x)$ can be extended (analytically or otherwise, cf. [B 6]) to the whole immediate influence domain of $x = 0$ and by taking into account excess antecedents, to the total influence domain. Combining several local solutions it is possible in principle to construct a global solution of (2-78). The equivalence between (2-80) and (2-80a) implies, however, that the constant c in (2-80a) plays a practically useful, but intrinsically an inessential role. In fact, without loss of generality it can be assumed that $c > 0$. Taking the logarithms of both sides of (2-80a), and dividing by log c, one obtain (2-78) with $v = \log w$. A proper choice of c, different from an eigenvalue of a fixed point or cycle of $x_{n+1} = f(x_n)$, may therefore lead directly to a global solution of (2-78). In other words, it may be more expedient to interprete (2-78) as an eigen-problem for $f(x)$ inside the complete (admissible) interval of x, and determine the value of c accordingly. Somé rather unexpected results can be obtained in this way (see Chapter VI). Moreover, it should be recalled that the eigenvalues $\lambda_{1,2}$ of a linear second order recurrence are only invariant with respect to inversible linear or inversible non linear almost-identity transformations. For more general inversible non linear transformations, the $\lambda_{1,2}$ change, even when the linearity of the recurrence is preserved. As an illustration consider the recurrence (2-19) when $\lambda_{1,2}$ are complex conjugate (eq. 2-32)). The solutions $w_1$, $w_2$ of the Schröder equations (2-49) are then also complex-conjugate, regardless of whether the f, g are taken from (2-19), (2-29) or from the real-valued canonical form (2-33). After a transformation to polar coordinates, however, the eigenvalues are real : $\lambda_1 = \sigma$, $\lambda_2 = 1$ (see eq. (2-33a)), and so are the corresponding solutions $w_1$, $w_2$ of the Schröder equations

(2-49a) $$w_1(\sigma\rho,\theta) = \sigma w_1(\rho,\theta),\ w_2(\sigma\rho,\theta) = w_2(\rho,\theta),\ w_1(0,\theta) = w_2(0,\theta) = 1$$

From the preceding argument it can be concluded that the solvability of the problem of fractional and continuous iterates is essentially conditioned by the extendability of the function $F(n, x_o)$ in (2-75) from n = positive integers to n ≠ positive integers, which depends only on the "intrinsic type" (or "order of transcendentality") of F. Particular cases of F are easily found where an integer → non-integer extension of n is not obvious at all. Consider, for example, the second order recurrence

(2-81) $$x_{n+1} = x_n, \quad y_{n+1} = (1+x_n)y_n, \qquad n = 0, 1, 2, \ldots .$$

When $x_o = y_o = 1$, a solution of (2-81) is given by

(2-82) $$x_n = n, \qquad y_n = n! ,$$

which is completely indeterminate for $n \neq 0, 1, 2, \ldots$ In principle, it is possible to extend n! to non-integer values in an arbitrary way. It is merely by convention that n! has been imbedded for all n into the Γ-function by means of the integral representation

$$\text{(2-83)} \qquad n! = \Gamma(n+1) = \int_0^\infty e^{-z}.z^n\,dz$$

The imbedding (2-83) does not follow from any analyticity considerations (in a real or complex domain), but from the continuity of the differentiation operator $D^n$, $D = \frac{d}{dx}$, for non-integer n (cf. [C 13], vol. 2, pp. 191-201). $D^n$ is defined by the symbolic identity

$$\text{(2-84)} \qquad D^n(\frac{1}{s}) = \int_0^\infty (D^n\, e^{-su})du = (-1)^n \int_0^\infty e^{-su}.u^n\,du = \frac{(-1)^n}{s^{n+1}}\,n! \qquad ,$$

which is valid without any restriction for $s > 0$ and $n = 0, 1, 2, \ldots$ Since the last integral in (2-84) converges not only for $n = 0, 1, 2, \ldots$ but also for any $n \neq -1, -2, \ldots$ it is permissible to cancel the factor $(-1)^n$ in the last two terms, and then let $s \to 1$. What distinguishes $\Gamma(n+1)$ "transcendentally" from all other extensions of n! is its logarithmic convexity (an intrinsic type of functions of a real variable). The definition (2-83) is entirely natural in a dynamic context because derivatives of a non-integer order arise in several physical problems. If possible, similar types of imbedding should be sought in other cases of non-unique extendability of $F(n,x_o)$.

Consider now the classical first-order example

$$\text{(2-85)} \qquad x_{n+1} = f(x_n) = e^{x_n}, \qquad -\infty < x_n < +\infty \; .$$

Since this recurrence has no point-singularities (except at infinity), it is dynamically uninteresting, because it can describe only unstable transients. The inverse recurrence $x_n = \log x_{n+1}$ exists only for positive $x_{n+1}$. The formal recurrence, written around the singularity at infinity, is

$$\text{(2-85a)} \qquad y_{n+1} = g(y_n) = e^{-1/y_n} \qquad , \quad -\infty < y_n < +\infty \; .$$

Its fixed point $y = 0$ is stable like if it had a (degenerate) eigenvalue $\lambda = 0$. The influence domain of $y = 0$ is of course the whole y-axis. Since the function $e^{-1/y}$ is discontinuous at $y = 0$, it has no Mac Laurin series, but since all its derivatives vanish for $y \to 0^+$, the study of the solutions of (2-85a) and (2-85) constitutes a highly interesting topic in the theory of functions of a real variable. In particular, (2-85) and (2-85a) may be used to characterize growth rates of functions as $y_n \to 0$ or $x_n \to \infty$, when the iterates $f_k(x)$, $g_k(y)$, $k = 0, 1, \ldots$ are used as a scale. If $x_n$ is assumed to be complex-valued in (2-85), then there exists a fixed point with finite coordinates, and a local solution can be found by means of analytical methods. This solution can be extended to the real $x_n$-asis not only for $n = 0, 1, 2, \ldots$ which is trivial, but also for $n = \frac{1}{2}$ (see for example [S 9] and the references therein). This $\frac{1}{2}$-iterate is by no means unique, and there is no evidence that it is priviledged in any mathematically fundamental sense. Since neither (2-85) nor (2-85a) describes a known natural process, no dynamic insight is available for the elimination of non-uniqueness. The problem of extending the functions $f_k$, $g_k$ to non-integer values of k is therefore still unsolved (cf. Chapter VI).

Another approach to the study of fractional and continuous iterates is based on the construction of certain formal power series [B 4]. The adjective formal means that the convergence of these series is not known a priori. As an illustration consider a particular form of the recurrence (2-41)

$$u_1 = \lambda u + \sum_{m+n=2}^{\infty} a_{mn} u^m v^n \quad ,$$

(2-86)

$$v_1 = \frac{1}{\lambda} v + \sum_{m+n=2}^{\infty} b_{mn} u^m v^n \quad ,$$

where $\lambda_1 = \lambda$, $\lambda_2 = 1/\lambda$, $(\lambda \neq 0, \pm 1)$, are the eigenvalues of the fixed point O = (u = 0, v = 0), and $a_{mn}$, $b_{mn}$ are known constants. The two series in (2-86) are assumed to be convergent. u, v, $u_1$, $v_1$ are written instead of $u_n$, $v_n$, $u_{n+1}$, $v_{n+1}$ for notational reasons. The method is constructed on the assumption that the k-th iterate of (2-86) has the form

$$u_k = \lambda^k u + \sum_{m+n=2}^{\infty} \varphi_{mn} u^m v^n \quad ,$$

(2-87)

$$v_k = \lambda^{-k} v + \sum_{m+n=2}^{\infty} \psi_{mn} u^m v^n \quad ,$$

where the coefficients $\varphi_{mn}$, $\psi_{mn}$ depend on k. These coefficients can be determined recursively in the following way. If the u, v, are replaced in the right-hand side of (2-87) by $u_1$, $v_1$, then the result is $u_{k+1}$, $v_{k+1}$. Inserting (2-87) into (2-86) and comparing the coefficients of $u^m v^n$, m+n = 2, yields six non homogeneous first order recurrences in which the index k is written inside parenthesis, like if it were an independent variable :

$$\varphi_{20}(k+1) - \lambda^2 \varphi_{20}(k) = a_{20} \lambda^k, \qquad \varphi_{11}(k+1) - \varphi_{11}(k) = a_{11} \lambda^k \quad ,$$

(2-88)

$$\varphi_{02}(k+1) - \lambda^{-2} \varphi_{02}(k) = a_{02} \lambda^k, \qquad \psi_{20}(k+1) - \lambda^2 \psi_{20}(k) = b_{20} \lambda^{-k} \quad ,$$

$$\psi_{11}(k+1) - \psi_{11}(k) = b_{11} \lambda^{-k}, \qquad \psi_{02}(k+1) - \lambda^{-2} \psi_{02}(k) = b_{02} \lambda^{-k} \, .$$

It can be verified by substitution that one solution of (2-88) is :

$$\varphi_{20}(k) = \frac{\lambda^k - \lambda^{2k}}{\lambda - \lambda^2} a_{20} \, , \qquad \varphi_{11}(k) = \frac{\lambda^k - 1}{\lambda - 1} a_{11} \quad ,$$

(2-89)

$$\varphi_{02}(k) = \frac{\lambda^k - \lambda^{-2k}}{\lambda - \lambda^{-2}} a_{02} \, , \qquad \psi_{20}(k) = \frac{\lambda^{-k} - \lambda^{2k}}{\lambda^{-1} - \lambda^2} b_{20} \quad ,$$

$$\psi_{11}(k) = \frac{\lambda^{-k} - 1}{\lambda^{-1} - 1} b_{11} \, , \qquad \psi_{02}(k) = \frac{\lambda^{-k} - \lambda^{-2k}}{\lambda^{-1} - \lambda^{-2}} b_{02} \, .$$

By induction it is shown that similar expressions exist for m+n = 3, 4, ... The coefficient of the series in (2-87) are therefore uniquely defined. All $\varphi_{mn}$, $\psi_{mn}$ are rational functions of $\lambda^k$. A similar result is obtained for other values of $\lambda_1$ and $\lambda_2$, real or complex.

If the validity of (2-87) is formally extended to fractional and continuous values of k, which is quite permissible when the series are convergent, then the equations in (2-87) represent a continuous group of transformations, with k an additive parameter for the group. Let $u_k = u(k)$, $v_k = v(k)$ and $U(u, v) = \frac{d\,u_k}{dk}\Big|_{k=0}$, $V(u, v) = \frac{d\,v_k}{dk}\Big|_{k=0}$, i.e. the functions U, V are defined by means of the derivatives of the series in (2-87), evaluated at k = 0. It follows then trivially that u(k), v(k) satisfy the differential equations

$$\text{(2-90)} \qquad \frac{du}{dk} = U(u, v), \quad \frac{dv}{dk} = V(u, v), \quad u(0) = u_0, \quad v(0), v_0.$$

The first terms of U, V are known in the following particular cases [B 4] :

$$\text{(2-91)} \qquad \begin{aligned} &\lambda_1 = \lambda_1, \lambda_2 = 1/\lambda, && U = u \log \lambda + \ldots, && V = -v \log \lambda + \ldots, \\ &\lambda_{1,2} = e^{\pm i\varphi}, && U = -\varphi v + \ldots, && V = \varphi u + \ldots, \\ &\lambda_{1,2} = \lim_{\varphi \to 0} e^{\pm i\varphi}, && U = a_{20} u^2 + \ldots, && V = b_{20} u^2 + \ldots. \end{aligned}$$

Unless the local phase portrait near (0,0) of the recurrence (2-41) is of a type which exists also in autonomous differential equations of order two, the convergence of series, like those in (2-87), is impossible in principle. There should be, hower, no obstacle to convergence when (0,0) is a focus, a node or a non-stochastic saddle (for a more precise meaning of the adjective non-stochastic, see Chapters III and IV). In the case of a centre ($\lambda_{1,2} = e^{\pm i\varphi}$) convergence is unlikely, except for special forms of F, G in (2-41).

As an illustration of the content of (2-90), consider the recurrences

$$\text{(2-92)} \qquad x_{n+1} = y_n, \quad y_{n+1} = -x_n + 2y_n/(a^2-y_n^2), \quad a \neq \pm 1,$$

and

$$\text{(2-93)} \qquad x_{n+1} = y_n, \quad y_{n+1} = -x_n + 2ay_n/(1+y_n^2) \quad a \neq +1,$$

admitting the fixed point (0,0). In the first case (0,0) is a saddle of type 1 when $|a| < 1$, and (in linear approximation) a centre when $|a| > 1$, whereas in the second case it is a centre (in linear approximation) when $|a| < 1$, a saddle of type 1 when $a > 1$, and a saddle of type 3 when $a < -1$. When $|a| > 1$ in (2-92), there exist two additional fixed points $(\pm\sqrt{a^2-1}, \pm\sqrt{a^2-1})$, which are centres (in linear approximation). When $a > 1$, the recurrence (2-93) possesses also two additional fixed points. They have the coordinates $(\pm\sqrt{a-1}, \pm\sqrt{a-1})$ and are centres (in linear approximation). The invariant curves of (2-92) and (2-93) are described explicitly by w(x, y) = C, C = positive "integration" constant, where

$$\text{(2-94)} \qquad w(x, y) = x^2y^2 - a^2(x^2 + y^2) + 2xy,$$

and

$$\text{(2-95)} \qquad w(x, y) = x^2y^2 + x^2 + y^2 - 2\,a\,x\,y,$$

respectively. It can be easily verified by substitution that the preceding $w(x, y)$ are solutions of the functional equation (2-25). The explicit expressions (2-94) and (2-95) have been found with the help of a rather ingenious educated guess [L 2]. Some arguments explaining this guess will be discussed in Chapter VI. An analysis of (2-94) and (2-95) shows readily that the saddles are quite normal and that the "linearized" centres of (2-92) and (2-93) are still centres after perturbation by the corresponding non linear terms. As it was already pointed out before, such a survival of centres is rather exceptional. The local phase plane portraits of (2-92) and (2-93) are shown in Fig. 2-16 and Fig. 2-17, respectively. In addition to their theoretical interest, the explicit solutions (2-94) and (2-95) provide a practical test object for an evaluation of efficiency and accuracy of computational algorithms.

Consider a one-parameter family of curves described by $w(x, y) = C$. It is obvious that this family can also be described implicitly by the differential equation $\frac{d}{dx} w(x, y) = 0$, where y is considered to be a dependent and x an independent variable. In fact, $w(x, y) = C$ happens to be the general solution of $\frac{d}{dx} w(x, y) = 0$. Applying this property to (2-94) and (2-95), one obtains the differential equations

$$\text{(2-96)} \qquad \frac{dy}{dx} = \frac{U(x,y)}{V(x,y)}, \qquad U = a^2x - y - xy^2, \qquad V = x - a^2y + x^2y \quad ,$$

and

$$\text{(2-97)} \qquad \frac{dy}{dx} = \frac{U(x,y)}{V(x,y)}, \qquad U = x - ay + xy^2, \qquad V = ax - y - x^2y \quad .$$

The trajectories of (2-96) and (2-97) coincide by construction with the invariant curves of the recurrences (2-92) and (2-93), respectively. The fixed points of (2-92) and (2-93) become simply positions of static equilibrium of (2-96) and (2-97). The latter are given by the real roots of the simultaneous algebraic equations $U(x,y) = 0$, $V(x,y) = 0$.

Consider now the two differential equations in (2-90). Dividing the first by the second yields

$$\text{(2-98)} \qquad \frac{du}{dv} = \frac{U(u,v)}{V(u,v)}$$

Choose a fixed point of (2-92) or (2-93) and transfer it to (0,0), if it is not already there. Since the Jacobian determinant of (2-92) and (2-93) is unity (except at $y = \pm a$ in (2-92), where it is undefined), the recurrences (2-92) and (2-93) can be expressed locally (i.e. near (0,0)) in a canonical form of the type (2-86). The functions U, V in (2-96) and (2-97) admit obviously convergent Mc Laurin developments near the corresponding fixed points of (2-92) and (2-93), hence so do the functions U, V in (2-98). Since the latter are derivatives of the series in (2-87), those series are also convergent. Their radius of convergence cannot, however, exceed the distance between (0,0) and the next closest phase plane singularity.

When the invariant curves of a recurrence are known, then their explicit expression $w(x,y) = C$ can be used to express y as a function of x (or vice-versa). After insertion into the recurrence (0-22a), two uncoupled first order recurrences are obtained. The solutions of the latter are functions of n, which represent $w(x,y) = C$ parametrically. $w(x, y) = C$ plays thus a role analoguous to that of an "integral of motion" of two simultaneous differential equations. In the particular case of (2-94), (2-95), and (2-92), (2-93), the general solution $x_n = G(x_o,y_o,n)$, $y_n = F(x_o,y_o,n)$, see eq. (0-23), can be expressed locally in the form of convergent Mc Laurin series in $x_o$, $y_o$.

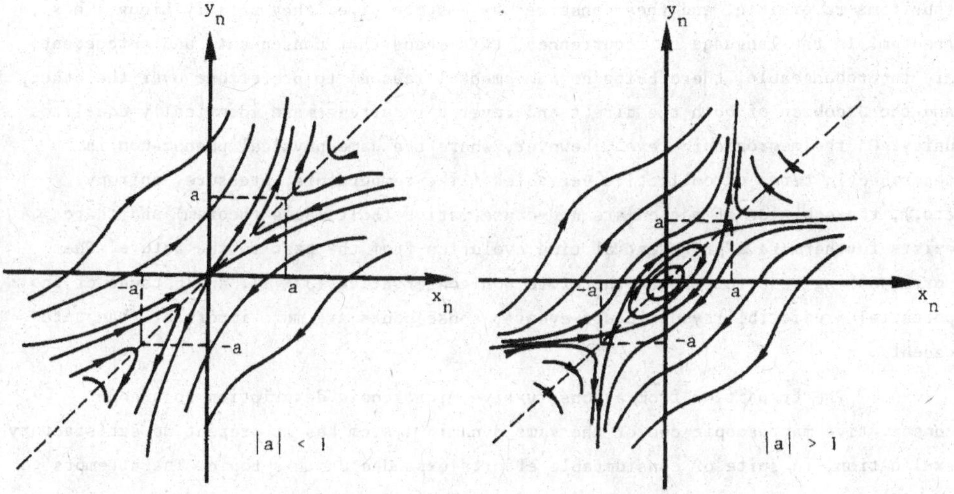


Fig. 2-16. Local phase plane portrait of the recurrence (2-92).

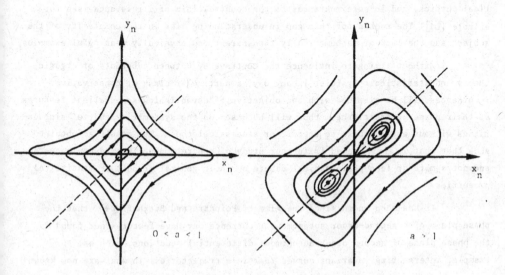

Fig. 2-18. Local phase plane portrait of the recurrence (2-93).

## Chapter III

## STOCHASTICITY IN CONSERVATIVE RECURRENCES

### 3.0 Introduction

Stochasticity in conservative dynamic system constitutes a source of a classical paradox. In microscopic physics, from the level of subnuclear particles to that of molecules, all equations of motion derive from Hamiltonian functions and are thus time-reversible, and they conserve phase space, i.e. they satisfy Liouville's theorem. In the language of recurrences, this means that consequents and antecedents are interchangeable, there being no fundamental reason to prefer one over the other, and the Jacobian of both the direct and inverse recurrences are identically equal to unity. On the macroscopic level, however, where the same physical phenomenon is described in terms of collective variables (like temperature, pressure, entropy, etc.), the equation of motion are non-conservative (Boltzman's theorem) and there exists furthermore a preferential time evolution from the past to the future. The corresponding recurrences are therefore non-conservative ($J \neq 1$), and because of the potential predictibility of future events, consequents are more important than antecedents.

The transition from a conservative microscopic description to a non-conservative macroscopic one of the same dynamic system has at present no satisfactory explanation, in spite of considerable efforts expanded on this topic. The attempts to provide an explanation within the framework of ergodic theory have been unconvincing, to say the least, leading in general to apathy on the part of most workers in statistical physics, and in some instances to the countre-claim of irrelevance (see for example [L1]).The reason for this gap in understanding lies in the complexity of the subject and the lack of mathematically transparent and physically meaningful examples.

Without trying to influence the controversy between adherents of ergodic theory and statistical physicist in any way, a particular class of conservative recurrences will be examined with the objective of establishing some salient features of their phase portrait. The method will be based on the systematic study of singularities without any recourse to ergodic or statistical tools. The result so obtained show that what is still needed is further study of concrete non linear dynamic system and not a more or less philosophical debate on their general ergodic or statistical properties.

It has been known since the time of Poincaré and Birkhoff [B4] that the phase plane of a second-order autonomous recurrence may have features not found in the phase plane of second order autonomous differential equations, such as, for example, intersecting invariant curves (particle trajectories) in what are now known as Birkhoff's instability rings. The pattern of intersecting trajectories is in general so complex that at first sight it appears to be stochastic. Of course there is no real stochasticity, the word stochastic (or chaotic) being used merely as a matter of convenience.

As stochastic solutions, i.e. deterministic solutions of a very complex nature, are described by correspondingly complex transcendental functions, it is understandable that in the last 80 years analytical progress has been rather slow. Some reasons for this slow progress will become clear in the light of concrete problems described later, which are solved by means of a mixed analytical-numerical approach. Analytical expressions are used to define the solutions, but numerical methods are used to determine the corresponding functions or constants. In contrast to the orderly non conservative case, a conservative recurrence is usually badly contitioned numerically. Systematic accuracy checks coupled with a certain amount of redundancy are essential. Detailed calculations are readily possible with present computers which have a high speed and a normal working word of about 15 digits. Eventual doubts about the influence of purely numerical phenomena can be resolved by recourse to double accuracy.

Similarly to differential equations, one may try to divide the phase plane of a second order conservative recurrence into cells, each cell filled with trajectories of the same qualitative type. But since there is no counterpart for recurrences of the Andronov-Pontryagin theorem on the number of phase cells [A4], such a procedure is not likely to be completely exhaustive. It is, however, reasonable to look first for singular solutions, which may turn out to be separatrices, as if a division into cells were possible.

The recurrence chosen for study, which possesses many of the properties of conservative recurrences in general, is

$$(3\text{-}1)\qquad \begin{aligned} x_{n+1} &= g(x_n, y_n) = y_n + F(x_n) , & F(0) &= 0 , \\ y_{n+1} &= f(x_n, y_n) = -x_n + F(x_{n+1}) , & n &= 0, \pm 1, \pm 2, \ldots , \end{aligned}$$

whose inverse, obtained by rearrangement, is :

$$(3\text{-}2)\qquad x_n = -y_{n+1} + F(x_{n+1}), \qquad y_n = x_{n+1} - F(x_n) .$$

In (3-1) and (3-2), $x_n$ and $y_n$ are real-valued and the function $F(x)$ is essentially arbitrary. In order to retain the applicability of the material on invariant curves described in Chapter II, it is necessary to assume that $F(x)$ possesses a certain number of continuous derivatives and admits, whenever required, an at least asymptotically convergent Mc Laurin series. Because of the presence of analytical and computational constraints, the non linearity of $F(x)$ will be chosen as follows :

$$
(3\text{-}3)\quad
\begin{array}{ll}
\text{a)} & x^2 \quad , \\
\text{b)} & x^3 \quad , \\
\text{c)} & x^2/(1+x^2) \quad , \\
\text{d)} & x^2 \,.\, e^{-(x^2-1)/4} \quad , \\
\text{e)} & \sin x - x \quad , \\
\text{f)} & \sin (bx+b_o) - bx - \sin b_o , \qquad b,\ b_o = \text{constants}
\end{array}
$$

The detailed motivation of the choice (3-3) is rather complex. The non linear functions in formulae (3-3) represent, however, four different types of behaviour for large x : unbounded algebraic, bounded algebraic, exponentially decreasing, and bounded periodic. They furnish a four-point scale for a rough classification of many reasonable smooth non linearities. For the sake of brevity the non linearities in (3-3) will be referred to as quadratic, cubic, bounded quadratic, quadratic exponential, sinusoïdal and sinusoïdal unsymmetric. These names are self-explanatory. In order to have as few parameters as possible the non linearities (3-3) will be incorporated into appropriate normalized forms. The special shapes chosen in (3-3) make it possible to obtain normalized forms with only one parameter, without significantly affecting the generality of the recurrences (3-1) and (3-2). If $F'_n = F'(x_n)$, $' = \frac{d}{dx_n}$, and

$$(3\text{-}4)\qquad a = F'_n,\ b = 1,\ c = F'_n . F'_{n+1} - 1,\ d = F'_{n+1}$$

then the Jacobian determinant of (3-1) is :

$$(3\text{-}5)\qquad J = J(x_n, y_n) = ab - cd \equiv 1 \ .$$

Because of (3-5), the eigenvalues of all fixed points and cycles satisfy $\lambda_1 . \lambda_2 = 1$. It is therefore sufficient to specify only one eigenvalue, the argument $\varphi$ when $\lambda_{1,2} = e^{\pm i\varphi}$, and the absolutely larger one, when $\lambda_{1,2}$ are real (i.e. $\lambda_1$ when $\lambda_1, \lambda_2 > 0$ and $\lambda_2$ when $\lambda_1, \lambda_2 < 0$).

## 3.1 Quadratic recurrence

### 3.1.1 General properties

The following combination of (3-1) and (3-3a) yields a convenient one-parameter form

$$
(3\text{-}6)\qquad
\begin{array}{ll}
x_{n+1} = y_n + F(x_n), & y_{n+1} = -x_n + F(x_{n+1}) \quad , \\
F(x) = \mu x + (1-\mu)x^2, & -1 < \mu < 1
\end{array}
$$

The fixed points (x, y) of (3-6) are given by y = 0 and the roots of

$$(3\text{-}7)\qquad F(x) - x = 0 \quad ,$$

which are (0,0) and (1,0), regardless of the value of $\mu$.

For (0,0), the eigenvalue is

$$(3\text{-}8)\qquad \lambda = \mu + i\sqrt{1-\mu^2} = e^{i\varphi}, \qquad \mu = \cos\varphi, \qquad 0 < \varphi < \pi .$$

In the linear approximation, the fixed point (0,0) is thus a centre, except possibly for

$$(3\text{-}9)\qquad \mu = \cos 2\pi p/q, \qquad p,\ q = \text{integers} .$$

The fixed point (1,0) is a saddle of type 1 ($\lambda_1 > 1$). The value of $\lambda_1$ depends monotonically on $\mu$ ; $\lambda_1 - 1$ increases as $\mu$ decreases. Since by definition $|\mu| < 1$, there are no cycles of order two.

After considerable albebraic manipulation of the iterated recurrence (0-25), it is found that the points of two distinct cycles of order k = 3 are given by y = 0 and

$$(3\text{-}10)\qquad x^2 + \frac{1+\mu}{1-\mu}x + \frac{1+2\mu}{2(1-\mu)^2} = 0 .$$

Equation (3-10) admits two real roots for $\mu < 1 - \sqrt{2} = \mu_o$, which merge into a double root

$$(3\text{-}11)\qquad x = -\frac{1}{2}\frac{1+\mu_o}{1-\mu_o}, \quad y = 0 ,$$

as $\mu$ approaches $\mu_o$ from below.

For k = 4, it is also possible to arrive at explicit expressions. There exist again two distinct cycles, one defined by

$$(3\text{-}12)\qquad x = \frac{1}{2(1-\mu)}(-\mu \pm \sqrt{\mu^2 - 2\mu}), \qquad y = \pm\frac{\sqrt{\mu(\mu-2)}}{2(1-\mu)} ,$$

and the other by

$$(3\text{-}13)\qquad x = \frac{1}{2(1-\mu)}(-\mu \pm \sqrt{\mu^2 - 2A}), \qquad y = 0 , \qquad A = \mu - \sqrt{\mu^2 - 2\mu} .$$

Both cycles of order k = 4 exist only for $\mu < 0$ and both merge with each other, and simultaneously with (0,0), when $\mu \to 0^-$. For k = 5 and k = 6 explicit expressions of x and y require the use of elliptic functions, which are awkward to manipulate and difficult to evaluate numerically. For k > 6 a purely analytical method becomes impossible in principle, the required transcendental functions being unstudied so far. The determination of the distribution of cycles in the phase plane thus unavoidably involves the use of numerical computation. As mentioned before, such an endeavour is far from being a trivial matter. By the combination of various methods and some trial and error, the cycles given in Tables 3-1 to 5 have been found. Cycles are classified in the same way as fixed points. The tables contain the order k, the coordinates of one point, the eigenvalue and, when real, the eigen-directions (cf. eq. (2-30)). Because of symmetry of (3-6), $p_1 = -p_2$ if y = 0. Consider the k points of a cycle and join the successive points (defined by consequents) by means of straight lines producing thus

Table 3-1. Cycles of the quadratic recurrence (3-6), $\mu = 0.8$, $k/r \geqslant (k/r)_{min}$

| | | Centres | | | Saddles | | | | |
|---|---|---|---|---|---|---|---|---|---|
| k | r | x | y | $\phi$ | x | y | $\lambda_1$ | $p_1$ | $p_2$ |
| 10 | 1 | 0.250525 | 0 | 0.00014 | 0.241022 | 0.036586 | 1.00014 | -1.900 | -1.893 |
| 11 | 1 | 0.541472 | 0.049656 | 0.0039 | 0.553787 | 0 | 1.0040 | 130.3 | |
| 12 | 1 | 0.701273 | 0 | 0.0108 | 0.690907 | 0.042711 | 1.0109 | -2.138 | -2.018 |
| 13 | 1 | 0.784273 | 0.033838 | 0.0199 | 0.792378 | 0 | 1.0201 | 46.02 | |
| 14 | 1 | 0.852882 | 0 | 0.0313 | 0.846716 | 0.025958 | 1.0318 | -2.278 | -2.017 |
| 15 | 1 | 0.889927 | 0.019591 | 0.0460 | 0.894553 | 0 | 1.0471 | 24.26 | |
| 16 | 1 | 0.923875 | 0 | 0.0653 | 0.920430 | 0.014648 | 1.0674 | -2.442 | -1.953 |
| 17 | 1 | 0.942232 | 0.010886 | 0.0910 | 0.944784 | 0 | 1.0053 | 13.39 | |
| 18 | 1 | 0.959826 | 0 | 0.1255 | 0.957941 | 0.008058 | 1.1338 | 1.823 | 2.746 |
| 19 | 1 | 0.969319 | 0.005948 | 0.1728 | 0.970709 | 0 | 1.1882 | 7.381 | |

Table 3-2. Cycles of the quadratic recurrence (3-6), $\mu = 0.6$, $k/r \geqslant (k/r)_{min}$

| | | Centres | | | Saddles | | | | |
|---|---|---|---|---|---|---|---|---|---|
| k | r | x | y | $\phi$ | x | y | $\lambda_1$ | $p_1$ | $p_2$ |
| 7 | 1 | 0.270826 | 0.078992 | 0.0202 | 0.291725 | 0 | 1.023 | 12.35 | |
| 8 | 1 | 0.636402 | 0 | 0.2323 | 0.612900 | 0.094901 | 1.259 | -4.303 | -1.402 |
| 9 | 1 | 0.768053 | 0.071259 | 0.5364 | 0.783907 | 0 | 1.686 | 2.570 | |
| 10 | 1 | 0.866626 | 0 | 0.9792 | 0.855053 | 0.049575 | 2.475 | 7.515 | -1.170 |
| 11 | 1 | 0.908344 | 0.033302 | 1.7514 | 0.915504 | 0 | 4.101 | 1.400 | |
| | | Centre-saddle bifurcation | | $\lambda_2$ | | | | | |
| 12 | 1 | 0.946208 | 0 | -3.618 | 0.941152 | 0.022154 | 7.729 | 1.014 | -1.600 |
| 13 | 1 | 0.962248 | 0.014531 | -12.25 | 0.965337 | 0 | 16.18 | 1.050 | |
| 14 | 1 | 0.977719 | 0 | -32.38 | 0.975562 | 0.009536 | 36.18 | 1.067 | -0.9718 |
| 15 | 1 | 0.984249 | 0.006201 | -80.22 | 0.985561 | 0 | 83.73 | 0.9816 | |
| 16 | 1 | 0.990688 | 0 | -194.0 | 0.989776 | 0.004048 | 196.9 | 0.9857 | -0.9712 |
| 17 | 1 | 0.993399 | 0.002623 | -464.9 | 0.993953 | 0 | 466.1 | 0.9751 | |
| 18 | 1 | 0.996095 | 0 | -1110 | 0.995711 | 0.001708 | 1107 | 0.9768 | -0.9752 |
| 19 | 1 | 0.997229 | 0.001105 | -2644 | 0.997462 | 0 | 2632 | 0.9768 | |
| | | Centres | | $\phi$ | | | | | |
| 15 | 2 | 0.503853 | 0.050169 | 0.0127 | 0.509777 | 0 | 1.013 | 101.62 | |
| 17 | 2 | 0.717118 | 0.041262 | 0.1367 | 0.723500 | 0 | 1.146 | 16.41 | |
| 19 | 2 | 0.827394 | 0.029627 | 0.5191 | 0.827460 | 0 | 1.664 | 5.995 | |

a polygone with k corners. An additional property of a cycle of order $k > 2$ (also listed in the tables) is its rotation number r, equal to the number of turns made around an interior point of the polygon when following successive consequents of (x, y), until one comes back to (x, y). The values chosen for $\mu$ are 0.8, 0.6, 0.5, 0.125 and 0. Since from an inspection of (3-6) it is obvious that the parameter $\delta = 1 - \mu$ is a sort of measure of the strength of the non linearity, the cycles saddles of type 1 of the same order and rotation number, have eigenvalues which increase simultaneously with $\delta$.

Table 3-3. Cycles of the quadratic recurrence (3-6), $\mu = 0.5$, $k/r \geqslant (k/r)_{min}$

| | | Centres | | | Saddles | | | | |
|---|---|---|---|---|---|---|---|---|---|
| k | r | x | y | $\varphi$ | x | y | $\lambda_1$ | $p_1$ | $p_2$ |
| 7 | 1 | 0.586 810 | 0.121 232 | 0.4641 | 0.612 089 | 0 | 1.569 | 2.220 | |
| 8 | 1 | 0.786 636 | 0 | 1.2129 | 0.763 337 | 0.090 327 | 2.920 | 3.689 | -1.061 |
| Centre - saddle bifurcation | | | | $\lambda_2$ | | | | | |
| 9 | 1 | 0.861 563 | 0.059 636 | -1.938 | 0.872 866 | 0 | 6.162 | 1.241 | |
| 10 | 1 | 0.924 574 | 0 | -10.80 | 0.915 075 | 0.038 856 | 14.45 | 1.326 | -1.050 |
| 11 | 1 | 0.948 864 | 0.024 260 | -33.07 | 0.953 394 | 0 | 36.06 | 1.363 | -1.039 |
| 12 | 1 | 0.971 841 | 0 | -91.35 | 0.968 149 | 0.015 418 | 92.65 | 1.081 | -1.113 |
| 13 | 1 | 0.980 667 | 0.009 480 | -244.0 | 0.982.430 | 0 | 240.9 | 1.097 | |
| 14 | 1 | 0.989 325 | 0 | -643.7 | 0.987 906 | 0.005 974 | 629.0 | 1.103 | -1.102 |
| 15 | 1 | 0.992 641 | 0.003 653 | -1690 | 0.993 319 | 0 | 1645 | 1.108 | |
| 16 | 1 | 0.995 933 | 0 | -4430 | | | | | |
| 17 | 1 | | | | 0.997 451 | 0 | 11.272 | 1.114 | |
| 18 | 1 | 0.998 448 | 0 | -30 381 | 0.998 240 | 0.000 878 | 29 508 | 1.1156 | -1.1148 |
| 19 | 1 | 0.998 928 | 0.000 535 | -79 545 | | | | | |
| Centres | | | | $\varphi$ | | | | | |
| 13 | 2 | 0.446 929 | 0.061 051 | 0.0357 | 0.455 747 | 0 | 1.036 | 32.29 | |
| 15 | 2 | 0.708 660 | 0.054 040 | 0.5992 | 0.707 000 | 0 | 1.793 | 4.830 | |
| Center - saddle bifurcation | | | | $\lambda_2$ | | | | | |
| 17 | 2 | 0.832 164 | 0.035 354 | -3.750 | 0.849 280 | 0 | 8.326 | 1.542 | |
| 19 | 2 | 0.899 523 | 0.024 282 | -55.05 | 0.884 444 | 0 | 67.30 | 1.296 | |
| Centres | | | | $\varphi$ | | | | | |
| 18 | 2 | 0.876 777 | 0.054 020 | 2.033 | | | | | |
| 20 | 2 | | | | 0.946 478 | 0 | 101.9 | 1.414 | |
| 19 | 3 | 0.372 605 | 0.038 109 | 0.0015 | 0.376 187 | 0 | 1.001 | 827.1 | |

Table 3-4. Cycles of the quadratic recurrence (3-6), $\mu = 1/8$, $k/r \geqslant (k/r)_{min}$

| | | Centres or type 3 - saddles | | | Saddles | | | | |
|---|---|---|---|---|---|---|---|---|---|
| k | r | x | y | $\varphi$ or $\lambda_2$ | x | y | $\lambda_1$ | $p_1$ | $p_2$ |
| 5 | 1 | 0.582 101 | 0.212 852 | -1.623 | 0.550 008 | 0 | 4.077 | 0.9716 | |
| 6 | 1 | 0.827 883 | 0 | -30.20 | 0.753 560 | 0.162 494 | 22.12 | 1.259 | -1.074 |
| 7 | 1 | 0.896 050 | 0.081 680 | -131.6 | 0.892 102 | 0 | 88.78 | 1.339 | |
| 8 | 1 | 0.953 811 | 0 | -485.3 | 0.932 781 | 0.054 863 | 322.5 | 1.416 | -1.442 |
| 9 | 1 | 0.970 650 | 0.024 927 | -1712 | 0.969 997 | 0 | 1134.2 | 1.512 | |
| 10 | 1 | 0.986.864 | 0 | -5961 | 0.980 828 | 0.016 453 | 39 46 | 1.545 | -1.532 |
| 11 | 1 | 0.991 565 | 0.007 318 | -20 671 | 0.991 413 | 0 | 13 680 | 1.564 | |
| 9 | 2 | 0.301 774 | 0.080 068 | 0.1669 | 0.321 761 | 0 | 1.179 | 3.664 | |
| 10 | 2 | 0.610 908 | 0.207 987 | 1.481 | none | none | | | |
| 11 | 2 | 0.708 786 | 0.119 505 | -109.6 | 0.617 273 | 0 | 84.53 | 1.200 | |
| 12 | 2 | 0.862 634 | 0.029 616 | -516.9 | 0.878 072 | 0 | 316.6 | 1.301 | |
| 13 | 2 | | | | 0.640 541 | 0 | 1119 | 1.328 | |
| 13 | 2 | | | | 0.888 954 | 0 | 1956 | 1.329 | |
| 14 | 2 | | | | 0.966 064 | 0 | 3904 | 1.501 | |
| 14 | 2 | | | | 0.955 037 | 0 | 13 898 | 1.462 | |
| 15 | 2 | | | | 0.642 399 | 0 | 13 556 | 1.340 | |
| 16 | 2 | | | | 0.642 518 | 0 | 46 953 | 1.341 | |
| 17 | 2 | | | | 0.642 552 | 0 | 162 568 | 1.341 | |
| 14 | 3 | 0.458 956 | 0 | 0.5589 | 0.442 893 | 0.050 955 | 1.768 | 1.175 | 5.046 |
| 15 | 3 | 0.673 006 | 0.192 560 | -16.71 | 0.615 285 | 0.191 886 | 12.89 | -4.290 | |
| 16 | 3 | 0.690 682 | 0.135 022 | -401.7 | 0.593.788 | 0.027 732 | 329.8 | 1.228 | -1.042 |
| 17 | 3 | 0.746 769 | 0.165 467 | -2469 | 0.619 716 | 0 | 1854 | 1.212 | |
| 17 | 3 | | | | 0.874 597 | 0 | 1145 | 1.293 | |
| 19 | 3 | | | | 0.957 144 | 0 | 33 875 | 1.468 | |
| 19 | 3 | | | | 0.965 058 | 0 | 13 407 | 1.499 | |
| 19 | 4 | 0.498 250 | 0.027 691 | 1.340 | 0.507 369 | 0 | 3.222 | 1.575 | |
| 20 | 4 | | | | 0.651 340 | 0.166 271 | 72.80 | -3.518 | |
| 21 | 4 | | | | 0.616 979 | 0.198 380 | 17.79 | 41.41 | |

For conciseness, a cycle saddle (or centre) of order k and rotation number r is designated by saddle (or centre) k/r or $k_{(r)}$. It is related to Poincaré's rotation number N by the relation N = r/k. Consider a family of two cycles, consisting of a cycle saddle k/r of type 1 and a cycle centre k/r. By tracing the two

Table 3-5. Cycles of the quadratic recurrence (3-6), $\mu = 0$, $4 < k/r \leq 5$

| Centres and saddles of type 3 | | | | | Saddles | | | |
|---|---|---|---|---|---|---|---|---|
| $k$ | $r$ | $x$ | $y$ | $\varphi$ or $\lambda_2$ ($< 0$) | $x$ | $y$ | $\lambda_1$ | $p$ |
| 5 | 1 | −0.537795 | 0 | −18.533 | 0.660586 | 0 | 11.79 | 21.031 |
| 9 | 2 | −0.455981 | 0 | −11.268 | 0.585191 | 0 | 11.669 | 0.8887 |
| 13 | 3 | −0.425889 | 0 | −4.378 | 0.536625 | 0 | 7.989 | 0.9126 |
| 17 | 4 | −0.405787 | 0 | 2.029 | 0.496506 | 0 | 4.665 | 1.190 |
| 21 | 5 | −0.389832 | 0 | 1.135 | 0.462456 | 0 | 2.804 | 1.980 |
| 25 | 6 | −0.376256 | 0 | 0.672 | 0.434919 | 0 | 1.913 | 3.644 |
| 29 | 7 | −0.364375 | 0 | 0.398 | 0.412971 | 0 | 1.481 | 6.711 |
| 33 | 8 | −0.353849 | 0 | 0.234 | 0.395143 | 0 | 1.262 | 12.13 |
| 37 | 9 | −0.344454 | 0 | 0.136 | 0.380283 | 0 | 1.145 | 21.65 |
| 41 | 10 | −0.336013 | 0 | 0.078 | 0.367615 | 0 | 1.082 | 38.39 |
| 45 | 11 | −0.328382 | 0 | 0.045 | 0.356618 | 0 | 1.046 | 67.89 |
| 49 | 12 | −0.321441 | 0 | 0.026 | 0.346933 | 0 | 1.026 | 119.9 |
| 53 | 13 | −0.315092 | 0 | 0.014 | 0.338303 | 0 | 1.014 | 211.7 |
| 57 | 14 | −0.309253 | 0 | 0.008 | 0.330539 | 0 | 1.008 | 373.7 |
| 61 | 15 | −0.303857 | 0 | 0.004 | 0.323498 | 0 | 1.004 | 659.7 |
| 65 | 16 | −0.298850 | 0 | 0.002 | 0.317068 | 0 | 1.002 | 1164 |
| 69 | 17 | −0.294183 | 0 | 0.001 | 0.311160 | 0 | 1.001 | 2054 |
| 73 | 18 | −0.289820 | 0 | 0.000 | 0.305703 | 0 | 1.000 | 3623 |
| 77 | 19 | −0.285726 | 0 | 0.000 | 0.300641 | 0 | 1.000 | 6387 |
| 81 | 20 | −0.281874 | 0 | 0.000 | 0.295924 | 0 | 1.000 | 11258 |
| 85 | 21 | −0.278239 | 0 | 0.000 | 0.291514 | 0 | 1.000 | 19831 |
| 89 | 22 | −0.274801 | 0 | 0.000 | 0.287376 | 0 | 1.000 | 34916 |
| 93 | 23 | −0,271542 | 0 | 0.000 | 0.283482 | 0 | 1.000 | 61439 |
| 97 | 24 | −0.268445 | 0 | 0.000 | 0.279808 | 0 | 1.000 | 108045 |

invariant curves (four segments) passing through each point of the cycle saddle k/r, one finds that these invariant curves either merge smoothly with the invariant curves passing through a neighbouring point of the cycle saddle k/r, or they intersect the latter curves, but remain inside an annulus containing all cycle points k/r. In the first case, the singularity configuration so obtained is called an island structure, and in the second an instability ring (in the sense of Birkhoff). In both cases the invariant curve-segments emanating from the saddles k/r surround the points of the cycle centre k/r. The intersection points of invariant curves have been named by Poincaré. Some of this nomenclature will be introduced in the following sections. For example, the intersections of consequent and antecedent curve segments originating from the same cycle k/r are called homoclinic points.

According to Birkhoff [B 4], the phase plane picture of a conservative recurrence like (3-6), around a reference fixed point centre (0,0) with $\varphi \neq 2\pi p/q$, is in general as shown in Fig. 3-1: the centre (0,0) is first surrounded by a set of smooth closed invariant curves (and non-singular ones, in the sense that they do not cross any saddles of a finite order), then by one or more island structures, and finally by one or more instability rings, each bounded by at least two smooth closed and non-singular invariant curves.

The main objective of this chapter consists in determining whether the solution structure shown in Fig. 3-1 is sufficient to account for all features of the recurrence (3-6). Even a superficial inspection of the Tables of known cycles suggests

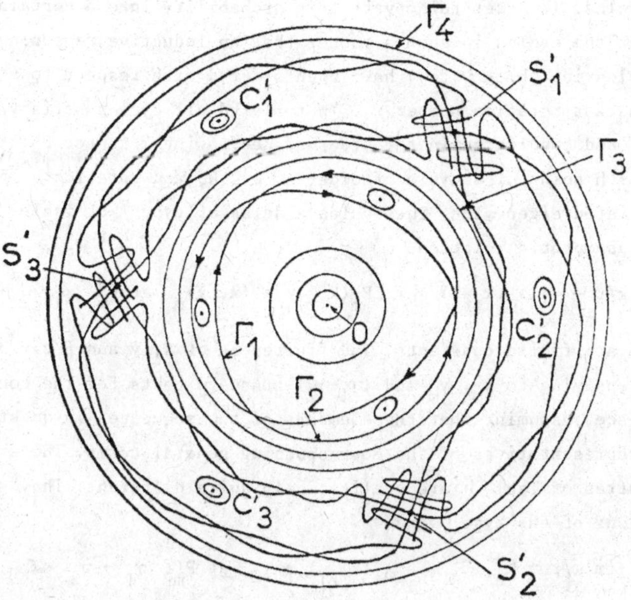


Fig. 3-1 : Phase plane picture according to Birkhoff

The center O is first surrounded by a family of regular invariant curves, then by one or more island structures, defined by the closed singular invariant curves $\Gamma_1$, $\Gamma_2$ passing through saddles of type 1, and finally by one or more instability rings, defined by the intersecting singular invariant curves emanating from the saddles of type 1 like $S'_1$, $S'_2$, $S'_3$. These intersecting invariant curves are limited from below and above by the closed regular invariant curves $\Gamma_3$ and $\Gamma_4$, respectively. Inside the instability ring there exist closed regular invariant curves surrounding centers $C'_1$, $C'_2$, $C'_3$.

a negative answer, because there exist, for example, cycles saddles of type 3 for which no place has been provided in the singularity configurations of Fig.3-1. Moreover, Fig.3-1 is essentially local in character and contains no indications about possible global properties of the phase portrait.

### 3.1.2 Ordering of cycles

The number of cycles listed in Tables 3-1 to 5 is not limited by any fundamental consideration, but merely by the saturation or exhaustion of the various methods employed to find them. An extension of the list encounters two essentially numerical constraints : increasingly slow convergence of iteration methods as one looks for saddles of type 1 with $\lambda-1 \ll 1$ (or centres with $\varphi \ll 1$), and an increasing strong sensitivity to rounding and truncations errors as one looks for saddles with max $|\lambda| \gg 1$. By means of standard methods it is possible to find a few additional cycles by using a computer with a longer working word (containing more than 15 digits) or by resorting to quadruple arithmetic. In both cases the extra gain is quite small, especially in view of the inescapable probability that the recurrence (3-6) admits an infinity of cycles. In order to convert this probability into a certainty it is advisable to order the cycles in such a manner that an inductive reasoning becomes possible. The cycles in Tables 3-1 to 4 have been ordered with respect to an increasing order k for a constant rotation number $r_o$. In table 3-5 all cycles satisfy the inequality $4 < k/r < 5$, and are listed in the order of decreasing k/r. Let $P_m(k_{(r)}) = (x_m, y_m)_k$ be the m-th point of a cycle of order k, $m \leq k$, and rotation number r, and $\lambda(k_{(r)})$ the associated eigenvalue. The cycles saddles of type 1 of Tables 3-1 to 4 satisfy then the asymptotic relations

(3-14) $$k/r \to \infty, \quad \lambda_1(k_{(r)}) \to \infty, \quad P_m(k_{(r)}) \to (\bar{x}, \bar{y})_m \quad \text{at } P_m, \; p_1 - p_2 \to \bar{p} \quad ,$$

where $\bar{x}$, $\bar{y}$, and $\bar{p}$ are finite constants. The finiteness of $\bar{x}$, $\bar{y}$ and $\bar{p}$ can be established directly by applying to $x_m$, $y_m$ and $(p_1 - p_2)$ standard tests for the convergence of a numerical sequence, assuming that the elements of the sequences in question are already typical representatives of the corresponding general terms. The cycles saddles of type 1 and centres of Table 3-5 are qualitatively quite different. They satisfy asymptotic relations of the type

(3-15) $$\lim_{\substack{k\to\infty \\ r\to\infty}} k/r = k_e, \qquad \lambda_1(k_{(r)}) \to 1 \quad \text{at } P_m, \; p_1 - p_2 \to 0 \quad ,$$
$$\varphi(k_{(r)}) \to 0, \qquad P_m(k_{(r)}) \to (\tilde{x}, \tilde{y})_m \quad ,$$

where $\tilde{x}$, $\tilde{y}$ and $k_e$ are finite constants. $k_e$ is usually known by construction of the table. In the particular case of Table 3-5 one has $k = 4r+1$ and $k_e = 4$. Moreover, $|\tilde{x}| \geqslant 0$ and $|\tilde{y}| \geqslant 0$, because the (Euclidian) distance between $P_m(k_{(r)})$ and (0,0) diminishes monotonically. The limits of $P_m(k_{(r)})$ and of $p_1 - p_2$ in (3-14) have to be found indirectly. This can be done by determining the invariant curves crossing the saddle ($x = 1$, $y = 0$), and then relating the positions of $P_m(k_{(r)})$ to these invariant curves. For conciseness, the saddle (1,0) will be called the main saddle, and the invariant curves crossing it, the main invariant curves of (3-6). For $\mu = 0.8$, 0.6, 0.5 and 0.125, these main invariant curves are shown in Fig.3-2 → 3-5, together with the positions of all known cycles.

A first conclusion from an inspection of Fig. 3-2 → 3-5 is that cycles exist only inside the area enclosed by the main invariant curves. The sequences $P_m(k_{(r)})$ in (3-14) are therefore bounded by points on the main invariant curves. Moreover, the points $P_m(k_{(r)}$ are not scattered all over the enclosed area, but are aligned on certain curves (cf. Fig. 3-2 to 5). Due to the symmetry of (3-6), one of these priviledged curves is given by $y_n = 0$ and an other by

(3-16) $$y_n = x_n - F(x_n) \quad .$$

The main invariant curves of (3-6) do not join smoothly, but intersect, forming homoclinic points of the same type as those accuring inside the instability rings of Fig. 3-1. Consider first the cycles of rotation number $r = 1$ and increasing order k. From an inspection of Tables 1 to 4 and Figures 3-2 to 5, it is obvious that the points

$P_m(k_{(1)})$ approach asymptotically homoclinic points on the main invariant curves. This conclusion is confirmed by a study of the limits of the sequences $(p_1-p_2)$ at $P_m(k_{(1)})$, $k \gg 1$, which are found to approach asymptotically the difference of slopes of the invariant curves at the homoclinic points in question. Homoclinic points can therefore be characterized explicitly as accumulation points of a set of cycles of a suitably chosen order.

Let $B_m = (\bar{x}, \bar{y})_m$ be the coordinates of the homoclinic points on the main invariant curves of (3-6), then

$$\lim_{k\to\infty} P_m(k_{(1)}) = \lim_{k\to\infty} (x_m, y_m)_k = B_m = (\bar{x}, \bar{y})_m \quad . \tag{3-17}$$

Main invariant curves and distribution of cycles. Recurrence (3-6).
ʘ centres, x saddles

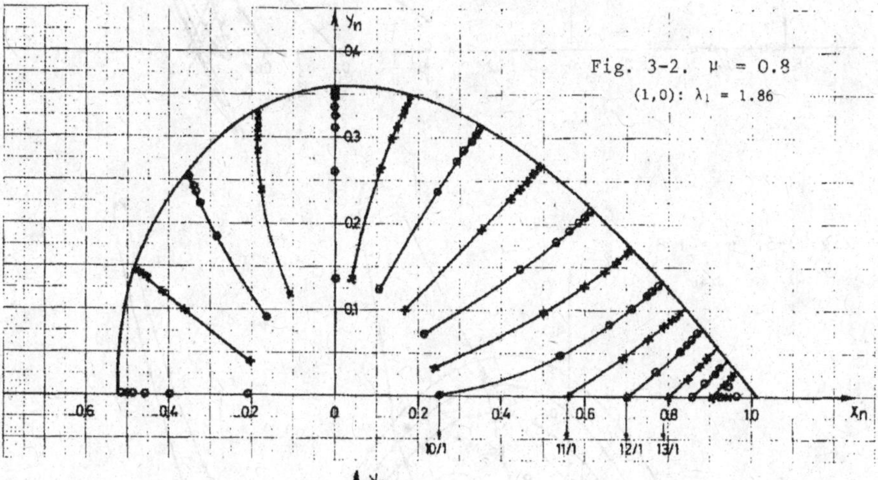


Fig. 3-2. $\mu = 0.8$
(1,0): $\lambda_1 = 1.86$

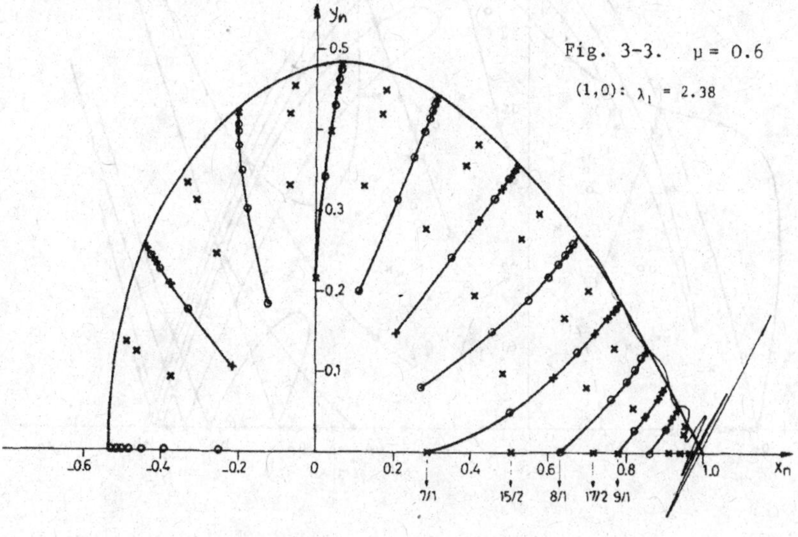


Fig. 3-3. $\mu = 0.6$
(1,0): $\lambda_1 = 2.38$

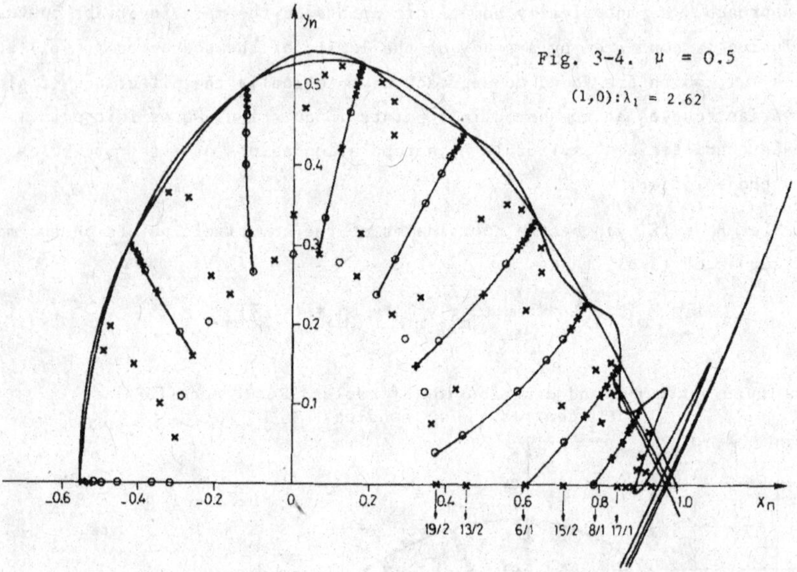


Fig. 3-4. $\mu$ = 0.5

(1,0): $\lambda_1$ = 2.62

Fig. 3-5. $\mu$ = 1/8

(1,0): $\lambda_1$ = 3.46

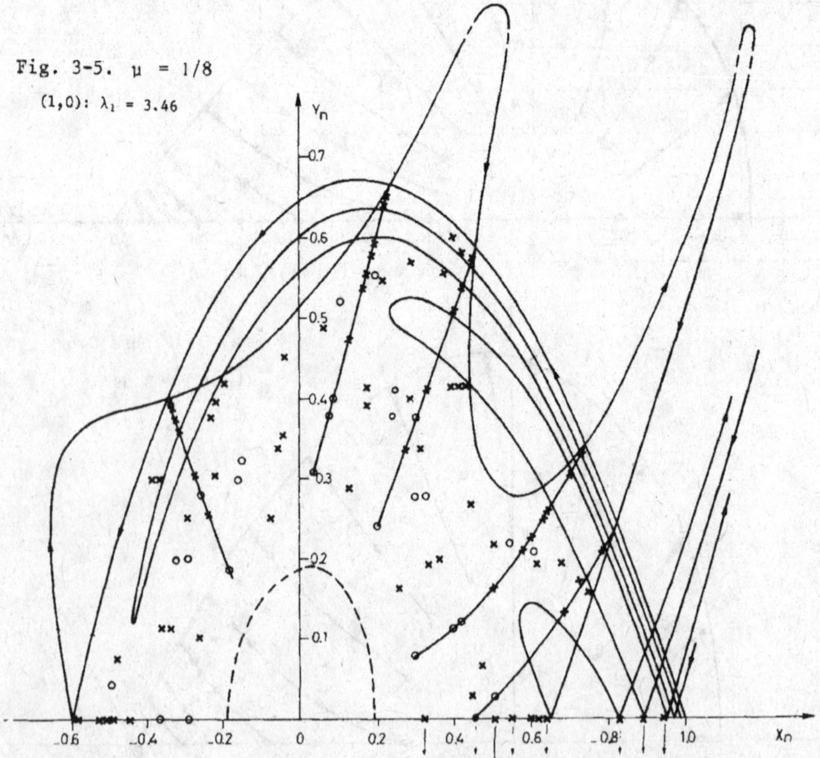

A relation of the same type holds for cycles of rotation number $r > 1$, except that the points $P_m(k_{(r)})$ converge to the homoclinic points $B_m$ in a less direct manner (cf. Fig.3-6. Homoclinic points appear thus as degenerate cycles of an infinite order, and are obviously enumerable.

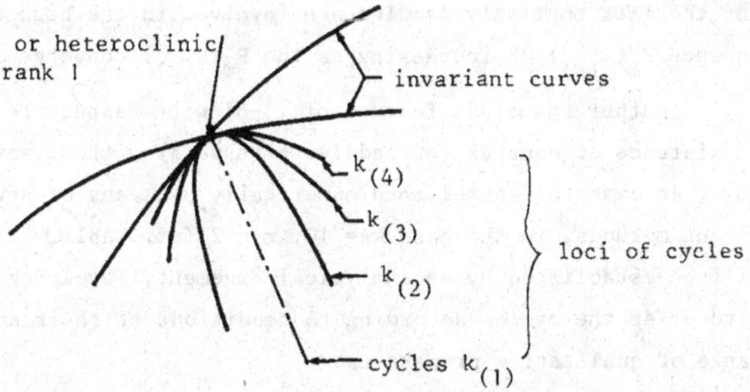


Fig. 3-6

A more detailed examination of the data leading to Tables 1 to 5 (listing all points of the cycles) shows that the distance between (0,0) and the position $P_m(k_{(r)})$ of the cycles of (3-6) is monotonically increasing as $k/r$ increases, and the positions of cycles $P_m(k_{(r)})$ populate the phase plane in a "natural" growth order. In order to express this idea more clearly, consider for example three cycles $k_{(1)}$, $k_{(r)}$ and $(k+1)_{(1)}$, satisfying the inequality $k < k/r < k+1$. Two neighbouring points of each of the three point sets $P(k)$, $P(k/r)$ and $P(k+1)$ can be joined by curvilinear segments, yielding three continuous curves $C(k)$, $C(k/r)$ and $C(k+1)$. The inequality $k < k/r < k+1$ implies that the curvilinear segments of each curve can be so chosen that these curves do not intersect, and form two "concentric" annuli in the phase plane. Moreover, for every value of $r$ and for a fixed $\mu$, there exists a cycle of a minimal order $k_{min}$. For example, when $r = 1$ and $\mu = 0.8$, 0.6, 0.5 and 0.125, then $k_{min} = 10$, 7, 7 and 5, respectively. An analytical explanation of this property will be given later.

If the tables of cycles are fairly complete, then after the study of individual cycles, it is possible to examine eventual cycle pairs of the same order and rotation number. For example, the pairing of a centre with a saddle of type 1 constitutes a basis of the island structures and instability rings present in Fig. 3-1. It was already noted that the existence of saddles of type 3 upsets things somewhat. A numerical study of the evolution of many centres $k_{(r)}$ as a function of decreasing $\mu$ (increasing strength of non linearity) has shown that the saddles of type 3 are former centres which have undergone a bifurcation, i.e. a gradual decrease of $\mu$ leads to a continuous evolution of the positions $P_m(k_{(r)})$ of the cycle, but a qualitatively discontinuous evolution of the associated eigenvalues. More specifically, at some $\mu = \mu_o$, a bifurcation takes place. For $\mu > \mu_o$ the eigenvalues are complex conjugate, and for $\mu < \mu_o$ they become real and negative. In other words, with an increase of the strength of non linearity $\delta = 1-\mu$ a center changes suddenly into a saddle of type 3.

The higher the order and the lower the rotation number, i.e. the larger the "effective" order k/r, the smaller the critical strength of the required non linearity for this bifurcation to take place (see for example the cycles $9_{(1)}$, $10_{(1)}$, $11_{(1)}$, $17_{(1)}$ and $19_{(1)}$ in Tables 3-2 and 3). The occurence of the centre → saddle of type 3 bifurcation explains the fact that only saddles are involved in the high effective-order part of the sequence $P_m(k_{(r)})$, k increasing, as the $P_m(k_{(r)})$ converge to homoclinic points.

Another anomalous feature of (3-6) with respect to the pairing of cycles is the existence of centres (or saddles of type 3) without any accompanying saddles of type 1. An example, established numerically by means of several algorithmically independent methods, is the case k = 10, r = 2 (cf. Table 3-4). The genuinness of this case will be established by an analytical argument, involving bifurcations which permit to order the cycles according to conditions of their appearence, disappearance and change of qualitative properties.

### 3.1.3 Properties of invariant curves

Let the homoclinic points $B_m$ be numbered as shown in Fig. 3-7. Such a numbering permits one to assign a measure A(m, m+1) to the "amplitude" of a "loop" formed by the invariant curves between $B_m$ and $B_{m+1}$. Fig.3-2 to 5 show that the amplitudes A(m, m+1) increase rapidly as the strength of the non linearity $\delta = 1-\mu$ increases.

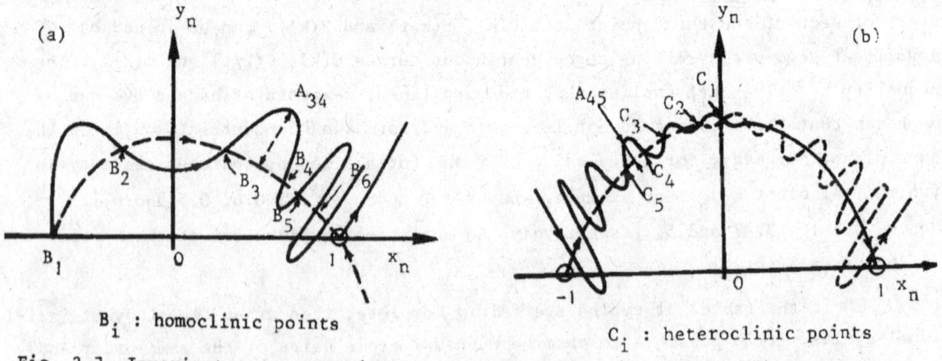


Fig. 3-7. Invariant curves passing through saddles.

In the study of the ordering of cycles it was found that the centres turn into saddles both with an increase of non-linearity, k being constant, and an increase of k, with $\delta = 1-\mu$ constant. It is worthwhile to examine what happens to the invariant curves in such a case. Consider for this purpose the cycle $9_{(1)}$ of (3-6) for $0.4 < \delta < 0.5, (0.6 > \mu > 0.5)$. The invariant curves from two adjacent saddles $9_{(1)}$ of type 1 are shown in Fig. 3-8 to 3-12.

An inspection of these figures shows that an increase of the strength of the non-linearity $\delta$ leads to a simultaneous increase of the eigenvalues $\lambda_1$ of the saddles of type 1, the angles between the eigenslopes $p_1$ and $p_2$, and the amplitudes

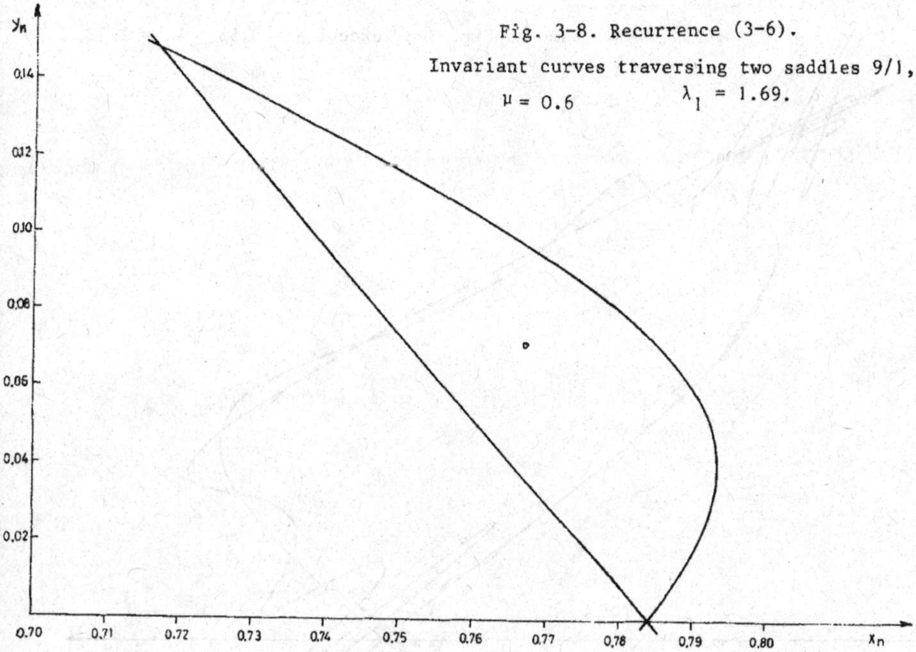


Fig. 3-8. Recurrence (3-6).
Invariant curves traversing two saddles 9/1, $\mu = 0.6$ $\lambda_1 = 1.69$.

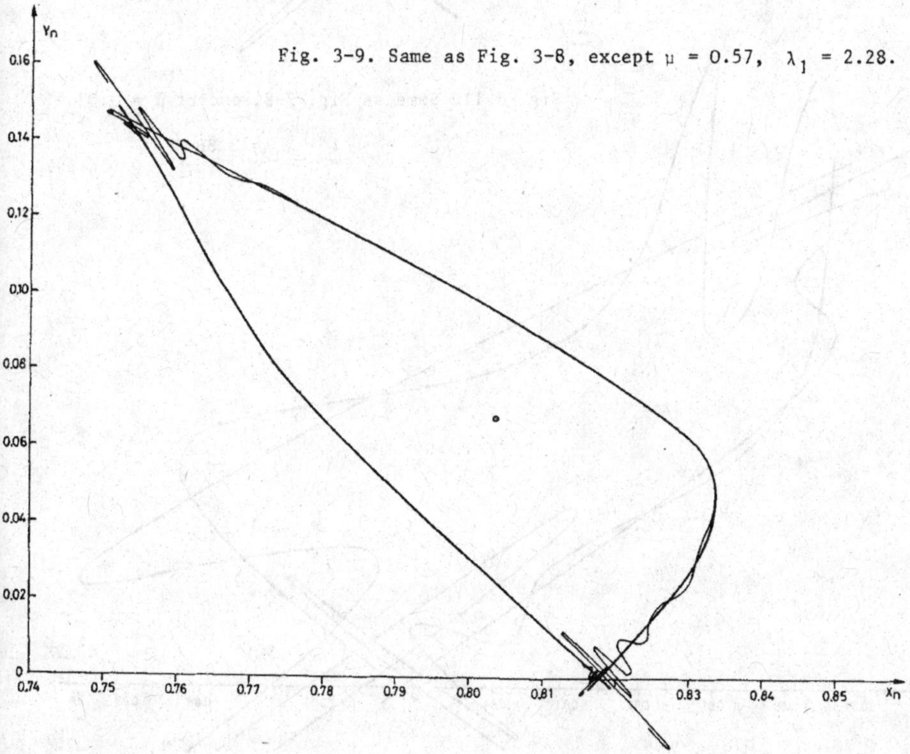


Fig. 3-9. Same as Fig. 3-8, except $\mu = 0.57$, $\lambda_1 = 2.28$.

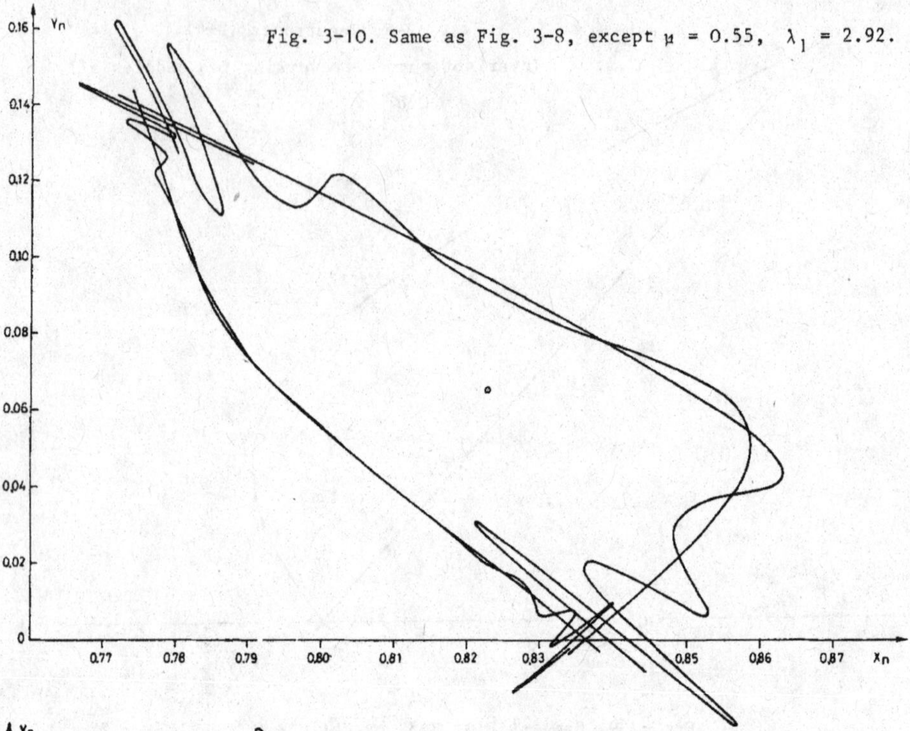


Fig. 3-10. Same as Fig. 3-8, except $\mu$ = 0.55, $\lambda_1$ = 2.92.

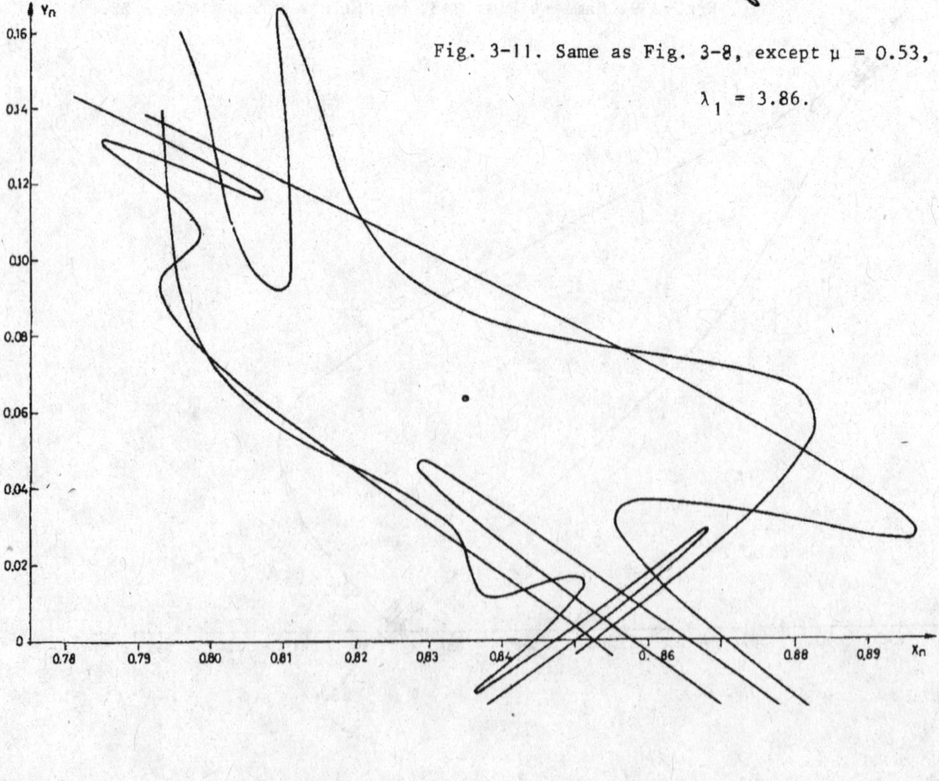


Fig. 3-11. Same as Fig. 3-8, except $\mu$ = 0.53, $\lambda_1$ = 3.86.

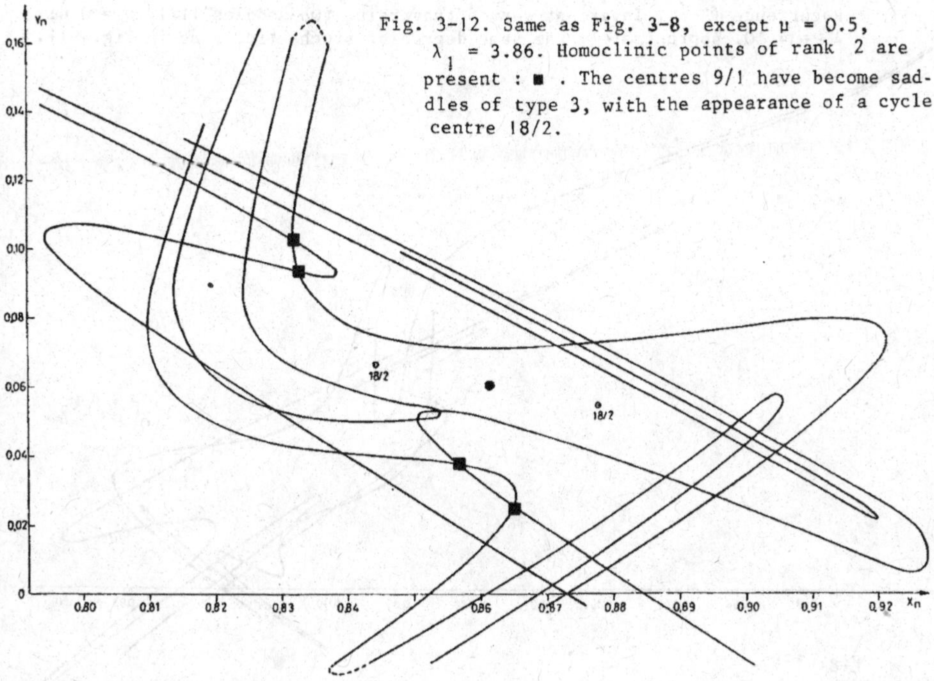


Fig. 3-12. Same as Fig. 3-8, except $\mu = 0.5$, $\lambda_1 = 3.86$. Homoclinic points of rank 2 are present : ■ . The centres 9/1 have become saddles of type 3, with the appearance of a cycle centre 18/2.

A(m, m+1) of the loops between two successive homoclinic points $B_m$, $B_{m+1}$. In Fig.3-12 the amplitudes of A(m, m+1) are already sufficiently large for the appearance on the first loop of secondary intersections and thus of secondary homoclinic points (identified by means of small black squares). This appears to happen simultaneously with the "centre→saddle of type 3" bifurcation of the centre-saddle of type 1" pair.

Fig. 3-13 shows the invariant curves of the cycle $11_{(1)}$ for $\mu = 0.6$, and Fig. 3-14 of the qualitatively similar cycle $19_{(4)}$ for $\mu = 0.125$. A comparison of the change of shape between Figs. 3-8 and 3-13, both having the same strength of the non-linearity and different k, to the change of shape between Figs. 3-8 and 3-12, which have the same k and different $\delta = 1-\mu$, shows that an increase of the distance of a cycle from the main centre (0,0) plays in (3-6) the same role as an increase of the strength of the non-linearity for a fixed distance. The qualitative equivalence between the strength of the non-linearity and the distance from (0,0) is even more apparent from a comparison of Figs. 3-11, 3-13 and 3-14, having roughly the same amplitude of loops for quite different values of $k_{(r)}$ and $\mu$.

Taking into account all results discussed so far for a constant strength of the non-linearity, the global phase portrait of (3-6) can be roughly described as follows : using the main centre (0,0) as a reference, the existence of island structures and instability rings is limited to cycles up to a critical effective order $k/r = k_{c1}$.

Recurrence (3-6). Invariant curves traversing two saddles 11/1, $\mu = 0.6$, $\lambda_1 = 4.10$. Approximately the same degree of stochasticity as in Fig.3-11.

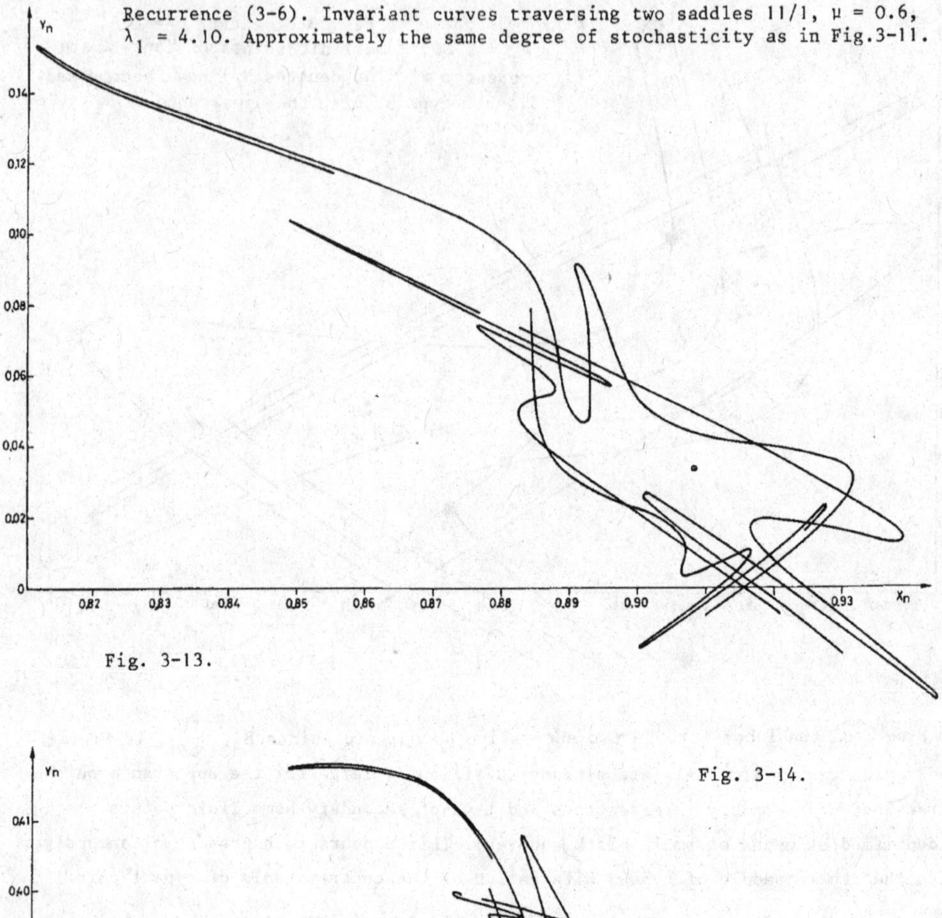


Fig. 3-13.

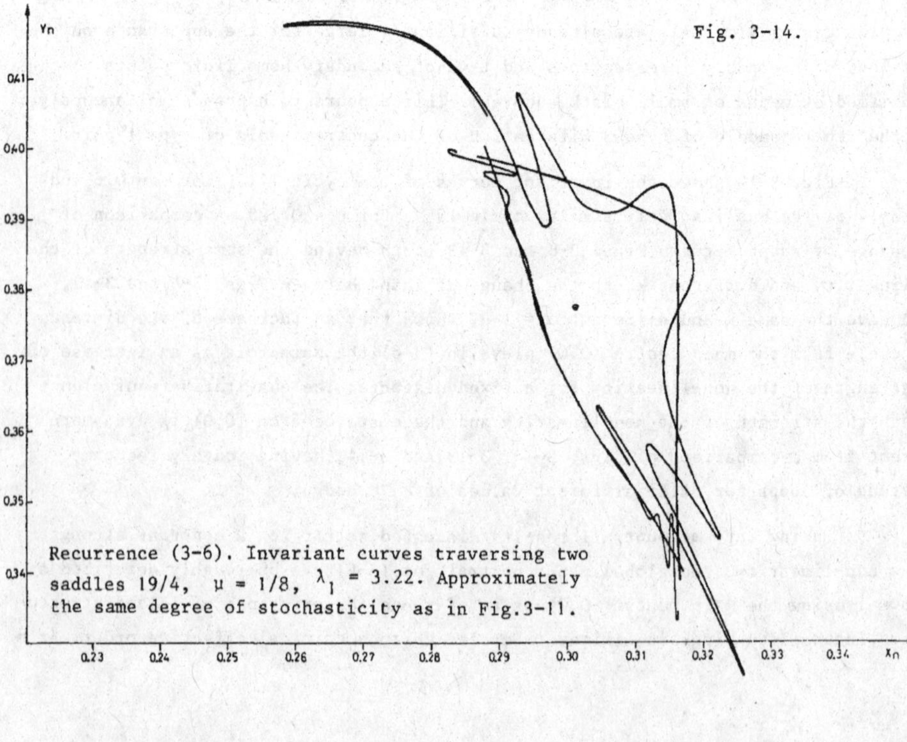


Fig. 3-14.

Recurrence (3-6). Invariant curves traversing two saddles 19/4, $\mu = 1/8$, $\lambda_1 = 3.22$. Approximately the same degree of stochasticity as in Fig.3-11.

Since the amplitudes A(m, m+1) of the loops formed by invariant curves increase while the distance between the cycles of an increasing effective order k/r diminishes indefinitely (these cycles converge to homoclinic points), a situation is reached when the invariant curves of two cycles of two different but close effective orders intersect. According to Poincaré, such intersections are called heteroclinic points. The appearance of heteroclinic points at the smallest possible distance from the main centre (0,0) may be used as an equivalent definition of $k_{c1}$.

If $k/r > k_{c1}$, a second critical value $k/r = k_{c2}$ is reached when the loops of the invariant curves of a saddle $k_{(r)}$ have sufficiently large amplitudes A(m,m+1) to fall into the influence domain of the main invariant curves. Since the latter invariant curves have two unbounded branches, the invariant curves emanating from the cycles $k_{(r)} > k_{c2}$ become also unbounded. A sufficient condition for the unboundedness of an invariant curve passing through a saddle $k_{(r)}$ of type 1 appears to be the existence of a companion saddle of type 3 produced by a bifurcation of a centre $k_{(r)}$, or equivalently the existence of secondary homoclinic points on the first loops of the invariant curves traversing the saddles $k_{(r)}$ of type 1 (see Fig. 3-12).

In the region defined by $k/r < k_{c1}$ the distance $s_n$ between (0,0) and the antecedents or consequents of an initial point $(x_o, y_o)$ is bounded. The discrete trajectories of (3-6) are therefore stable in the sense of orbital stability of the theory of differential equations. For $k_{c1} < k/r < k_{c2}$, the distance $s_n$ grows in the mean, but at a less than exponential rate. This region may be described qualitatively

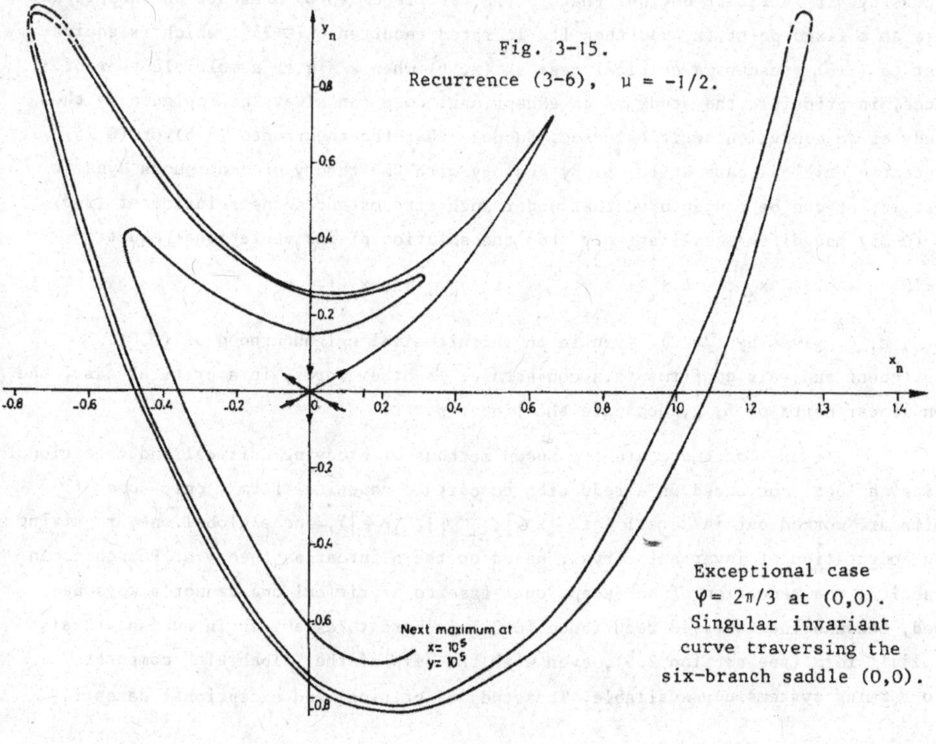


Fig. 3-15.
Recurrence (3-6), $\mu = -1/2$.

Exceptional case
$\varphi = 2\pi/3$ at (0,0).
Singular invariant curve traversing the six-branch saddle (0,0).

as a region of diffusion. For $k/r > k_{c2}$ the mean of the distance $s_n$ grows exponentially. Since the antecedents or consequents of an initial point $(x_o, y_0)$ form geometrically a complicated sequence of points, this region is well described by the name stochastic instability, already used in a similar context [C 2]. Finally, the total region containing intersecting invariant curves, Birkhoff's instability rings included, may be described as the region of stochasticity. As the strength $\delta = 1-\mu$ of the non linearity increases, the region of stochasticity may coincide with the whole phase plane. For the recurrence (3-6), this appears to happen only for $\mu = -1/2$.

### 3.1.4 Critical and exceptional cases

Unless they are saddles, all cycles (and fixed points) of a conservative recurrence constitute critical cases in the sense of Liapunov. The preceding analysis of the properties of the recurrence (3-6), leading to the phase portrait of Fig. 3-1, was carried out with the assumption that the main fixed point (0,0) does not have an eigenvalue of the form $\lambda = \pm 1$ (critical case in a narrow sense) or $\lambda = e^{i\varphi}$ with $\varphi = 2\pi(p/q)$, p, q being integers (exceptional case in a narrow sense). The first assumption is obviously verified in the interval $-1 < \mu < +1$, but the second is not. In fact, there exists an infinite set of p, q for which it fails.

The distinction between critical cases ($\lambda = \pm 1$) and exceptional ones ($\varphi = 2\pi p/q$) introduced above is a matter of convenience and not one of mathematical necessity. It is rather obvious that if the recurrence (3-6) presents an exceptional case at a fixed point (x, y), then its iterated recurrence (0-25), which is equivalent to (3-6), presents a critical case at (x, y) when k = q or a multiple thereof. Hence, in principle the study of an exceptional case can always be replaced by the study of an equivalent critical case. Suppose that the recurrence (3-6) or (0-25) present a critical case at (0,0). By analogy with the theory of continuous dynamic systems, it can be conjectured that under such circumstances the solution of (3-6) or (0-25) may differ qualitatively from the solution of its variational equation

$$x_{n+1} = a\,x_n + b\,y_n, \qquad y_{n+1} = c\,x_n + d\,y_n \quad , \tag{2-19}$$

a, b, c, d, given by (2-22), even in an infinitesimal neighbourhood of (0,0). The subsequent analysis confirms this conjecture. In other words, in a critical case, the non linear parts of f, g, dominate the linear parts.

At present there are two known methods of studying critical and exceptional cases, a local one based on a reduction to certain canonical forms, only some of which are worked out in detail (cf. [L 6], [C 4], [B 4]), and a global one, requiring the computation of invariant curves, based on the bifurcation theory of Poincaré. In practice, the reduction of an exceptional case to a critical one cannot always be used, because the iterated recurrence (0-25) is hard to establish in an analytical explicit form (see section 2.5), even with the help of the "albebraic" computer programming systems now available. The study of critical and exceptional cases is therefore lengthy and laborious. In order to preserve some transparency in the link-up of ideas, it is necessary to omit many theoretically straightforward, but practically tedious and difficult intermediate steps.

The essence of the method of reduction to a canonical form consists in identifying and isolating "dominating terms" of the non linear parts of f and g. One tries to attain this objective by means of a sequence of non linear transformations, which successively remove non-dominating terms. If by means of the successive transformations one has managed to remove completely the non linear parts of f and g, i.e. no dominating terms have appeared, the recurrences (3-6) and (2-19) become qualitatively equivalent. Since the recurrence (2-19) is readily solved explicitly for an arbitrary initial point $(x_o, y_o)$, the sequence of inverse transformations, if it converges, yields then a general explicit solution of (3-6). Such an occurence is of course globally impossible in principle, because it would mean that the solution of (3-6) is completely defined by a single phase cell. A local reduction to (2-19) may be, however, possible under some conditions.

Practically the method of successive transformations is only successful when the dominating terms happen to be of a sufficiently low order, because in general the complexity as well as the notational bulk of the successive transformation increases very rapidly. The "order" or "degree" of a dominating term can be defined in various ways. In what follows, because of the continuity and differentiability assumptions already made, it will be defined as the degree m of a term in a Mc Laurin series. For a two-dimensional Mc Laurin series the term of order m is actually a sum

$$\sum_m a_{ij} x^i y^j, \qquad m = i+j ,$$

with constant coefficients $a_{ij}$. Such a sum is called an "m-th order form".

Consider now the method of successive transformations mentioned above. To be more specific, suppose that the recurrence (0-25) or (3-6) has an exceptional case at (0,0). If not, it is always possible to transform a point (x, y) to (0,0) by means of a linear transformation. As a second preparatory step, the recurrence to be studied is transformed into a complex-conjugate form by means of

$$x = u+v, \quad y = p_1 u + p_2 v, \quad \lambda_{1,2} = e^{\pm i\varphi}, \quad p_{1,2} = (\lambda_{1,2} - a)/b \tag{3-18}$$

In the new variables the recurrence takes the form :

$$\left\{ \begin{aligned} u_{n+1} &= e^{i\varphi}.u_n + \sum_{j=m}^{\infty} P_j(u_n, v_n) \\ v_{n+1} &= e^{-i\varphi}.v_n + \sum_{j=m}^{\infty} Q_j(u_n, v_n) \end{aligned} \right\}, \quad m \geqslant 2 , \tag{3-19}$$

where

$$P_j(u, v) = \sum_{\gamma=0}^{j} \binom{j}{\gamma} C_{j\gamma} . u^{j-\gamma} . v^{\gamma} \quad ,$$

(3-20)

$$Q_j(u, v) = \sum_{\gamma=0}^{j} \binom{j}{\gamma} C^{*}_{j\gamma} . u^{\gamma} . v^{j-\gamma} \quad .$$

The asterisk represents the complex conjugate, and the binomial coefficients $\binom{j}{\gamma}$ are introduced for convenience. They permit simplifications in some later algebra. The constants $C_j$ are completely defined by equations (3-18) and the original recurrence. The successive transformations permitting dominating terms to be identified are of the almost-identity type

$$\xi = u + \varphi_m(u, v), \qquad \eta = u + \psi_m(u, v) \quad ,$$

(3-21)

$$\varphi_m(u, v) = \sum_{\gamma=0}^{m} \binom{m}{\gamma} \alpha_{\gamma} . \xi^{m-\gamma} . \eta^{\gamma} \quad ,$$

$$\psi_m(u, v) = \sum_{\gamma=0}^{m} \binom{m}{\gamma} \alpha^{*}_{\gamma} . \xi^{\gamma} . \eta^{m-\gamma} \quad ,$$

the free undetermined constants $\alpha_\gamma$ to be so chosen that in the new variables $\xi$, $\eta$ the recurrence (3-19) contains as few terms as possible in the corresponding m-th order forms. This condition implies the following values for $\alpha_\gamma$ :

(3-22) $$\alpha_\gamma = \frac{e^{-i\varphi}}{1-e^{-i(m-2\gamma-1)\varphi}} C_{m\gamma}, \qquad \gamma = 0, 1, \ldots, m \quad .$$

The transformation (3-21) will thus succeed only if none of the denominators in equation (3-22) vanishes, i.e. it will succeed if one does not have $(m-2\gamma-1)\varphi = 2\pi$, or a multiple of $2\pi$, which can only happen if

(3-23) $$\varphi = 2\pi\, p/q, \qquad p, q = \text{integers} \quad .$$

The exceptional case (3-23) is therefore defined as a case where the removal of all terms in the m-th order form cannot be attained. In other words, if equation (3-23) holds and the integers p, q, are such that not all $a_\gamma$ in (3-21) can be chosen according to equations (3-22), then the recurrence (3-20) contains some m-th order terms which will subsist after the transformation (3-21). The constants $\alpha_\gamma$ inconsistent with the equations (3-21) can be chosen arbitrarily (for example, set to unity).

Let the transformed recurrence (3-19) be

$$\xi_{n+1} = e^{i\varphi} . \xi_n + U_m(\xi_n, \eta_n) + \sum_{j=m+1}^{\infty} U_j(\xi_n, \eta_n) \quad ,$$

(3-24)

$$\eta_{n+1} = e^{-i\varphi} . \eta_n + V_m(\xi_n, \eta_n) + \sum_{j=m+1}^{\infty} V_j(\xi_n, \eta_n) \quad ,$$

where $U_j$, $V_j$, are j-th order forms. The "canonical" forms $U_m$, $V_m$ define the dominating terms. The advantage of equations (3-24) compared to the original recurrence is twofold :

i) If the critical case is such that one or more invariant curves traverse the point (0,0), then at least in principle it is possible to determine the slope of these invariant curves at (0,0) from the knowkedge of $U_m$ and $V_m$ alone (i.e. there is no need to study the $U_j$, $V_j$, $j > m$).

ii) If no invariant curves traverse (0,0), then (3-24) is analytically more convenient if one wishes to determine the approximate slope of invariant curves near (0,0).

In the case i), the angles between the invariant curves can be found by transforming the recurrence (3-24) into polar coordinates

(3-25) $$\xi = \rho . e^{i\theta}, \qquad \eta = \rho . e^{-i\theta} \quad .$$

Since $\xi . \eta = \rho^2$, one readily finds

(3-26) $$\rho^2_{n+1} = \rho^2_n + A_m(\theta_n, ). \rho^4_n + \dots \quad ,$$

$A_m$ defined by $U_m$ and $V_m$, and after some algebraic manipulations

(3-27) $$\theta_{n+1} = \theta_n + \Theta_m(\rho_n, \theta_n, \varphi) + \dots \quad .$$

As a matter of convenience, the dependence of $A_m$ and $\Theta_m$ on the "unperturbed" rotation angle $\varphi$ has been indicated explicitly. The condition for the existence of invariant angles $\bar{\theta}$ is obviously

(3-28) $$\theta_{n+1} = \theta_n + \varphi \quad .$$

The elimination of $\theta_{n+1}$ between equations (3-27) and (3-28) leads in the limit $\rho_n \to 0$ to an algebraic equation for the unknown real constants $\theta_n = \bar{\theta}$.

Once the angles $\bar{\theta}_j$, j = 1, 2, ... are known $\bar{\theta}_n = \bar{\theta}_j + \dots$ defines the initial shope of an invariant curve segment emanating from the fixed point $\rho_n = 0$. The relative positions of successive consequents on each invariant curve segment is given near $\rho_n = 0$ by equation (3-26), i.e. one knows then whether these consequents converge to or diverge from $\rho_n = 0$. Once one is in possession of this knowkedge, there remains only the straightforward but tedious task of transforming the angles $\bar{\theta}_j$ into slopes $p_{ij}$ expressed in the variables x, y, and then computing the invariant curves by means of the series expansion (2-45).

Consider now an application of the preceding method to the study of the exceptional cases $\varphi = 2\pi/3$ and $\varphi = 2\pi/4$ at (0,0) of the recurrence (3-6). Since

(3-29) $$\cos\varphi = \mu, \qquad p_{1,2} = \pm i\sqrt{1-\mu^2} = \pm i \sin\varphi,$$

these exceptional cases occur for $\mu = -\frac{1}{2}$ and $\mu = 0$, respectively.

The use of the transformation (3-18) leads to the recurrence (3-19), (3-20) with

$$C_{20} = \frac{1}{2}\left[(1+2\cos\varphi)\ \sin\varphi + i(1+\cos\varphi)(1-2\cos\varphi)\right]\ \mathrm{tg}\ \frac{\varphi}{2}$$

$$C_{21} = \frac{1}{2}\left[\sin\varphi - i(1+\cos\varphi)\right]\ \mathrm{tg}\ \frac{\varphi}{2}$$

$$C_{22} = \frac{1}{2}\left[1-2\cos\varphi\right]\left[\sin\varphi + i(1+\cos\varphi)\right]\ \mathrm{tg}\ \frac{\varphi}{2}$$

(3-30) $$C_{30} = (1 - \cos\varphi)(\sin\varphi - i\cos\varphi)\ \mathrm{tg}\ \frac{\varphi}{2}$$

$$C_{31} = (1 - \cos\varphi)(\frac{1}{3}\sin\varphi - i\cos\varphi)\ \mathrm{tg}\ \frac{\varphi}{2}$$

$$C_{32} = -C^{*}_{31}$$

$$C_{33} = -C^{*}_{30}$$

$$C_{4j} = -\frac{i}{2}(1-\cos\varphi)\ \mathrm{tg}\ \frac{\varphi}{2}, \quad j = 0, 1, \ldots, 4$$

$$C_{kj} = 0, \ k > 4 \quad .$$

Since from equations (3-20) one finds $m = 2$ in equation (3-19), the first of the successive transformations of type (3-21) is

(3-31) $$\begin{aligned} \xi &= x + (\alpha_o\, x^2 + 2\alpha_1\, xy + \alpha_2\, y^2) \\ \eta &= y + (\alpha^{*}_2\, x^2 + 2\alpha^{*}_1\, xy + \alpha^{*}_0\, y^2) \end{aligned} \quad ,$$

and the objective $U_2 \equiv 0$, $V_2 \equiv 0$ in the recurrence (3-24) requires

(3-32) $$\begin{aligned} \alpha_0 &= e^{-i\varphi}.(1-e^{i\varphi})^{-1}.C_{20} \\ \alpha_1 &= e^{-i\varphi}.(1-e^{-i\varphi})^{-1}.C_{21} \\ \alpha_2 &= e^{-i\varphi}.(1-e^{-i3\varphi})^{-1}.C_{22} \end{aligned} \quad .$$

The three constants $\alpha_j$ are well defined if $\varphi = 2\pi/4$, and all quadratic terms can therefore be eliminated. If $\varphi = 2\pi/3$, the constant $\alpha_2$ is undefined, and one quadratic term turns out to be dominating. The resulting canonical form is

(3-33) $$\begin{aligned} \xi_{n+1} &= e^{i\varphi}.\xi_n + C_{22}\,\eta_n^2 + \ldots \\ \eta_{n+1} &= e^{-i\varphi}.\eta_n + C^{*}_{22}\,\xi_n^2 + \ldots \end{aligned}$$

In order to determine the possible existence of invariant angles, the transformation to polar coordinates is carried out :

(3-34) $$\begin{aligned} \rho_{n+1}^2 &= \rho_n^2 + \left[C_{22}.e^{-i(\varphi+3\theta_n)} + C^{*}_{22}\ e^{i(\varphi+3\theta_n)}\right|\ \rho_n^4 + \ldots \\ \theta_{n+1} &= \theta_n + \rho_n + \arg\left[1 + C_{22}.\rho_n.e^{-i(\varphi+3\theta_n)}\right| + \ldots \end{aligned} \quad .$$

For $|\rho_n| \ll 1$, the $\theta$-recurrence simplifies into :

(3-35) $$\theta_{n+1} = \theta_n + \ - \left[|C_{22}|\sin(\varphi+3\theta_n + \arg C^{*}_{22})\right]\ \rho_n + \ldots \quad .$$

Continuing (3-35) and (3-28), one finds six values for the invariant angle $\bar{\theta}$ :

$$\bar{\theta}_k = \frac{1}{3}(k\pi - \varphi - \arg C_{22}^{*}), \quad k = 0, 1, \ldots, 5, \tag{3-36}$$

or explicitly after substitution of $C_{22}$,

$$\theta = -\frac{\pi}{6}, \frac{\pi}{6}, \frac{\pi}{2}, \frac{5\pi}{6}, \frac{7\pi}{6} \text{ and } \frac{3\pi}{2} \quad . \tag{3-37}$$

In the x, y variables of the recurrence (3-6) these six values of $\bar{\theta}$ correspond to three invariant curves crossing (0,0), one vertically and two with slopes $\beta_1 = \pm\frac{1}{2}$ in equation (2-45). For $\varphi = 2\pi/3$ the fixed point (0,0) is thus not a centre, but a six-branch saddle. One invariant curve segment crossing this saddle is shown in Fig. 3-15, (p. 121).

For $\varphi = 2\pi/4$ it is necessary to repeat the transformation (3-19) with m = 3. The dominating terms turn out to be of order three

$$\begin{aligned} \xi_{n+1} &= i\xi_n + 2\xi_n^2\eta_n - 2\eta_n^3 + \ldots \\ \eta_{n+1} &= -i\eta_n - 2\xi_n^3 + 2\xi_n\eta_n^2 + \ldots \quad . \end{aligned} \tag{3-38}$$

The transformation to polar coordinates leads, however, to the result that there exist no (real) invariant angles $\bar{\theta}$. There are thus no invariant curves crossing (0,0). Close to (0,0), i.e. for $|x_n| + |y_n| \ll 1$ an approximate solution of (3-38) and (2-28) is

$$x_n^2\, y_n^2 = \text{const.} \tag{3-39}$$

A small segment of an invariant curve given by (3-39) can be continued numerically. Such a mixed method shows that the hyperbolic slope deforms gradually near $x_n = 0$ and $y_n = 0$ and leads to a closed invariant curve (Fig. 3-16). This result can be

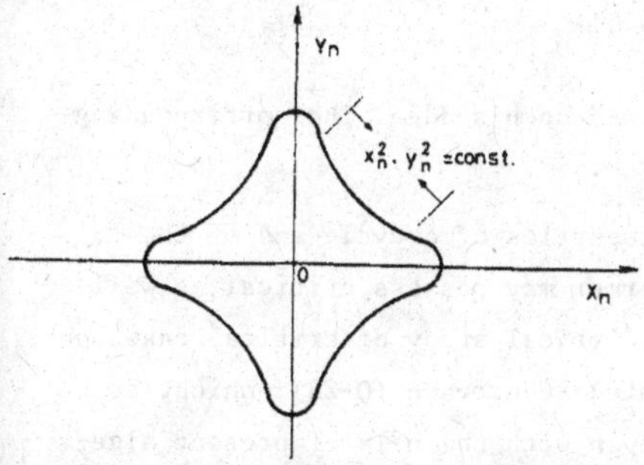


Fig. 3-16. Recurrence (3-6), $\mu = 0$. Exceptional case $\varphi = 2\pi/4$ at (0,0). Approximate closed invariant curve or unresolved island structure $k/r \simeq 4$, $k \gg 1$, $r \gg 1$, surrounding the centre (0,0).

confirmed analytically by adding higher order terms to (3-38). The corresponding fourth-order forms are :

$$\begin{aligned} U_4(\xi, \eta) &= \frac{1}{2}(1-i)(\xi-\eta)^4 + 2(\xi^2-\eta^2)\left[i(\xi+\eta)^2 - (\xi-\eta)^2\right] \\ V_4(\xi, \eta) &= \frac{1}{2}(1+i)(\xi-\eta)^4 + 2(\xi^2-\eta^2)\left[i(\xi+\eta)^2 + (\xi-\eta)^2\right] \quad . \end{aligned} \tag{3-40}$$

With the use of equations (3-40), the polar form of the recurrence (3-38) becomes

$$\text{(3-41)}\qquad \begin{aligned} \rho_{n+1}^2 &= \rho_n^2 + (4 \sin 4\theta_n)\, \rho_n^4 + \ldots \\ \theta_{n+1} &= \theta_n + \frac{\pi}{2} - 2(1 - 4 \cos 4\theta_n)\, \rho_n^2 + \ldots \quad , \end{aligned}$$

which leads to the iterated recurrence

$$\text{(3-42)}\qquad \begin{aligned} \rho_{n+4}^2 &= \rho_n^2 + (16 \sin 4\theta_n)\, \rho_n^4 + \ldots \\ \theta_{n+4} &= \theta_n - 8(1 - \cos 4\theta_n)\, \rho_n^2 + \ldots \quad . \end{aligned}$$

From the equation (3-42), it is easily seen that close to the origin $\rho_n$ is stationary for $\theta_n = 0, \frac{\pi}{2}, \pi$ and $3\pi/2$.

Using the same method no (real) invariant angles were found for $\varphi = 2\pi/5$ and $\varphi = 2\pi/6$. This result was verified numerically by computing some discrete trajectories near (0,0). The numerical computations confirm the theoretical deductions (on the basis of equations like (3-42)) that $\varphi = 2\pi/3$ and $\varphi = 2\pi/4$ are the only exceptional cases of the recurrence (3-6), for which the invariant curves near (0,0) differ substantially from a family of concentric ovals (similar to ellipses).

The successive transformation method can of course also be applied to exceptional cases arising in connection with cycles. Let (x, y) be the coordinates of one point of a cycle k/r, which is a centre in linear approximation. In order to arrive at the starting point of the transformation method, it is first necessary to translate (x, y) to (0,0), i.e. it is necessary to apply a non-homogeneous linear transformation to the iterated recurrence. Such a transformation is rather time-consuming, because the low-order terms of the transformed recurrence depend both on low <u>and</u> high-order terms of the starting recurrence. For this reason, it was carried out only for the centre 3/1 of (3-6), which attains $\varphi = 2\pi/3$ at (cf. eq. 3-10)

$$\text{(3-43)}\qquad \mu \simeq -0.47865, \qquad x \simeq -0.32229, \qquad y = 0 \quad .$$

It was found that this centre also turns into a six-branch saddle. The corresponding slopes in x, y - coordinates are $p_{1,2} = \pm 2$ and $p_3 = \infty$.

The main difference between the local properties of a cycle and of the fixed point of the recurrence (3-6) is that the former may possess critical as well as exceptional cases. The starting point for an analytical study of critical cases on a cycle is normally the explicit form of the iterated recurrence (0-25), which, as pointed out before, is extremely hard to obtain, even with the help of present algebraic computer programs. By applying suitable almost-identity transformations to the properly prepared recurrence (0-25) it is possible to define dominating terms and canonical forms in the same manner as for fixed points. A partial classification of dominating terms was already carried out by Birkhoff, but as a rule the solutions, i.e. the knowledge of invariant curves near (0,0) are still lacking. This is unfortunate because solutions of critical cases appear to be the key to a profound understanding of large-amplitude behaviour of non linear dynamic systems.

As an illustration of the present limitations of canonical transformation analysis, which moreover is merely local in character, consider the critical case $\lambda = -1$ occuring during the transition "centre → saddle of type 3". As can be seen from Tables 3-1 to 3-5, such a transition is frequent for cycles of the recurrence (3-6). Since cycles of order $k = 2$ do not occur in this recurrence, the simplest possible case is $k = 3$. One has then $\lambda = -1$ for $\mu = -1/2$, the coordinates of one point of the cycle 3/1 being $(-\frac{1}{3}, 0)$. The first terms of the iterated recurrence (0-25), with $(-\frac{1}{3}, 0)$ moved to (0,0), are

$$\text{(3-43)}\quad \begin{aligned} x_{n+3} &= -x_n + 6y_n^2 - 24x_ny_n + \frac{51}{2}x_n^2 + \dots \\ y_{n+3} &= -y_n + 4x_n - 12y_n^2 + 45x_ny_n - 45x_n^2 + \dots \end{aligned}$$

Although the quadratic terms appear to be dominating, the qualitative nature of the general solution and the shape of invariant curves are still unknown. Other methods of studying critical cases are therefore indispensable. One such method consists in the implementation of the bifurcation theory of Poincaré.

### 3.1.5 Bifurcations

Consider a family of functions G(x, c), depending continuously on the (real) parameter c in a given interval $c_1 < c < c_2$. A value $c_0$ is said to be a bifurcation value of G(x, c) (in a generalized sense) if $c_1 < c_0 < c_2$ and if G(x, c) undergoes a qualitative change at $c = c_0$, i.e. if the qualitative properties of $G(x, c_0)$ are different from the qualitative properties of $G(x, c_0+\Delta c)$, $|\Delta c| << |c_0|$. G(x, c) is called a bifurcation function. The nature of the qualitative properties can be most diverse.

The properties of the recurrence (3-6) will now be examined from the bifurcation point of view. The particular solutions of (3-6) are assumed to play the role of the functions G(x, c), whereas the parameter c will be, according to need, either $\mu$, or an eigenvalue $\lambda$, or the rotation angle $\varphi$ when $\lambda = e^{i\varphi}$. The simplest qualitative property of G(x, c) to be considered is the number of real roots of equations (0-24) which exist in a specified part of the phase plane. An additional qualitative property, if required, might be the nature of these roots. The simplest bifurcation values of $\mu$ will thus be the values $\mu_o$ for which cycles appear, disappear, or change their properties.

The first bifurcation uncountered in the study of the recurrence (3-6) happens to be the transition centre → saddle of type 3, the bifurcation value being $\lambda = -1$. While the bifurcation solution of (3-6) is still unknown (see, for example, (3-43)), in principle it is possible to obtain the non-bifurcated solution just "before" and just "after" the bifurcation, i.e. for a slightly weaker and a slightly stronger non-linearity. The properties of the bifurcation solution can then be roughly estimated by means of a continuity argument. If there were no practical limitations on the analytical or numerical operations, the choice of the specific cycle to be studied in connection with the $\lambda = -1$ bifurcation would be a matter of indifference. But since there exist several practical constraints, this choice is best made by experience. Extensive data on cycles, such as those shown in Tables 3-1 to 3-5 are of course quite helpful. Although all computations were carrier out with at least ten significant digits, for convenience only the first digits will be given in the subsequent discussions.

For the recurrence (3-6) and the cycle 4/1 the relevant bifurcation data are

$$\text{(3-44)}\qquad \begin{array}{llll} \mu = -0.103 & x = 0.532268 & y = 0 & \varphi = 2.98835 \\ \mu = -0.104 & x = 0.533703 & y = 0 & \lambda_2 = -1.22558 \end{array},$$

the eigenslopes in the latter case being $p_1 = -0.0545$, $p_2 = -p_1$. The bifurcation $\lambda = -1$ thus takes place in the interval $-0.104 < \mu < -0.103$. Since the saddle of type 3 has real eigenslopes, it is possible to compute the invariant curves passing through one point of this cycle. A representative part of the invariant curves traversing the point given in equation (3-44) is shown in Fig. 3-17. The corresponding saddles 4/1

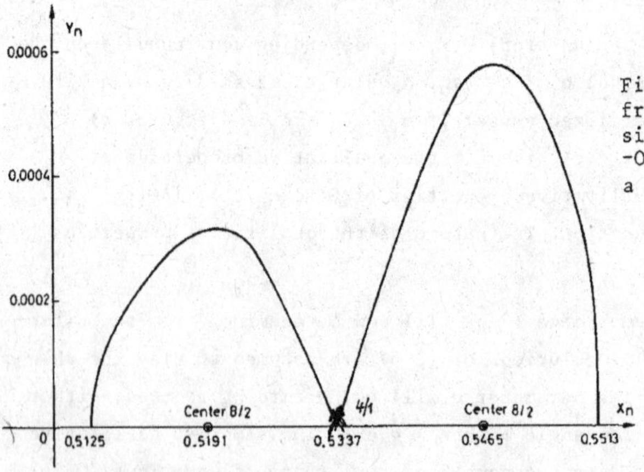


Fig. 3-17. Centres 8/2 bifurcated from a cycle point 4/1 at the transition $\varphi = \pi$. Recurrence (3-6), $\mu = -0.0014$. Invariant curves crossing a saddle 4/1, $\lambda_2 = -1.2226$.

of type 1 are far removed from the invariant curve loopes of Fig. 3-17. The closest points of the cycle saddle 4/1 of type 1 are

$$\text{(3-45)}\qquad \mu = -0.104 \qquad x = 0.164754 \qquad y = \pm\, 0.211856 \qquad \lambda_1 = 2.33845\,.$$

Thus a natural question is whether there exist points of another cycle inside the main loops of the invariant curves of Fig. 3-17. The answer turns out to be in the affirmative : for $\mu = -0.104$ two points of the cycle centre 8/2 with $\varphi = 0.583397$ are located at ($x = 0.519141$, $y = 0$) and ($x = 0.546543$, $y = 0$), respectively.

For intermediate values of $\mu$ one finds for the cycle 4/1

$$\mu = -0.1036 \qquad x = 0.533130 \qquad y = 0 \qquad \lambda_2 = -1.13268$$

$$(3\text{-}46) \qquad \mu = -0.1034 \qquad x = 0.532844 \qquad y = 0 \qquad \lambda_2 = -1.05210$$

$$\mu = -0.1032 \qquad x = 0.532556 \qquad y = 0 \qquad \varphi = 3.03975 \ ,$$

and for the cycle 8/2

$$\mu = -0.1036 \qquad \left\{ \begin{matrix} x = 0.524416 & y = 0 \\ x = 0.541198 & y = 0 \end{matrix} \right\} \qquad \varphi = 0.354186$$

$$(3\text{-}47) \qquad \mu = -0.1034 \qquad \left\{ \begin{matrix} x = 0.529369 & y = 0 \\ x = 0.536210 & y = 0 \end{matrix} \right\} \qquad \varphi = 0.143779$$

$$\mu = -0.1032 \qquad x = \text{none} \qquad y = \text{none} \ .$$

From Fig. 3-17, and the data in equation (3-46) and (3-47), it appears that the centres 8/2 approach the saddles 4/1 of type 3 as $\mu$ approaches $\mu_o$ from below. Since for $\mu > \mu_o$ the centres 8/2 cease to exist, it is natural to conclude that they merge with the saddles 4/1 of type 3 as $\mu \to \mu_o$. This conclusion is confirmed by the behaviour of the invariant curves, whose loops get flatter as $\mu \to \mu_o$. For $\mu = -0.1036$, the eigenslopes are $p_{1,2} = 0.0334$ and for $\mu = -0.1034$ they are already $p_{1,2} = \pm 0.0136$. One can therefore say that as $\mu$ traverses $\mu_o$ in the direction of increasing non-linearity, the centres 4/1 bifurcate at $\mu = \mu_o$, $(\varphi = \pi)$, into saddles of type 3, $(\lambda_2 = -1)$, and into centres 8/2, $(\varphi = 0)$.

In order to verify whether the preceding conclusion holds also for the previously established equivalence between the strength of the non-linearity $\delta = 1 - \mu$, and the distance $s_n$ of a point $(x_n, y_n)$ of a cycle from (0,0), the singular invariant curves defined by the saddle of type 1 → saddle of type 3 configuration were

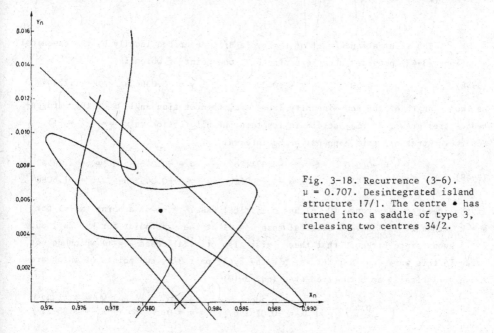

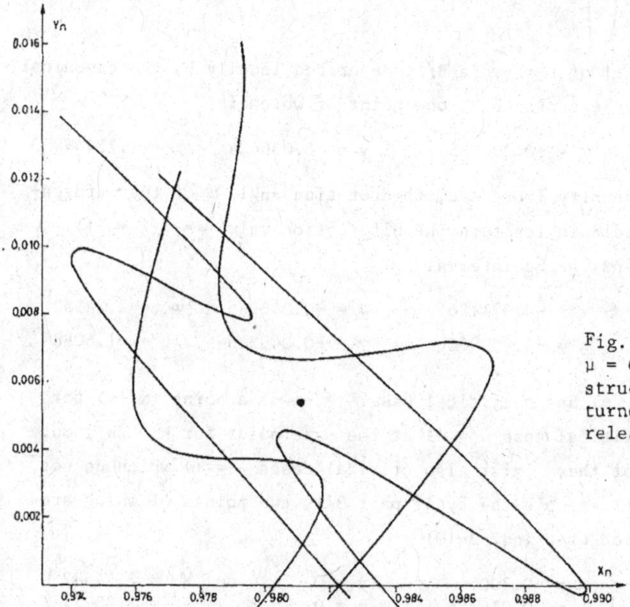


Fig. 3-18. Recurrence (3-6). $\mu = 0.707$. Desintegrated island structure 17/1. The centre • has turned into a saddle of type 3, releasing two centres 34/2.

computed for the rather weakly non linear case $\mu = 0.707$, $k = 17$, $r = 1$. These invariant curves are shown in Fig. 3-18 and 3-19. Inside the first loops of Fig. 3-19 were found two points of the cycle centre 34/2. For a moderate non-linearity each "island" of the cycle 9/1 with $\mu = 0.5$ (Fig. 3-12) was found to contain in its interior two points of the cycle centre 18/2.

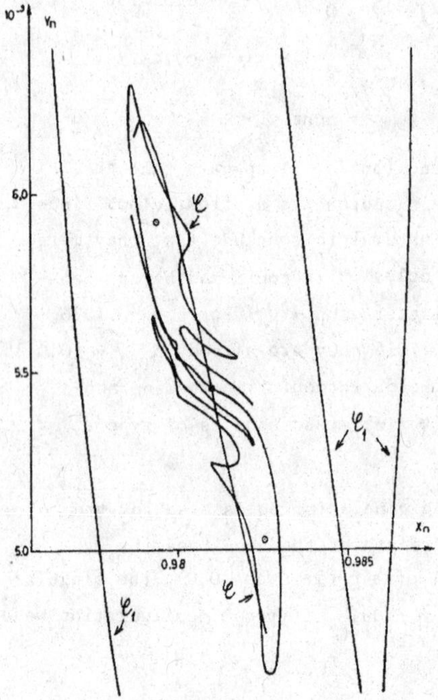


Fig. 3-19. Invariant curves $\mathcal{C}$ crossing the saddle • 17/1 of type 3 shown in Fig. 3-18. Loops of $\mathcal{C}$ surround the centres ⊙ 34/2 with the formation of numerous homoclinic points. $\mathcal{C}_1$ : invariant curves originating at the saddles 17/1 of type 1.

The bifurcation $\lambda = -1$ of the cycle 3/1, described locally by the canonical equations (3-43) produces a cycle centre 6/2, one point of which is

(3-48) $\mu = -0.506$ $\quad x = -0.332005$ $\quad y = -0.036424$ $\quad = 1.72444$ .

As the strength of the non-linearity increases, the rotation angle $\varphi$ of the bifurcated centres grows, and may attain in its turn the bifurcation value $\varphi = \pi, (\lambda = -1)$. For the centres 6/2 this happens in the interval

(3-49)
$$\begin{array}{llll} \mu = -0.510 & x = -0.331126 & y = -0.046945 & \varphi = 2.74582 \\ \mu = -0.511 & x = -0.330906 & y = -0.049216 & \lambda_2 = -1.60884 . \end{array}$$

If the iterated recurrence (0-25) has a critical case $\lambda = -1$ at a point $(x, y)$ for $k = k_o$, then it will have a critical case $\lambda = +1$ at the same point for $k = 2k_o$. But it is known since Birkhoff that there exist also critical cases $\lambda = +1$ which do not arise in this way. Consider for example the cycle pair 3/1, two points of which are known analytically as a function of $\mu$ (eq. 3-10) :

(3-50)
$$\mu = -0.45 \quad \left\{ \begin{array}{ll} x = -0.30054 & y = 0 \\ x = -0.079256 & y = 0 \end{array} \right. \quad \left. \begin{array}{l} \varphi = 2.74582 \\ \lambda_1 = 1.33221 \end{array} \right\} .$$

As was already pointed out before, these two points merge for $\mu = \mu_o = 1-\sqrt{2}$ into a double root of (0-24):

(3-51) $$x = -0.207107 \qquad y = 0 \qquad \lambda = +1 \quad .$$

For $\mu > \mu_o$ the roots of equation (3-10) are complex, i.e. the cycles 3/1 no longer exist.

By straightforward but practically tedious algebraic manipulations, it is possible to express the iterated recurrence (0-25) at the double root (3-51) in canonical form (i.e. (x, y) moved to (0,0) and only dominating terms retained explicitly)

(3-52) $$\begin{aligned} x_{n+3} &= x_n - y_n + (1-\mu_o)\, x_n^2 + \ldots \\ y_{n+3} &= y_n - 2(1-\mu_o)\, x_n^2 + 2(1-\mu_o) x_n y_n - (1-y_o)\, y_n^2 + \ldots \end{aligned}$$

The invariant curves of the critical case (3-52) cannot be expressed in the form (2-43). By means of computer-programmed algebraic operations it was found that two invariant curves passing through the point given by eq. (3-51) can be described by an at least asymptotically convergent series in fractional powers of $x_n$, the first term of which is

(3-53) $$y_{n+3} = \pm 2\sqrt{\frac{1-\mu_o}{3}\, x^3_{n+3}} \quad .$$

The invariant curves therefore form a cusp at the double point (3-51). Longer segments of the invariant curve defined "initially" by eq. (3-53) are shown in Fig. 3-20. These segments do not join smoothly, but form homoclinic points.

It is obvious that the cycle 3/1 has also something to do with the exceptional case $\varphi = 2\pi/3$, because contrary to what happens to the saddles k/r, k > 3 in Tables 3-1 to 3-5, the saddles 3/1 move in-wards (i.e. towards (0,0)) as the strength $\delta = 1-\mu$ of the non-linearity increases. More closely to $\mu_o = -\frac{1}{2}$, one finds :

(3-54) $$\begin{cases} \mu = -0.49 & x = -0.013709 & y = 0 & \lambda_1 = 1.06158 \\ \mu = -0.5 & x = 0 & y = 0 & \lambda_1 = 1 \text{ and } \varphi = 2\pi/3 \text{ at } (0,0) \\ \mu = -0.51 & x = 0.012995 & y = 0 & \lambda_1 = 1.06201 \\ \mu = -0.55 & x = 0.059493 & y = 0 & \lambda_1 = 1.35216 \end{cases}$$

The unstable exceptional case $\varphi = 2\pi/3$ at (0,0) is therefore produced by the coalescence of a centre with three saddles of type 1 of the cycle 3/1. The invariant curves corresponding to the crossing of this bifurcation are shown in Fig. 3-21 ($\mu > \mu_o$), Fig. 3-15 ($\mu = \mu_o$), and Fig. 3-22 ($\mu < \mu_o$). It should be noted that just as in the case of the cusp (Fig. 3-20), from which these invariant curves have evolved, the invariant curves emanating from the saddles 3/1 of type 1 do not join smoothly. The loops on the interior invariant curves (those closer to the origin) are much less pronounced than those on the external ones. The iterates on the latter diverge very rapidly.

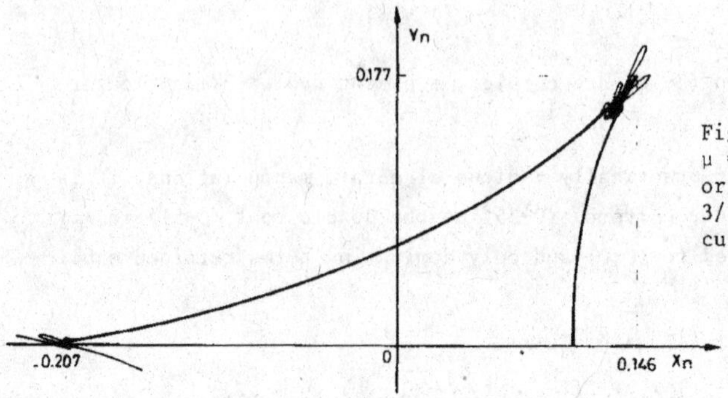


Fig. 3-20. Recurrence (3-6). $\mu = \mu_o = 1-\sqrt{2}$. Invariant curves originating at the double cycle 3/1 and forming a stochastic cusp. Critical case $\lambda = +1$.

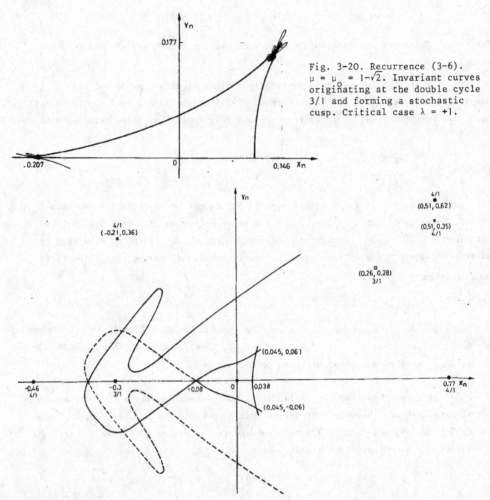

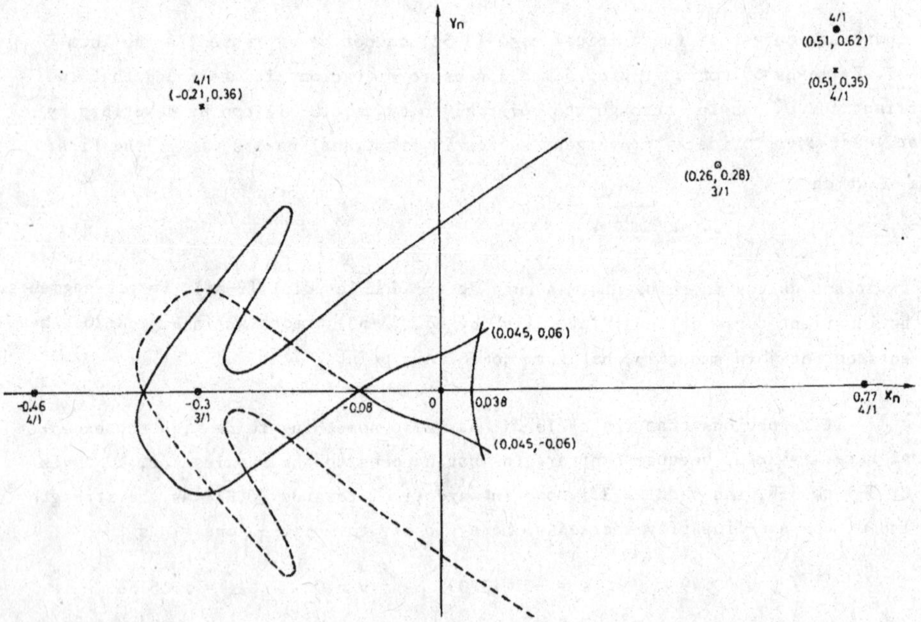


Fig. 3-21. Recurrence (3-6). $\mu = -0.45 < \mu_o$. Invariant curves (not to scale) crossing the saddles 3/1 and surrounding the 3/1 centres before merging with (0,0). x saddle of type 1. • saddle of type 3.

The exceptional case $\varphi = 2\pi/3$ occuring on the centres 3/1, cf. eq. (3-43), is a bifurcation involving the saddles 9/3 of type 1. For $\mu = -0.47870$ one such saddle is at ($x = -0.322341$, $y = 0$) with $\lambda_1 = 1.00263$ and the eigenslopes $p_{1,2} = \pm 1.973$.

Since the correlation of the cycles 3/1 with the exceptional case $\varphi = 2\pi/3$ at (0,0) turned out to be successful, it is worth while to examine the relationship between the cycles 4/1 and the exceptional case $\varphi = 2\pi/4$. The locations of the points of the cycles 4/1 are given explicitly by the equation (3-12) and (3-13). As was

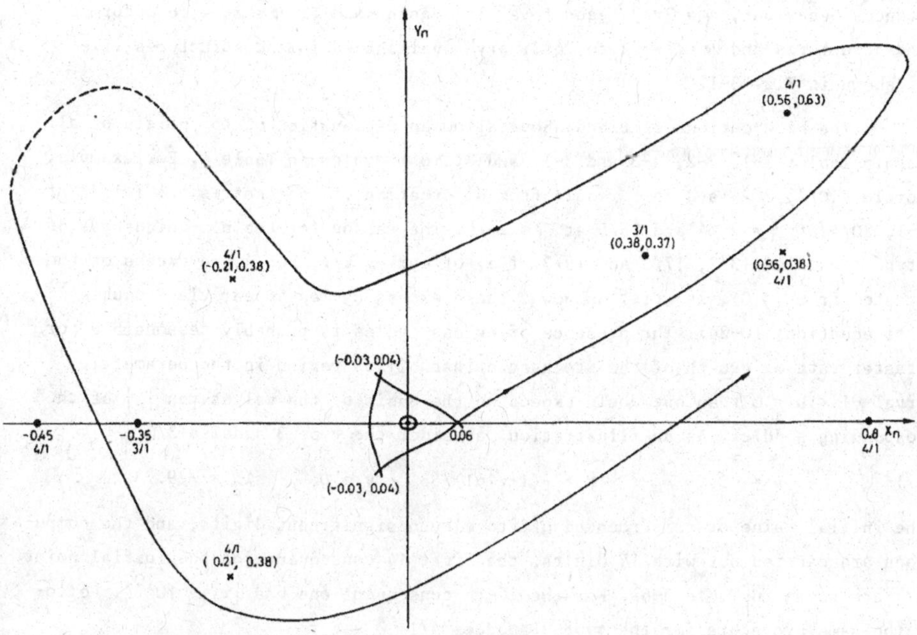


Fig. 3-22. Recurrence (3-6), $\mu = -0.55$. Invariant curves (not to scale) crossing the saddles 3/1 after they have separated from (0,0).

already pointed out, both cycles exist only for $\mu < 0$ and both merge with (0,0) as $\mu \to 0^-$. The rate of approach of the saddles and centres 4/1 to (0,0) is, however, unequal. Sample locations are

$$
(3\text{-}55)\quad \left\{
\begin{array}{llll}
\mu = -0.0.0001 & x = -0.084329 & y = 0 \quad , & \varphi = 0.009548 \\
\mu = -0.001 & x = -0.150466 & y = 0 \quad , & \varphi = 0.0544121 \\
\mu = -0.005 & \left\{ \begin{array}{l} x = -0.225650 \\ x = -0.047326 \end{array} \right. & \begin{array}{l} y = 0 \quad , \\ y = 0.049813 \; , \end{array} & \left. \begin{array}{l} \varphi = 0.183911 \\ \lambda_1 = 1.04091 \end{array} \right\} \\
\mu = -0.1 & \left\{ \begin{array}{l} x = -0.267843 \\ x = -0.065235 \end{array} \right. & \begin{array}{l} y = 0 \quad , \\ y = 0.070185 \; , \end{array} & \left. \begin{array}{l} \varphi = 0.313781 \\ \lambda_1 = 1.0837 \end{array} \right\}
\end{array}
\right.
$$

The bifurcation of the exceptional case $\varphi = 2\pi/4$ therefore releases an island structure 4/1 when $\mu = \mu_o = 0$ is crossed in the direction of increasing non-linearity. The difference of the rate at which centres and saddles 4/1 leave the vicinity of (0,0) explains the deformation of the regular invariant curves at $\mu = \mu_o = 0$, described approximately by eq. (3-39), because in order to preserve continuity the shapes of these invariant curves (Fig. 3-16) must "preadjust" to the shape of the singular invariant curves traversing the 4/1 saddles, which are about to appear.

All higher exceptional cases $\varphi = 2\pi r/k$, $k > 4$ at (0,0) examined so far show the same behaviour : the island structure k/r is released at

$$
(3\text{-}56) \qquad \mu_o = \cos\,(2\pi r/k)
$$

and exists only for $\mu < \mu_o$. The exceptional cases on centres of a cycle have an analoguous behaviour, $\varphi = 2\pi/3$ leads to a six-branch saddle, $\varphi = 2\pi/4$ to deformed island structures and $\varphi = 2\pi r/k$ to "ordinary" oval-shaped island structures like those shown in Fig. 3-1.

The bifurcations described above allow an explanation of the origin of all cycles in Tables 3-1, 3-2, 3-3 and 3-5, and of most cycles in Table 4. For example, the cycles 10/2, 20/4 and 25/5 result from bifurcation of the centres 5/1 : 10/2 at $\lambda = -1$, 20/4 at $\varphi = 2\pi/4$ and 25/5 at $\varphi = 2\pi/5$. The reason for the non-uniqueness of the type 1 saddles 13/2, 17/3 and 19/3, i.e. of cycles k/r, k > 4, in excess of those bifurcated from (0,0), is still unknown. These excess cycles appear via a double root of equations (0-24). The presence of excess cycles is probably responsible for the faster rate of growth of the stochastic instability region in the parametric interval $-1 < \mu < 0$ than one would expect on the basis of the values max $|\lambda|$ of the corresponding saddles. As an illustration, consider the type 3 saddles 3/1 :

(3-57) $\mu = -0.9 \qquad x = -(1+\sqrt{161})/38 \ , \ y = 0 \ , \quad \lambda_2 = -29.73 \ .$

If the initial value of x is rounded off to eleven significant digits, and the computations are carried out with 15 digits, the first 14 consequents of the initial point (3-57) are shown in Table 3-6a. For the 20-th consequent one has $|x| > 10^{100}$. Following the same procedure for the type 3 saddles 4/1

(3-58) $\mu = -0.045 \qquad x = (9 - \sqrt{1281})/58 \ , y = 0 \ , \quad \lambda_2 = -54.47 \ ,$

but rounding off the initial value of x to nine significant digits, one finds the first 14 consequents of Table 3-6b. At the 20-th consequent the value of $|x|$ is still below one. For the saddles 4/1 both the rounding error and $|\lambda_2|$ are larger, but the divergence of consequents is much weaker. This more pronounced stochasticity is the

| | (a) Type 3 saddle 3/1 $\mu = -9/10$, $\lambda_2 = -29.73$ | | (b) Type 3 saddle 4/1 $\mu = -9/20$, $\lambda_2 = -54.47$ | |
|---|---|---|---|---|
| n | $x_n$ | $y_n$ | $x_n$ | $y_n$ |
| 0 | -0.360228 | 0 | -0.461915 | 0 |
| 1 | 0.570757 | 0.465497 | 0.517241 | 0.617087 |
| 2 | 0.570756 | -0.465476 | 0.772260 | 0.000000 |
| 3 | -0.360196 | 0.000081 | 0.517241 | -0.617087 |
| 4 | 0.570603 | 0.465271 | -0.461915 | 0.000000 |
| 5 | 0.570345 | -0.465856 | 0.517241 | 0.617087 |
| 6 | -0.361108 | 0.002411 | 0.772260 | 0.000000 |
| 7 | 0.575168 | 0.472012 | 0.517242 | -0.617087 |
| 8 | -0.582915 | -0.454190 | -0.461914 | -0.000001 |
| 9 | -0.333213 | -0.072065 | 0.517239 | 0.617084 |
| 10 | 0.438786 | 0.304118 | 0.772254 | -0.000007 |
| 11 | 0.275024 | -0.542295 | 0.517224 | -0.617100 |
| 12 | -0.646404 | 1.10063 | -0.461946 | 0.000073 |
| 13 | 2.47629 | 10.0685 | 0.517370 | 0.617254 |
| 14 | 19.4907 | 701.767 | 0.772562 | 0.000411 |
| | $\|x_{20}\| > 10^{100}$ | | $\|x_{20}\| < 0.55$ | $\|y_{24}\| > 7 \cdot 10^4$ |

Table 3-6. Recurrence (3-6). Stochastic instability. Growth rate of consequents.

reason why, for example, the cycle 3/1 was not chosen for a detailed numerical study of the $\lambda = -1$ bifurcation.

### 3.2 Cubic recurrence

The one-parameter form analoguous to (3-6) is :

$$x_{n+1} = y_n + F(x_n) \quad , \quad y_{n+1} = -x_n + F(x_{n+1}) \quad , \tag{3-59}$$
$$F(x) = \mu x + (1-\mu)x^3 \quad , \quad -1 < \mu < 1 \quad .$$

Similarly to (3-6), the fixed-point equation (3-7) is readily solved : there exist three main fixed points, the centre (0,0) and the saddles of type 1 (1,0) and (-1,0). The eigenvalues of (0,0) are still given by equation (3-8), and the eigenvalue $\lambda_1$ of the main saddle increases as $\mu$ decreases. For the same reason as before there are no cycles of order k = 2.

After some tedious algebraic manipulations the positions of some cycles can be obtained in an analytically explicit form. The first case of interest is k = 3, associated with the exceptional case $\varphi = 2\pi/3$ at (0,0). As one might expect, there exist centres and saddles 3/1. One centre is at

$$x = 0 \quad , \quad y = \sqrt{\tfrac{1}{2}(1+2\mu)/(\mu-1)} \quad , \tag{3-60}$$

and one saddle at

$$x = \sqrt{(z-\mu)/(1-\mu)} \quad , \quad y = 0 \quad , \tag{3-61}$$

where z is a real root of

$$z^3 + (1-\mu)z^2 + (1-\mu)z + \frac{1}{2} = 0 \quad .$$

Both cycles exist only for $\mu < -\frac{1}{2}$ and coalesce with (0,0) as $\mu \to -\frac{1}{2}$. For the centres 3/1 this conclusion is obvious from an inspection of eq. (3-60). For the saddles 3/1 it can be arrived at by examining the well-known algebraic inequalities for cubic polynomials, but it is simplest to inspect the following numerical results

(3-62)

| | | | |
|---|---|---|---|
| $\mu = -0.5001$ | $x = 0.009428$ | $y = 0$ | $\lambda_1 = 1.00001$ |
| $\mu = -0.501$ | $x = 0.029804$ | $y = 0$ | $\lambda_1 = 1.00018$ |
| $\mu = -0.55$ | $x = 0.207352$ | $y = 0$ | $\lambda_1 = 1.06457$ |

One point of the cycle centre 4/1 is at

$$x = \sqrt{-\mu/(1-\mu)}, \quad y = \sqrt{-\mu/(1-\mu)} \quad , \tag{3-63}$$

whereas one point of the saddles is at

$$x = \sqrt{-\mu/(1-\mu)} \quad , \quad y = 0 \tag{3-64}$$

The island structure 4/1 exists only for $\mu < \mu_o = 0$, and is generated at (0,0) when $\varphi = 2\pi/4$. No explicit expressions in terms of elementary functions are known for the locations of the saddles 5/1 and 5/2.

One point of the centres 5/1 and centres 5/2 is at

$$(3\text{-}65) \qquad x = 0 \quad , \quad y = \sqrt{(z-\mu)/(1-\mu)} \quad ,$$

where z is a real root of

$$(3\text{-}66) \qquad 16z^4 - 16\mu z^3 + (4\mu+z)z - 1 = 0 \quad .$$

The nature of the roots of (3-66) can also be characterized in terms of algebraic inequalities, and z can be expressed in terms of square roots or cube roots. The centres 5/1 exist only for $\mu < \mu_o = \cos(2\pi/5) \simeq 0.309$, and they merge with (0,0) as $\mu \to \mu_o$. The same property holds for the centres 5/2, except that $\mu_o = \cos(4\pi/5) \approx 0.809$.

One saddle 6/1 is at

$$(3\text{-}67) \qquad x = 0 \quad , \quad y = \sqrt{\tfrac{1}{2}(1-2\mu)/(1-\mu)} \quad ,$$

and one centre, and later a bifurcated saddle of type 3, at

$$(3\text{-}68) \qquad x = \sqrt{(z-\mu)/(1-\mu)}, \quad y = 0 \quad ,$$

where

$$z^3 - (1+\mu)z^2 + (1+\mu)z - \frac{1}{2} = 0 \quad .$$

Both cycles 6/1 exist only for $\mu < \mu_o = \cos(2\pi/6) = \frac{1}{2}$ and merge with (0,0) as $\mu \to \frac{1}{2}$. For the centres this is easy to see from

$$(3\text{-}69) \qquad \begin{array}{llll} \mu = 0.499 & x = 0.051588 & y = 0 & \varphi = 0.000358 \\ \mu = 0.4999 & x = 0.016328 & y = 0 & \varphi = 0.000011 \\ \mu = \frac{1}{2} & x = 0 & y = 0 & \varphi = 0 \end{array} \quad .$$

The cycles of other orders have been found numerically and are given in Tables 3-7 to 3-10. For symmetry reasons all cycles of an odd order consist of two "complementary" cycles of the same order.

A local analysis of the recurrence (3-59) does not disclose anything qualitatively new. The exceptional case $\varphi = 2\pi/3$ is trivial because there are no quadratic

Table 3-7. Cycles of the cubic recurrence (3-59), $\mu = 0.8$, $k/r \geqslant (k/r)_{min}$

| | | Centres | | | Saddles | | | | |
|---|---|---|---|---|---|---|---|---|---|
| k | r | x | y | $\phi$ | x | y | $\lambda_1$ | $p_1$ | $p_2$ |
| 10 | 1 | 0.244854 | 0 | 0.00034 | 0.233105 | 0.044088 | 1.0003 | 1.840 | -1.826 |
| 12 | 1 | 0.637054 | +0.075703 | 0.0289 | 0.655575 | 0 | 1.0293 | 49.85 | |
| 14 | 1 | 0.802387 | 0 | 0.0807 | 0.788431 | 0.059665 | 1.0841 | -2.339 | -2.030 |
| 16 | 1 | 0.870022 | 0.042294 | 0.1525 | 0.879650 | 0 | 1.1647 | 18.89 | |
| 18 | 1 | 0.924653 | 0 | 0.2553 | 0.918194 | 0.028817 | 1.2904 | -2.749 | -1.999 |
| 11 | 1 | 0.516110 | 0.037849 | 0.00003 | 0.520909 | 0 | 1.00003 | 33095 | |
| 13 | 1 | 0.737916 | 0.034151 | 0.00069 | 0.741984 | 0 | 1.0007 | 2766 | |
| 15 | 1 | 0.843619 | 0.024989 | 0.0032 | 0.846531 | 0 | 1.0032 | 812.1 | |
| 17 | 1 | 0.903045 | 0.017232 | 0.0099 | 0.904969 | 0 | 1.0100 | 312.0 | |
| 19 | 1 | 0.938721 | 0.011578 | 0.0262 | 0.940079 | 0 | 1.0265 | 129.2 | |

Table 3-8. Cycles of the cubic recurrence (3-59), $\mu = 0.6$, $k/r \geqslant (k/r)_{min}$

| Centres | | | | | Saddles | | | | |
|---|---|---|---|---|---|---|---|---|---|
| $k$ | $r$ | $x$ | $y$ | $\varphi$ | $x$ | $y$ | $\lambda_1$ | $p_1$ | $p_2$ |
| 8 | 1 | 0.553 125 | 0.153 559 | 0.2495 | 0.590 354 | 0 | | 6.688 | |
| 10 | 1 | 0.813 451 | 0 | 0.9222 | 0.786 546 | 0.119 978 | 2.476 | 5.39 | −1.59 |
| 12 | 1 | 0.889 315 | 0.074 389 | 2.0371 | 0.904 285 | 0 | 6.512 | 2.659 | |
| 14 | 1 | 0.949 266 | 0 | 1.5132 | 0.940 350 | 0.043 535 | 27.44 | 4.476 | −1.498 |
| Bifurcation centre - saddle | | | | $\lambda_1$ | | | | | |
| 16 | 1 | 0.967 668 | 0.024 625 | −93.05 | 0.972 417 | 0 | 184.7 | 1.620 | |
| 18 | 1 | 0.975 081 | 0.032 603 | −1 472 | 0.991 167 | 0 | 1 562 | 1.491 | |
| 9 | 1 | 0.721 785 | 0.071 478 | 0.0752 | 0.728 748 | 0 | 1.078 | 36.89 | |
| 11 | 1 | 0.862 461 | 0.046 432 | 0.4703 | 0.870 566 | 0 | 1.589 | 8.615 | |
| 13 | 1 | 0.927 654 | 0.027 835 | 2.3109 | 0.924 416 | 0 | 5.353 | 3.574 | |
| Bifurcation centre - saddle | | | | $\lambda_2$ | | | | | |
| 15 | 1 | 0.961 070 | 0.015 697 | −38.61 | 0.968 778 | 0 | 49.57 | 1.692 | |
| 17 | 1 | | | | 0.973 770 | 0 | 518.79 | 1.662 | |
| 17 | 1 | | | | 0.986 155 | 0 | 281.84 | 1.369 | |
| 17 | 1 | | | | 0.989 760 | 0 | 159.70 | 1.522 | |
| 19 | 1 | | | | 0.974 386 | 0 | 4979,1 | 1.629 | |
| 19 | 1 | | | | 0.991 600 | 0 | 5283,4 | 1.490 | |
| Centres | | | | $\varphi$ | | | | | |
| 17 | 2 | 0 | 0.481 665 | 0.0021 | 0.669 270 | 0 | 1.0021 | 2059.4 | |
| 19 | 2 | 0 | 0.533 124 | 0.0191 | 0.773 804 | 0 | 1.019 | 377.0 | |

Table 3-9. Cycles of the cubic recurrence (3-59), $\mu = 0.5$, $k/r \geqslant (k/r)_{min}$

| Centres | | | | | Saddles | | | |
|---|---|---|---|---|---|---|---|---|
| $k$ | $r$ | $x$ | $y$ | $\varphi$ | $x$ | $y$ | $\lambda_1$ | $p_1$ |
| 8 | 1 | 0.688 931 | 0.180 973 | 0.9608 | 0.726 901 | 0 | 2.540 | 3.378 |
| 10 | 1 | 0.876 382 | | 2.678 | 0 | 0.616 666 | 10.19 | −0.0496 |
| 12 | 1 | 0.946 784 | 0.049 044 | 1.309 | none | none | | |
| | | Bifurcation | | $\lambda_1$ | | | | |
| 12 | 1 | 0.926 513 | 0.065 584 | 19.79 | 0 | 0.631 078 | 70.68 | −0.0352 |
| 14 | 1 | | | | 0.982 162 | 0 | 624.7 | 1.700 |
| 14 | 1 | | | | 0.969 197 | 0 | 573.1 | 1.473 |
| 16 | 1 | | | | 0.945 823 | 0 | 8 039.2 | 1.936 |
| 16 | 1 | | | | 0.984 037 | 0 | 9 609.9 | 1.699 |
| Centres | | | | $\varphi$ | | | | |
| 7 | 1 | 0.555 585 | 0.095 388 | 0.0403 | 0.568 726 | 0 | 1.041 | 41.77 |
| 9 | 1 | 0.810 851 | 0.074 465 | 0.6616 | 0.811 357 | 0 | 1.899 | 7.174 |
| Bifurcation centre - saddle | | | | $\lambda_2$ | | | | |
| 11 | 1 | 0 | 0.626 372 | −8.402 | 0.924 473 | 0 | 13.595 | 2.157 |
| 13 | 1 | 0 | 0.633 557 | −169.7 | 0.943 500 | 0 | 199.27 | 1.896 |
| 15 | 1 | | | | 0.945 687 | 0 | 2 192.7 | 1.933 |
| 15 | 1 | | | | 0.933 610 | 0 | 2 641.3 | 1.698 |
| 17 | 1 | | | | 0.945 860 | 0 | 29857.6 | 1.937 |
| Centres | | | | $\varphi$ | | | | |
| 13 | 2 | 0 | 0.353 688 | 0.00007 | 0.423 887 | 0 | 1.000007 | 26100 |
| 15 | 2 | 0 | 0.520 267 | 0.0132 | 0.664 642 | 0 | 1.013 | 329.3 |
| 17 | 2 | 0 | 0.578 281 | 0.2314 | 0.771 603 | 0 | 1.259 | 26.77 |
| 19 | 2 | 0 | 0.611 442 | 1.2407 | 0.835 656 | 0 | 3.043 | 19.81 |
| 19 | 3 | 0 | 0.297 748 | 0.000008 | 0,352 597 | 0 | 1.000008 | 3.104 |
| 20 | 3 | | | | 0 | 0.396 481 | 1.004 | −0.00062 |

Table 3-10. Cycles of the cubic recurrence (3-59), $\mu = 1/8$, $k/r \geqslant (k/r)_{min}$

| | | Centres and bifurcated saddles | | | Saddles | | | | |
|---|---|---|---|---|---|---|---|---|---|
| k | r | x | y | $\phi$ or $\lambda$ | x | y | $\lambda_1$ | $p_1$ | $p_2$ |
| 5 | 1 | 0 | 0.499598 | $\phi$ = 0.2693 | 0.506708 | 0 | 1.3059 | 8.862 | |
| 7 | 1 | 0 | 0.719507 | $\lambda_2$ = -33.64 | 0.872819 | 0 | 38.938 | 2.116 | |
| 9 | 1 | 0.900774 | 0 | $\lambda_1$ = 1398 | 0.962584 | 0 | 1368.4 | 2.271 | |
| 9 | 1 | 0.976489 | 0 | $\lambda_1$ = 678.8 | | | | | |
| 11 | 1 | 0.902004 | 0 | $\lambda_1$ = 39000 | | | | | |
| 11 | 2 | 0 | 0.617970 | $\phi$ = 0.7666 | 0.606154 | 0 | 2.0807 | 2.744 | |
| 13 | 2 | 0 | 0.674263 | $\lambda_2$ = -251.1 | 0.863099 | 0 | 320.45 | 2.139 | |
| 15 | 2 | 0.875734 | 0 | $\lambda_1$ = 4999 | 0.894221 | 0 | 9990.5 | 2.150 | |
| 15 | 2 | 0.972147 | 0 | $\lambda_1$ = 263.3 | | | | | |
| 17 | 3 | 0 | 0.643999 | $\lambda_2$ = - -2.952 | | | | | |
| 19 | 3 | 0 | 0.660568 | $\lambda_2$ = -2667 | 0.862151 | 0 | 3490.7 | 2.144 | |
| 6 | 1 | 0.749219 | 0 | $\phi$ = 3.0860 | 0 | 0.654654 | 11.511 | -0.2950 | |
| 8 | 1 | 0.895574 | 0 | $\lambda_1$ = 272.2 | | | | | |
| 10 | 1 | 0.980539 | 0 | $\lambda_1$ = 7353 | 0.960505 | 0 | 9477.6 | 2.255 | |
| 12 | 1 | 0.902042 | 0 | $\lambda_1$ = 20713 | | | | | |
| 12 | 2 | 0.799854 | 0 | $\phi$ = 0.7856 | 0.762687 | 0.016234 | 1.2590 | 0.0053 | -6.091 |
| 14 | 2 | 0 | 0.682554 | $\lambda_1$ = 1499.6 | | | | | |
| 16 | 2 | 0.977099 | 0.005240 | $\lambda_2$ = -22271 | | | | | |
| 16 | 3 | 0.584123 | 0 | $\lambda_1$ = 2.496 | | | | | |
| 18 | 3 | 0.633322 | 0 | $\lambda_2$ = -40.20 | | | | | |
| 20 | 3 | 0.865464 | 0 | $\lambda_1$ = 11012 | | | | | |

terms to be eliminated from the canonical forms. For $\mu = -\frac{1}{2}$, the origin is therefore a centre, surrounded for $\mu = -\frac{1}{2} - \varepsilon$, $0 < \varepsilon \ll 1$, by a double island structure 3/1. The exceptional case $\varphi = 2\pi/4$ has cubic dominating terms, but they are such that no invariant angles $\bar{\theta}$ exist. Similarly to eq. (3-39), for $\mu = 0$, one finds near the origin the invariant curve approximation

$$x_n^4 + y_n^4 = \text{const.} \tag{3-70}$$

The deformation of the invariant curves with respect to an elliptical shape (Fig. 3-23) is therefore not the same as in the case of a quadratic non-linearity.

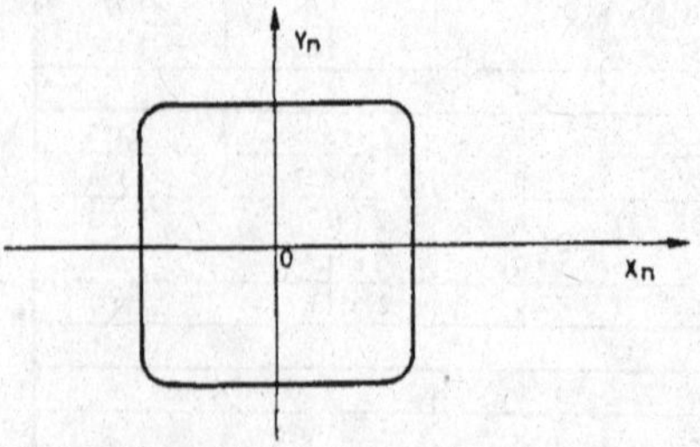


Fig. 3-23. Recurrence (3-59), $\mu = 0$. Exceptional case $\varphi = 2\pi/4$ at (0,0). Approximate closed invariant curve or unresolved island structure $k/r \simeq 4$, $k \gg 1$, $r \gg 1$, surrounding the centre (0,0).

The cycles of Tables 3-7 to 3-10 are shown in Fig. 3-24a to d, together with the main invariant curves. The conclusion drawn from an inspection of these figures is the same as in the case of a quadratic non-linearity. The positions of cycles form sequences $P_m(k_{(r)})$ which converge from the inside to intersection points of the main invariant curves. The centre → saddle of type 3 bifurcation takes place in the same manner, and the saddles satisfy the asymptotic relations (3-14) with the

Main invariant curves and distribution of cycles.

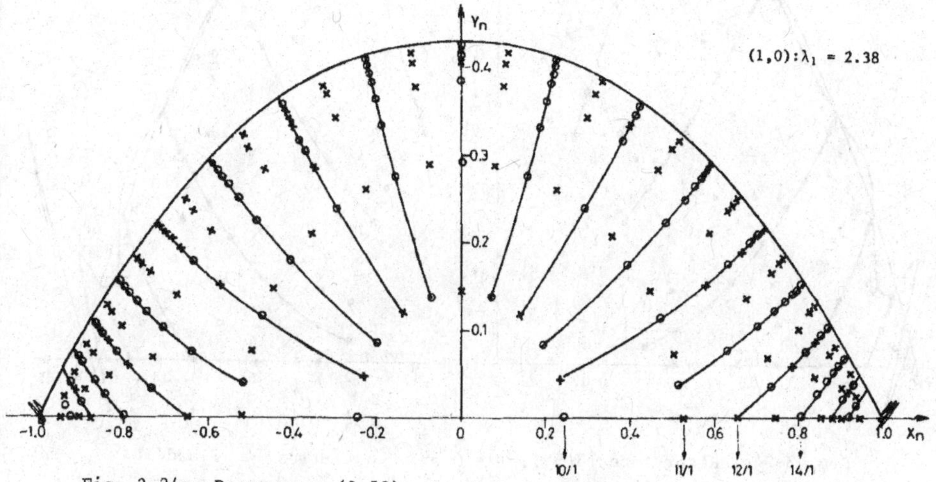


Fig. 3-24a. Recurrence (3-59), $\mu$ = 0.8. o centres x saddles

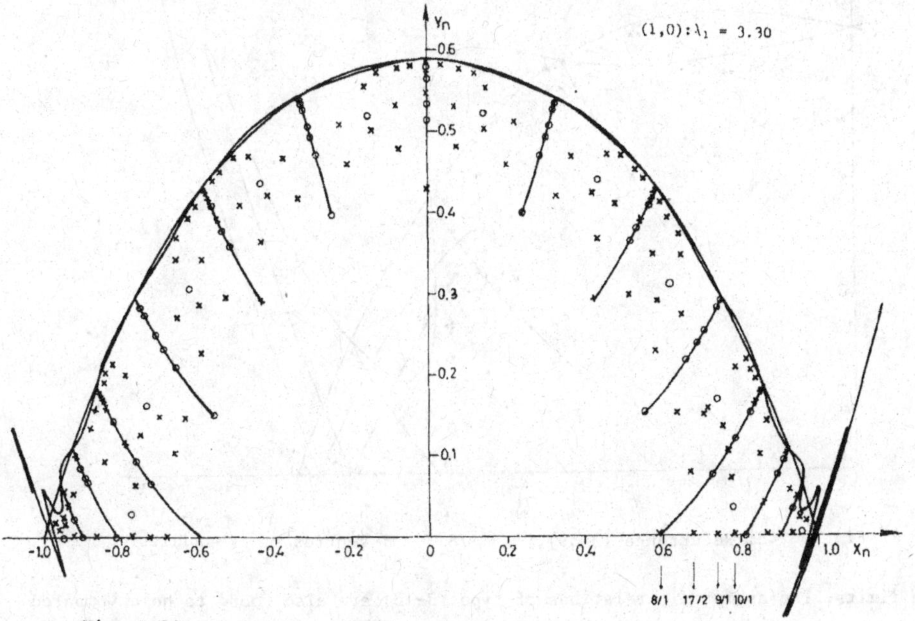


Fig. 3-24b. Recurrence (3-59), $\mu$ = 0.6 o centres x saddles

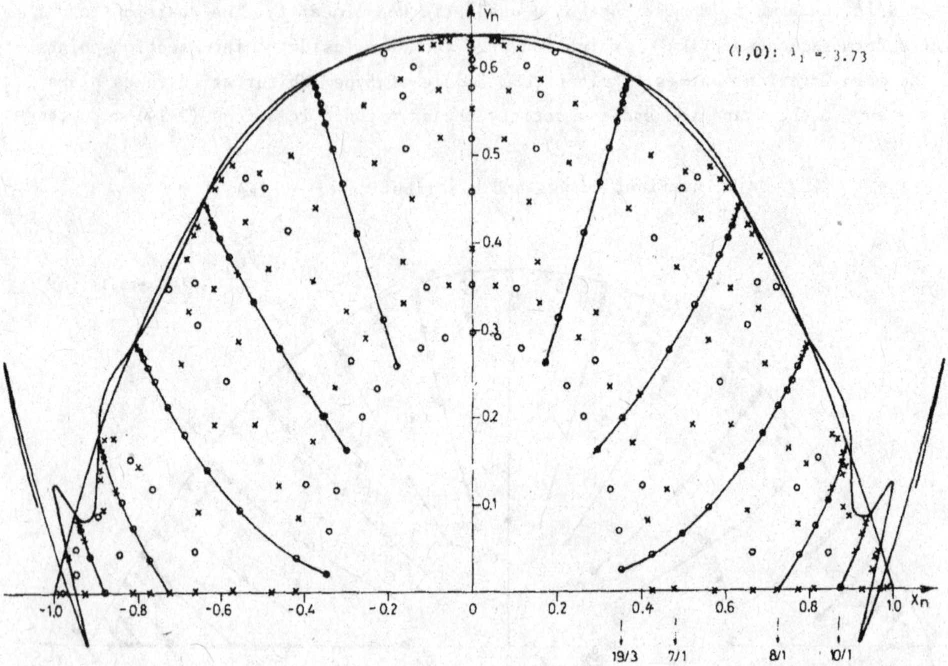


Fig. 3-24c. Recurrence (3-59), $\mu$ = 0.5   o centres   x saddles

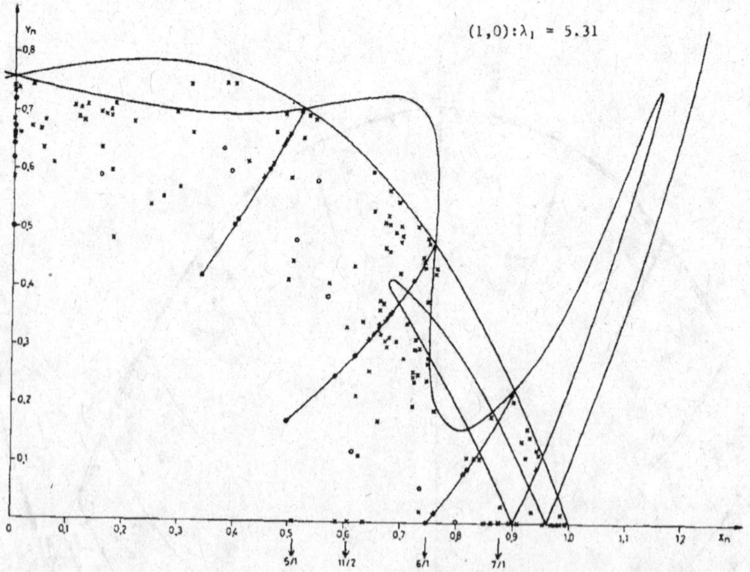


Fig. 3-24d. Recurrence (3-59), $\mu$ = 1/8   o centres   x saddles

same limits. The asymptotic relations of type (3-15) are also found to hold. Compared to the quadratic case, the major features of the phase portrait appear therefore to differ only by the effect of the $F(x) = -F(-x)$ symmetry, which leads to the doubling of cycles of an odd order k (the value k = 1 included), and the absence of quadratic forms responsible for the instability of the exceptional case $\varphi = 2\pi/3$.

Stochasticity is also generated with an increase of the strength of non-linearity $\delta = 1-\mu$ for a fixed distance in the phase plane from (0,0), or for a fixed $\mu$ and an increase of the distance. The cycle pair 12/1 constitutes a typical illustration. It comes into existence at (0,0) when $\mu$ traverses $\mu_o = \cos(2\pi/12) \simeq 0.8660$. When $\mu = 0.63$, two saddles of type I ($\lambda_1 = 3.772$) are at (0.88888,0) and (0.81986, 0.16848), and one centre ($\varphi = 1.3828$) at (0.87296, 0.07686). The configuration of invariant curves is shown in Fig. 3-25a.Because of the presence of small loops on the

Disintegration of an island structure with an increase of non linearity.

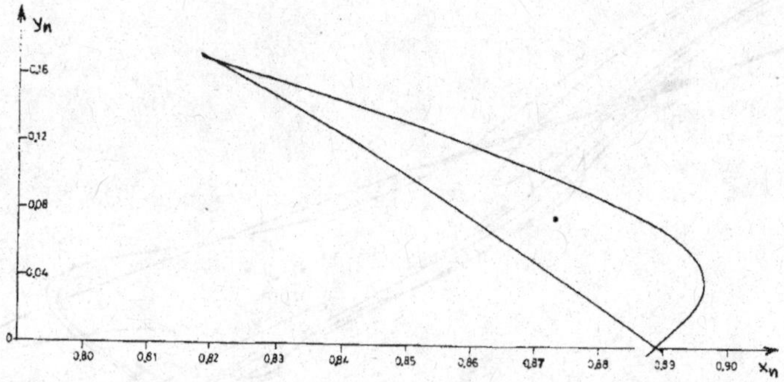


Fig. 3-25a. Recurrence (3-59). Invariant curves traversing two saddles 12/1. $\mu = 0.63$, $\lambda_1 = 3.772$.

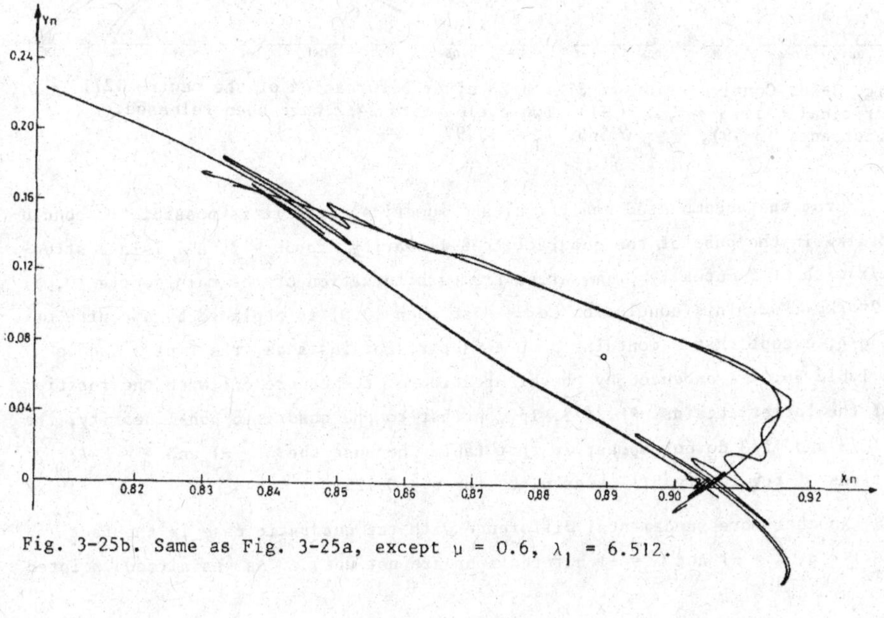


Fig. 3-25b. Same as Fig. 3-25a, except $\mu = 0.6$, $\lambda_1 = 6.512$.

invariant curves, too small and too narrow to be visible at the scale of reproduction, it is already an instability ring an no longer an island structure. When $\mu = 0.6$ the saddles ($\lambda_1 = 6.512$) are at (0.90428, 0) and (0.83836, 0.16558), and the centre ($\varphi = 2.0371$) at (0.88932, 0.07439). As expected the loops on the invariant curves are much more pronounced (Fig. 3-25b). A degenerated exceptional case $\varphi = \pi$ (critical case $\lambda = -1$) to be discussed later, occurs between $\mu = 0.566$ and $\mu = 0.565$. For $\mu = 0.565$, the invariant curve structure is shown in Fig. 3-26. The loops of the invariant curves have grown sufficiently to present secondary homoclinic points, similarly to what happens in the case of a quadratic non-linearity (cf. Fig. 3-12).

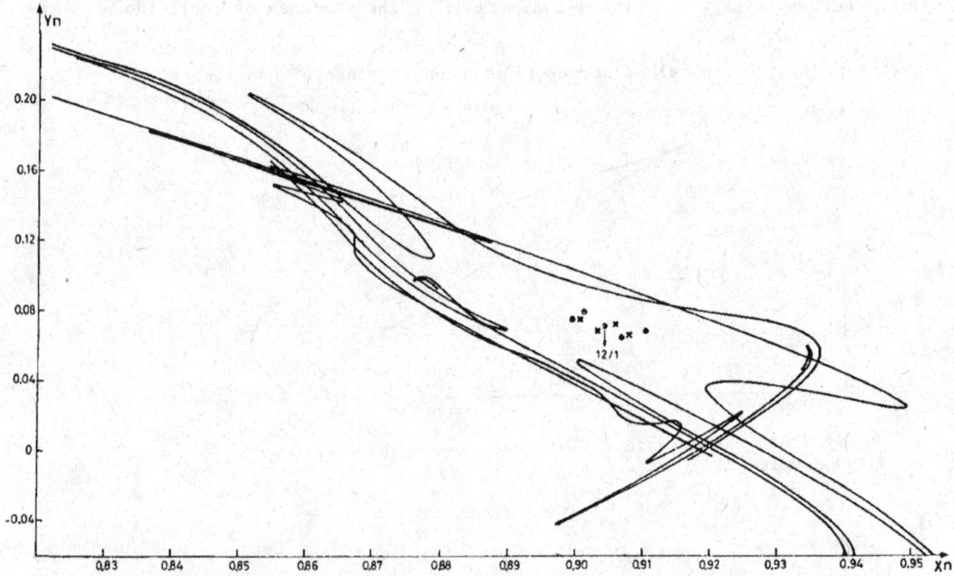


Fig. 3-26. Configuration of Fig. 3-25 after bifurcation of the centre 12/1 (critical case $\varphi = \pi$, $\lambda = -1$). Two cycle pairs 24/2 have been released. Recurrence (3-59), $\mu = 0.565$, $\lambda_1 = 13.9$.

From the accumulated analytical and numerical data it is possible to conclude that like in the case of the quadratic non-linearity, for $k > 3$, all island structures $k/r$ with $k$, $r$ mutually prime arise from a bifurcation of the main centre (0,0) at $\varphi = 2\pi r/k$, $k > 2$. This conclusion holds also when (0,0) is replaced by a centre of a cycle $q/p$, except that $q$ contains $p$ as a factor. For instance, the type 3 saddle 18/3 in Table 3-10 is produced by the bifurcation of the centre 6/1 when the rotation angle of the latter attains $\varphi = 2\pi/3$. In contrast to the quadratic non-linearity, the cycles 10/2 and 15/3 do not appear in that table, because the $\lambda = -1$ and $\varphi = 2\pi/3$ bifurcations of the centre 5/1 takes place for $\mu < 0.125$.

Another more fundamental difference with the quadratic case is the fact that the $\varphi = \pi$, $\lambda = -1$ and $\lambda = +1$ bifurcations are not unique. As was already pointed out, first there exists the centre-k/r → type 3 saddle-k/r bifurcation, which gives off a cycle centre 2k/2r. An example is : k = 12, r = 1 ,

$$(3\text{-}71)\qquad \begin{array}{llll} \mu = 0.5206 & x = 0.933734 & y = 0.057360 & \varphi = 2.90028 \\ \mu = 0.5204 & x = 0.933918 & y = 0.057241 & \lambda_2 = -1.05826 \end{array}$$

the centres produced being k = 24, r = 2, with one point at (0.93402, 0.05695) for $\mu = 0.5204$.

The second $\varphi = \pi$ bifurcation is of the type centre k/r → centre k/r. It produces one (or two) cycle pairs centre – type 1 saddle 2k/2r. An example is the cycle 12/1 :

$$(3\text{-}72)\qquad \begin{array}{llll} \mu = 0.5660 & x = 0.904441 & y = 0.071435 & \varphi = 3.12009 \\ \mu = 0.5655 & x = 0.904641 & y = 0.071390 & \varphi = 3.14028 \\ \mu = 0.5650 & x = 0.904840\,. & y = 0.071346 & \varphi = 3.12253 \end{array} ,$$

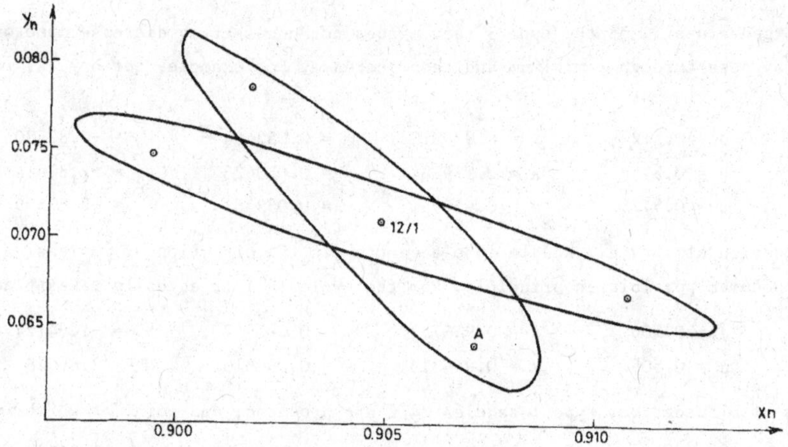


Fig. 3-27. Secondary island structure formed by the bifurcated cycle pair 24/2 shown in Fig. 3-26. Invariant curves crossing the saddles 24/2, $\mu = 0.565$, $\lambda_1 = 1.066$. A : one centre 24/2.

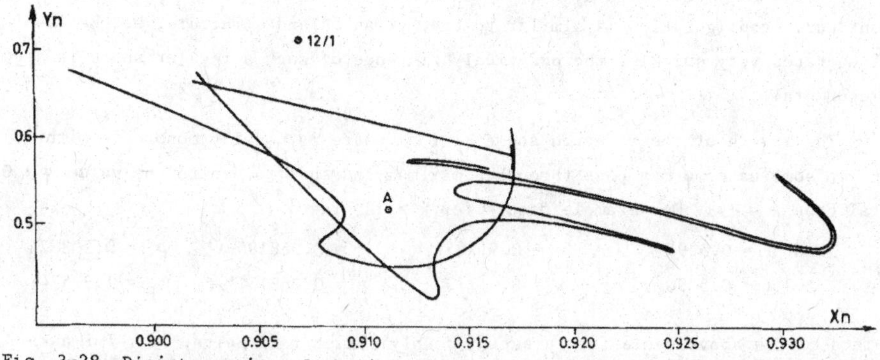


Fig. 3-28. Disintegration of the island structure of Fig. 3-27. Invariant curves crossing one saddle 24/2, $\mu = 0.56$, $\lambda_1 = 2.11$.

with the positions of the various points shown in Fig. 3-26. The invariant curves passing through the bifurcated saddles 24/2 form an island structure around each centre 12/1 (Fig. 3-27). The existence of such an island structure is theoretically significant, but practically rather ephemeral, as can be seen from Fig. 3-28, which shows a part of Fig. 3-27 when $\mu = 0.560$. The island structure 24/2 has completely disintegrated.

Another example is the cycle 6/1 :

| | | | | |
|---|---|---|---|---|
| | $\mu = 0.135$ | $x = 0.743725$ | $y = 0$ | $\varphi = 3.0269$ |
| (3-73) | $\mu = 0.130$ | $x = 0.746496$ | $y = 0$ | $\varphi = 3.1108$ |
| | $\mu = 0.125$ | $x = 0.749219$ | $y = 0$ | $\varphi = 3.0860$ |

with the bifurcated cycle pair 12/2

| | | | | |
|---|---|---|---|---|
| (3-74) | $\mu = 0.125$ | $x = 0.799854$ | $y = 0$ | $\varphi = 0.785630$ |
| | | $x = 0.762687$ | $y = 0.016234$ | $\lambda_1 = 1.25904$ |

Since the eigenvalues $\lambda_2$ of the type 3 saddle does not necessarily decrease monotonically, it may pass through a minimum and then increase. For example, for $n = 12$, $r = 1$, one has

| | | | | |
|---|---|---|---|---|
| | $\mu = 0.5127$ | $x = 0.939818$ | $y = 0.053464$ | $\lambda_2 = -2.87580$ |
| (3-75) | $\mu = 0.5126$ | $x = 0.939884$ | $y = 0.053423$ | $\lambda_2 = -2.87614$ |
| | $\mu = 0.5125$ | $x = 0.939950$ | $y = 0.053381$ | $\lambda_2 = -2.87599$ ; |

and a bifurcation of a type 3 saddle into a centre (in the direction of increasing non linearity) becomes possible in principle. For the cycle 12/1 it actually takes place :

| | | | | |
|---|---|---|---|---|
| (3-76) | $\mu = 0.506$ | $x = 0.943782$ | $y = 0.050947$ | $\lambda_2 = -1.64124$ |
| | $\mu = 0.505$ | $x = 0.944313$ | $y = 0.050610$ | $\varphi = 2.86028$ . |

At this $\lambda = -1$ bifurcation, type 1 saddles 24/2 are produced, one point of which is :

| | | | | |
|---|---|---|---|---|
| (3-77) | $\mu = 0.505$ | $x = 0.943231$ | $y = 0.051506$ | $\lambda_1 = 2.22857$ . |

Since two cycles saddle 12/1 of type 3 traverse simultaneously this bifurcation, the invariant curves passing through each saddle 24/2 encircle one centre 12/1. The invariant curve configuration is similar to that of an island structure, but because it disintegrates vety quickly, the practical existence of such a regular shape is also rather ephemeral.

The growth of the rotation angle $\varphi$ of a centre is usually monotonic with $\delta = 1-\mu$. In some cases $\varphi$ may pass through a maximum and then return to the value $\varphi = 0$ (critical case $\lambda = +1$). One example occurs for $k = 12$, $r = 1$ :

| | | | | |
|---|---|---|---|---|
| (3-78) | $\mu = 0.5295$ | $x = 0.917631$ | $y = 0.068195$ | $\varphi = 0.208874$ |
| | $\mu = 0.5290$ | $x = 0.917794$ | $y = 0.068151$ | $\lambda_1 = 1.37574$ |

The points of the above cycle form a set of doubly symmetrical pairs, i.e. for each point $(x, y)$ there exists beto a point $(-x, y)$, $x \neq 0$, and a point $(x, -y)$, $y \neq 0$.

Such a symmetry is to be expected from the general form of the recurrence (3-59). The bifurcation centre k/r → type 1 saddle k/r generates two distinct cycles centre k/r, with complementary symmetry and the same eigenvalue. Each newly generated cycle is only symmetrical in y, i.e. for every point (x, y), y ≠ 0, there exists a point (x, -y). The missing x-symmetrical points, required for over-all symmetry, are supplied by the other newly generated cycle, so that the combined points of the two cycles form a complete set of doubly symmetrical pairs. The two unsymmetrical cycles 12/1, generated by the bifurcation (3-78), are defined by :

$$(3\text{-}79) \qquad \mu = 0.525 \qquad \left\{ \begin{array}{ll} x = 0.907622 & y = 0.075973 \\ x = 0.0928938 & y = 0.060483 \end{array} \right\} \qquad \varphi = 1.69965 \,.$$

Their points are located between the newly created saddles, as if they formed a cycle 24/2, conserving however the periodicity k = 12 in two independent sets. Like in the case of the bifurcation saddle of type 3 → centre, two cycles 12/1 cross $\varphi = 0$ ($\lambda = +1$) simultaneously. The invariant curve configuration after this bifurcation is shown in Fig. 3-29. The first loops of the invariant curves passing through the symmetrical saddles, which result from the bifurcated symmetrical centres, wind around the newly generated unsymmetrical centres. This configuration is in principle similar to the centre → saddle of type 3 bifurcation with a quadratic non-linearity (cf. Fig. 3-19), except that the role of the cycle centre 2k/2r is now taken over by the two unsymmetrical centres k/r. The bifurcations described above are sufficient to explain the presence of all cycles listed in Tables 3-7 to 3-10.

Fig. 3-29. Recurrence (3-59), $\mu = 0.525$. Invariant curve configuration after the bifurcation $\varphi = 0$, $\lambda = +1$ of the centres 12/1. The newly bifurcated cycles are saddles 12/1 of type 1.

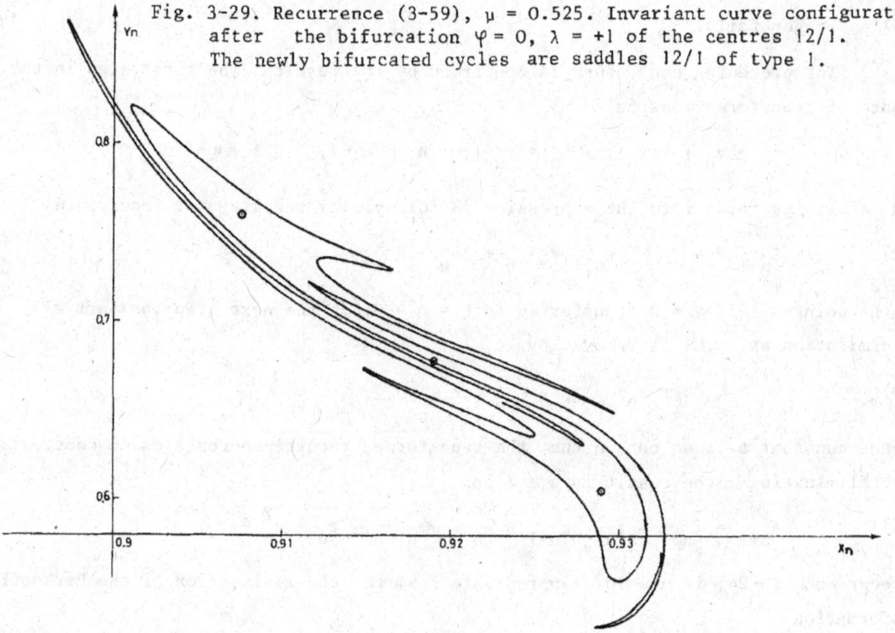

## 3.3 Bounded quadratic recurrence

The normalized one-parameter form analoguous to (3-6) and (3-59) is :

$$\text{(3-80)}\qquad \begin{aligned} &x_{n+1} = y_n + F(x_n), \qquad y_{n+1} = -x_n + F(x_{n+1}) \quad , \\ &F(x) = \mu x + 2(1-\mu)x^2 \,/\, (1+x^2), \qquad -1 < \mu < +1 \quad . \end{aligned}$$

The above form differs from the quadratic recurrence only by the presence of higher order terms in the Mc Laurin development of F(x). The principal similarity is the existence of the same main fixed points (0,0) and (1,0). Whereas (0,0) is still a centre, the point (1,0) is a degenerate singular point with the eigenvalues $\lambda_1 = \lambda_2 = +1$ regardless of the value of $\mu$. An exploratory analytic study of the canonical form of (3-80) near (1,0) has shown that this critical case $\lambda = 1$ differs essentially from that of a saddle.

From bifurcation theory of conservative differential equations it is known that the simplest critical case, analoguous to $\lambda = 1$ in a recurrence, involves the coalescence of a centre with a saddle. The resulting singular point is traversed by two trajectories forming a cusp. In such a case the series representation (2-43) of an invariant curve must fail in principle, as indeed it does at the point (1,0) of (3-80). On the other hand, it is obvious that a cusp on a trajectory can be removed by means of an appropriate algebraic transformation. An example of such a transformation, performed directly on a recurrence, was given already by Birkhoff. This transformation, when applied to the singular point (1,0) of the recurrence (3-80), should yield a recurrence centred on (0,0) for which the expansion (2-43) no longer fails.

The preceding conjecture is confirmed by the results. The first step in the sequence of transformations is

$$\text{(3-81)}\qquad u_n = y_n + F(x_n), \quad \xi_n = x_n - 1, \quad \eta_n = u_n - 1, \quad \beta = \mu - 1 \quad ,$$

which after insertion into the expression (3-80), yields the diagonal recurrence

$$\text{(3-82)}\qquad \xi_{n+1} = \eta_n, \qquad \eta_{n+1} = -\xi_n + 2\eta_n + \beta\eta_n^2 + \ldots \quad ,$$

with the point x = 1, y = 0 transferred to $\xi = \eta = 0$. In the next step the partial indetermination at (0,0) is removed by

$$\text{(3-83)}\qquad \eta_n = \xi_n(a + z_n) \quad ,$$

where the constant a is so chosen that the transformed recurrence contains no constant terms. Eliminating $\eta$ the result is a = 1 and

$$\text{(3-84)}\qquad \xi_{n+1} = \xi_n + \xi_n z_n, \quad z_{n+1} = z_n + \beta\xi_n + \beta\xi_n z_n - z_n^2 + \ldots$$

The recurrence (3-84) is now in a appropriate form for the application of the Birkhoff transformation

(3-85) $\xi_n = z_n \cdot v_n$

Inserting eq. (3-85) into the first equation of (3-84) and dividing by the second yields

$$v_{n+1} = v_n + 2\, z_n v_n - \beta v_n^2 + \ldots$$
(3-86)
$$z_{n+1} = z_n + \beta z_n v_n - z_n^2 + \ldots$$

The recurrence (3-86) should now admit an invariant curve of the form

(3-87) $z = a_1 v + a_2 v^2 + \ldots$

Inserting eq. (3-87) into eqs. (2-28), (3-86) and equating the coefficients of equal powers of v, one obtains an identity for $v^1$. From the coefficients of $v^2$ one finds

(3-88) $a_1 = \frac{2}{3} \beta$

and then recursively the $a_m$ from the coefficients of $v^{m+1}$, $m > 1$. Carrying out the inverse transformation the series (3-87) becomes

(3-89) $y_n = \pm\sqrt{\frac{2}{3} (\mu-1)(x_n-1)^3} + \ldots$ ,

which clearly shows a cusp with a horizontal asymptote. The same method was used to establish the equation of the invariant curve (3-53), passing through one point of the composite cycle 3/1 of the quadratic recurrence, except that the algebra involved in an iterated recurrence is too long to be reproduced in detail.

No cycles of the recurrence (3-80) are known analytically. They were sought numerically using the property that many of them are located on the curves $y_n = 0$ and $y_n = x_n - F(x_n)$. The fundamental difference with the quadratic non-linearity is the fact that cycles of the recurrence (3-80) exist also for $x > 1$, i.e. to the right of the cusp of the main invariant curves. It is thus necessary to classify cycles into "internal" (all points $P_m(k_{(r)})$ inside the main invariant curves) and "external" ones (all points $P_m(k_{(r)})$ outside the main invariant curves). A list of the cycles found so far are given in Tables 3-11 to 3-16, for a weak ($\mu = 0.8$), medium ($\mu = 0.5$) and

Table 3-11. Internal cycles of the recurrence (3-80), $\mu = 0.8$.

| | | Centres or Saddles of type 3 | | | | | Saddles of type 1 | | | | |
|---|---|---|---|---|---|---|---|---|---|---|---|
| k | r | x | y | $\phi$ or $\lambda_2$ (> 0) (< 0) | $p_1$ | $p_2$ | x | y | $\lambda_1$ | $p_1$ | $p_2$ |
| 11 | 1 | 0.292039 | 0.026974 | 0.0050 | | | 0.298778 | 0 | 1.0050 | 88.3 | |
| 12 | 1 | 0.399464 | 0 | 0.0145 | | | 0.393281 | 0.025076 | 1.0146 | -2.142 | -1.940 |
| 13 | 1 | 0.468378 | 0.021712 | 0.0267 | | | 0.473700 | 0 | 1.0270 | 24.80 | |
| 14 | 1 | 0.532349 | 0 | 0.0407 | | | 0.527844 | 0.018406 | 1.0416 | -2.333 | -1.843 |
| 15 | 1 | 0.576584 | 0.015516 | 0.0564 | | | 0.580377 | 0 | 1.0580 | 12.52 | |
| 16 | 1 | 0.620622 | 0 | 0.0735 | | | 0.617420 | 0.013086 | 1.0763 | -2.606 | -1.705 |
| 17 | 1 | 0.652184 | 0.011071 | 0.0922 | | | 0.654893 | 0 | 1.0966 | 7.536 | |
| 18 | 1 | 0.684443 | 0 | 0.1127 | | | 0.682136 | 0.009407 | 1.1192 | -3.081 | -1.547 |
| 19 | 1 | 0.708199 | 0.008032 | 0.1350 | | | 0.710168 | 0 | 1.1443 | 4.939 | |
| 20 | 1 | 0.732752 | 0 | 0.1593 | | | | | | | |
| 21 | 1 | | | | | | 0.752710 | 0 | 1.2035 | 3.411 | |

Table 3-12. External cycles of the recurrence (3-80), $\mu = 0.8$.

| | | Centres or Saddles of type 3 | | | | | Saddles of type 1 | | | | |
|---|---|---|---|---|---|---|---|---|---|---|---|
| k | r | x | y | $\phi$ or $\lambda_2$ (> 0) (< 0) | $p_1$ | $p_2$ | x | y | $\lambda_1$ | $p_1$ | $p_2$ |
| 17 | 1 | 2.055558 | 0 | 0.2186 | | | | | | | |
| 18 | 1 | 1.938042 | 0.071713 | 0.1865 | | | 0.439751 | 0.349019 | 1.204 | -0.104 | -0.145 |
| 19 | 1 | 1.873977 | 0 | 0.1642 | | | 0.696347 | 0.286989 | 1.178 | -0.047 | -0.086 |
| 20 | 1 | 1.789121 | 0.053041 | 0.1487 | | | 0.370208 | 0.298242 | 1.160 | -0.148 | -0.173 |
| 21 | 1 | 1.741434 | 0 | 0.1381 | | | 0.346101 | 0.280784 | 1.148 | -0.164 | -0.184 |
| 23 | 1 | 1.639532 | 0 | 0.1273 | | | 0.311401 | 0.255778 | 1.136 | -0.187 | -0.200 |
| 34 | 1 | 1.332010 | 0.010585 | 0.1860 | | | | | | | |
| 35 | 1 | 1.318555 | 0 | 0.1982 | | | | | | | |

Table 3-13. Internal cycles of the recurrence (3-80), $\mu = 0.5$.

| | | Centres or Saddles of type 3 | | | | | Saddles of type 1 | | | | |
|---|---|---|---|---|---|---|---|---|---|---|---|
| k | r | x | y | $\phi$ or $\lambda_2$ (> 0) (< 0) | $p_1$ | $p_2$ | x | y | $\lambda_1$ | $p_1$ | $p_2$ |
| 7 | 1 | 0.322040 | 0.067055 | 0.2784 | | | 0.337426 | 0 | 1.3170 | 2.969 | |
| 8 | 1 | 0.468721 | 0 | 0.6428 | | | 0.454143 | 0.056090 | 1.8574 | 9.715 | -1.032 |
| 9 | 1 | 0.546961 | 0.043205 | 1.080 | | | 0.556136 | 0 | 2.6788 | 1.244 | |
| 10 | 1 | 0.622972 | 0 | 1.669 | | | 0.614215 | 0.033187 | 3.9290 | 1.460 | -0.7641 |
| 11 | 1 | 0.668076 | 0.025445 | -1.101 | -0.1633 | -0.1135 | 0.673385 | 0 | 5.8174 | 0.7835 | |
| 12 | 1 | 0.715063 | 0 | -4.549 | -0.3404 | | 0.709696 | 0.019888 | 8.6249 | 0.7980 | -0.6085 |
| 13 | 1 | 0.744759 | 0.015604 | -8.720 | -0.4375 | 0.3569 | 0.747979 | 0 | 12.717 | 0.5969 | |
| 14 | 1 | 0.776245 | 0 | -14.60 | -0.4240 | | 0.772800 | 0.012488 | 18.562 | 0.5885 | -0.5157 |
| 15 | 1 | 0.797144 | 0.010029 | -22.81 | -0.4335 | 0.4133 | 0.799190 | 0 | 26.750 | 0.5007 | |
| 16 | 1 | 0.819377 | 0 | -34.09 | -0.4179 | | 0.817070 | 0.008198 | 38.017 | 0.4899 | -0.4537 |
| 17 | 1 | 0.834733 | 0.006718 | -49.37 | -0.4071 | 0.4066 | 0.808576 | 0.029857 | 53.273 | 0.5755 | -0.4274 |
| 18 | 1 | 0.828339 | 0.024472 | -69.74 | -0.4015 | 0.3951 | 0.849444 | 0.005592 | 73.631 | 0.4301 | -0.4080 |
| 19 | 1 | 0.862693 | 0.004662 | -96.56 | -0.3779 | 0.3839 | 0.859018 | 0.009513 | 100.4 | 0.4228 | -0.3853 |
| 13 | 2 | 0.234885 | 0.032203 | 0.0136 | | | 0.239247 | 0 | 1.0137 | 77.46 | |
| 15 | 2 | 0.406588 | 0.031394 | 0.2005 | | | 0.408328 | 0 | 1.2213 | 10.08 | |
| 17 | 2 | 0.513146 | 0.024174 | 0.7121 | | | 0.520519 | 0 | 1.9457 | 3.169 | |
| 19 | 2 | 0.589506 | 0.019036 | 1.773 | | | 0.583959 | 0 | 4.2621 | 1.792 | |
| 21 | 2 | 0.401584 | 0.219454 | -6.804 | -0.4624 | -0.9948 | 0.369878 | 0.239408 | 11.728 | -0.4991 | -0.6919 |
| 22 | 2 | 0.669789 | 0.025208 | 0.2726 | | | none | none | | | |
| 19 | 3 | 0.192616 | 0.019684 | 0.0004 | | | 0.194385 | 0 | 1.004 | 3477 | |

Table 3-14. External cycles of the recurrence (3-80), $\mu = 0.5$.

| | | Centres or Saddles of type 3 | | | | | Saddles of type 1 | | | | |
|---|---|---|---|---|---|---|---|---|---|---|---|
| k | r | x | y | $\phi$ or $\lambda_2$ (> 0) (< 0) | $p_1$ | $p_2$ | x | y | $\lambda_1$ | $p_1$ | $p_2$ |
| 7 | 1 | | | | | | 2.710985 | 1.331931 | 1.998 | -1.888 | -1.247 |
| 9 | 1 | 2.420416 | 0 | 1.215 | | | | | | | |
| 13 | 1 | | | | | | 1.495338 | 0.282701 | 3.650 | -0.146 | -0.680 |
| 14 | 1 | | | | | | 1.668985 | 0 | 3.919 | -2.075 | |
| 15 | 1 | 1.600740 | 0 | 1.841 | | | 0.678707 | 0.280761 | 4.291 | 0.100 | -0.314 |
| 16 | 1 | 1.524098 | 0.062993 | 2.100 | | | 0.355699 | 0.351313 | 4.794 | -0.151 | -0.393 |
| 17 | 1 | 0.488499 | 0.283163 | 2.540 | | | 0.342306 | 0.338394 | 5.460 | -0.182 | -0.398 |
| 18 | 1 | 0.474248 | 0.272473 | -1.987 | -0.495 | -0.596 | 0.332381 | 0.328864 | 6.328 | -0.206 | -0.401 |
| 19 | 1 | 0.463619 | 0.264553 | -3.279 | -0.487 | -0.644 | 0.324957 | 0.321760 | 7.446 | -0.224 | -0.402 |
| 20 | 1 | 0.141086 | 0.365554 | -4.780 | -0.151 | -0.216 | 0.319354 | 0.316412 | 8.873 | -0.238 | -0.404 |
| 21 | 1 | 0.138739 | 0.361275 | -6.625 | -0.147 | -0.217 | 0.315089 | 0.312349 | 10.68 | -0.249 | -0.405 |
| 17 | 2 | | | | | | 0.753226 | 0.976755 | 3.152 | 0.134 | -0.271 |
| 19 | 2 | | | | | | 1.962155 | 0.478486 | 3.950 | -0.425 | -1.013 |
| 21 | 2 | 2.057983 | 0.100515 | 1.893 | | | | | | | |
| 19 | 3 | | | | | | 5.129405 | 1.219153 | 16.42 | -0.159 | 1.75 |

Table 3-15. Internal cycles of the recurrence (3-80), $\mu = 1/8$.

| | | Centres or Saddles of type 3 | | | | | Saddles of type 1 | | | | |
|---|---|---|---|---|---|---|---|---|---|---|---|
| k | r | x | y | $\phi$ or $\lambda_2$ (> 0) (< 0) | $p_1$ | $p_2$ | x | y | $\lambda_1$ | $p_1$ | $p_2$ |
| 5 | 1 | 0.322420 | 0.117328 | 2.424 | | | 0.296695 | 0 | 3.8041 | 0.8635 | |
| 6 | 1 | 0.510336 | 0 | -17.40 | -0.6594 | | 0.447637 | 0.099555 | 15.059 | 0.7588 | -0.9928 |
| 7 | 1 | 0.593803 | 0.063380 | -50.64 | -0.7116 | 0.7540 | 0.579227 | 0 | 42.957 | 0.8194 | |
| 8 | 1 | 0.677231 | 0 | -111.7 | -0.7319 | | 0.643863 | 0.050514 | 94.756 | 0.8109 | -0.7663 |
| 9 | 1 | 0.721215 | 0.032264 | -216.6 | -0.698 | 0.708 | 0.711891 | 0 | 185.9 | 0.732 | |
| 10 | 1 | 0.768441 | 0 | -388.4 | -0.660 | | 0.749057 | 0.026439 | 337.8 | 0.705 | -0.676 |
| 11 | 1 | 0.795732 | 0.017788 | -658.6 | -0.615 | 0.636 | 0.789316 | 0 | 580.0 | 0.641 | |
| 12 | 1 | 0.812223 | 0.028204 | -1070 | -0.579 | 0.620 | 0.813159 | 0.014952 | 953.3 | 0.618 | -0.593 |
| 9 | 2 | -0.228948 | 0 | 0.1609 | | | 0.164703 | 0 | 1.1687 | 3.6998 | |
| 11 | 2 | -0.303135 | 0 | -61.69 | 3.040 | | 0.346987 | 0 | 54.046 | 1.0727 | |
| 12 | 2 | 0.546788 | 0.016689 | -181.2 | -0.807 | 0.696 | 0.559092 | 0 | 117.8 | 0.816 | |
| 13 | 2 | 0.522141 | 0.184241 | -350.5 | -0.755 | 0.980 | 0.627122 | 0.054759 | 215.4 | 0.819 | -0.774 |
| 13 | 2 | -0.341112 | 0 | -835.7 | 2.316 | | 0.504417 | 0.155184 | 669.5 | 0.994 | -0.765 |
| 14 | 2 | 0.651563 | 0.092360 | -617.7 | -0.742 | 0.716 | 0.699934 | 0 | 371.4 | 0.743 | |
| 14 | 2 | 0.626646 | 0.040098 | -1992 | -0.750 | 0.813 | | | | | |
| 15 | 2 | 0.683616 | 0.094739 | -1046 | -0.708 | 0.790 | 0.738802 | 0.028531 | 621.6 | 0.718 | -0.688 |
| 14 | 3 | 0.240858 | 0 | 0.5310 | | | 0.232131 | 0.026743 | 1.6772 | -0.4522 | -1.134 |
| 15 | 3 | 0.151723 | 0.317812 | -6.580 | -0.086 | $-1.5 \times 10^4$ | | | | | |
| 16 | 3 | | | | | | 0.329896 | 0.018083 | 189.6 | 0.917 | -1.143 |
| 17 | 3 | -0.304797 | 0 | -1002 | 2.98 | | 0.430598 | 0.111496 | 839.7 | 0.965 | -0.742 |
| 18 | 3 | 0.615437 | 0.061299 | -356.9 | -0.781 | 0.684 | | | | | |
| 18 | 3 | -0.338607 | 0 | -2358 | 2.35 | | | | | | |
| 19 | 3 | 0.654600 | 0.086658 | -9.414 | -0.7373 | 0.0475 | | | | | |
| 19 | 3 | -0.308866 | 0 | -5812 | 2.84 | | | | | | |
| 19 | 3 | -0.330884 | 0 | $-1.36 \times 10^4$ | 2.49 | | | | | | |
| 22 | 3 | -0.310131 | 0 | $-3.42 \times 10^4$ | 2.79 | | | | | | |
| 19 | 4 | -0.201826 | 0 | 1.143 | | | 0.269438 | 0 | 2.8082 | 1.5653 | |
| 20 | 4 | 0.397294 | 0.109064 | -28.77 | -0.677 | 0.900 | -0.288666 | 0 | 33.77 | -3.822 | |
| 21 | 4 | -0.297366 | 0 | -724.1 | 3.27 | | 0.325862 | 0.022511 | 698.8 | 0.883 | -1.168 |
| 22 | 4 | | | | | | -0.303135 | 0 | 3805 | -3.04 | |
| 23 | 5 | | | | | | 0.213209 | 0 | 1.1193 | 17.427 | |

Table 3-16. External cycles of the recurrence (3-80), $\mu = 1/8$.

| | | Centres or Saddles of type 3 | | | | | Saddles of type 1 | | | | |
|---|---|---|---|---|---|---|---|---|---|---|---|
| k | r | x | y | $\phi$ or $\lambda_2$ (> 0) (< 0) | $p_1$ | $p_2$ | x | y | $\lambda_1$ | $p_1$ | $p_2$ |
| 5 | 1 | 4.339865 | 0 | -4.386 | 0.484 | | -1.223643 | 0 | 9.590 | -0.999 | |
| 7 | 1 | -0.381711 | -0.556551 | -23.25 | 0.458 | -1.409 | -0.638332 | 0 | 22.09 | -1.592 | |
| 8 | 1 | 1.305709 | -0.702068 | -27.40 | -0.457 | 0.751 | -0.548362 | 0 | 27.77 | -1.69 | |
| 9 | 1 | 1.035614 | 0.564160 | -32.66 | -0.501 | 0.684 | -0.492612 | 0 | 34.68 | -1.76 | |
| 10 | 1 | 1.684247 | 0.179839 | -40.50 | 281 | -0.522 | -0.456131 | 0 | 44.03 | -1.82 | |
| 11 | 1 | -0.243740 | 0.311409 | -52.16 | 1.237 | -0.101 | -0.431622 | 0 | 57.24 | -1.87 | |
| 12 | 1 | 0.355936 | 0.474702 | -69.22 | -0.516 | 2.59 | -0.414918 | 0 | 76.03 | -1.91 | |
| 13 | 1 | -0.226439 | 0.283488 | -93.72 | 1.219 | -0.040 | 0.579640 | -0.318045 | 102.6 | 0.596 | -0.725 |
| 14 | 1 | -0.221637 | 0.275872 | -128.3 | 1.214 | -0.022 | -0.395423 | 0 | 139.8 | -1.97 | |
| 15 | 1 | 1.380839 | 0 | -176.6 | 1.33 | | -0.389815 | 0 | 191.2 | -1.99 | |
| 16 | 1 | -0.215955 | 0.266938 | -242.7 | 1.21 | -0.001 | -0.385840 | 0 | 261.4 | -2.00 | |
| 17 | 1 | | | | | | -0.382991 | 0 | 355.9 | -2.02 | |
| 18 | 1 | | | | | | -0.380926 | 0 | 481.6 | -2.02 | |
| 19 | 1 | | | | | | -0.379413 | 0 | 647.0 | -2.03 | |
| 20 | 1 | | | | | | -0.378293 | 0 | 862.0 | -2.03 | |
| 21 | 1 | | | | | | -0.377454 | 0 | 1139 | -2.04 | |
| 22 | 1 | | | | | | -0.376820 | 0 | 1492 | -2.04 | |
| 9 | 2 | | | | | | -3.372304 | 0 | 1.470 | -1.487 | |
| 10 | 2 | 5.615846 | 0 | 2.148 | | | | | | | |
| 11 | 2 | 1.208474 | 1.734202 | -189.8 | -1.03 | 1.44 | 0.955621 | -1.940308 | 149.8 | 1.13 | -1.53 |

Table 3-16. continued.

| | | Centres or Saddles of type 3 | | | | | Saddles of type 1 | | | | |
|---|---|---|---|---|---|---|---|---|---|---|---|
| k | r | x | y | $\phi$ or $\lambda_2$ (> 0) (< 0) | $p_1$ | $p_2$ | x | y | $\lambda_1$ | $p_1$ | $p_2$ |
| 12 | 2 | -0.613292 | -0.723919 | -234.2 | 0.545 | -1.52 | 2.840561 | -0.928460 | 168.8 | 12.0 | 0.164 |
| 16 | 2 | | | | | | 2.426940 | 0 | 342.7 | -1.24 | |
| 17 | 2 | -0.531285 | 0 | -920.3 | 1.70 | | | | | | |
| 19 | 2 | -0.481030 | 0 | -1411 | 1.77 | | | | | | |
| 21 | 2 | 1.444178 | 0.007099 | -86.19 | 1.89 | -0.996 | | | | | |
| 21 | 2 | -0.448108 | 0 | -2295 | 1.83 | | | | | | |
| 14 | 3 | 3.392015 | 1.357949 | -8.684 | 3.62 | 0.061 | 1.309334 | 3.153667 | 8.637 | 1.79 | -0.998 |
| 14 | 3 | 3.216959 | -2.174155 | -2.784 | -2.04 | -11.7 | | | | | |
| 15 | 3 | 4.828657 | 0.411528 | -26.02 | 0.383 | -0.466 | 3.312140 | 0.495458 | 24.66 | -0.310 | 1.94 |
| 18 | 3 | 0.890631 | 1.822365 | -3705 | -1.17 | 1.56 | | | | | |
| 21 | 3 | | | | | | -0.595739 | 0 | 2989 | -1.61 | |
| 21 | 3 | | | | | | -0.571335 | 0 | 1903 | -1.68 | |
| 22 | 3 | | | | | | -0.556874 | 0 | 9938 | -1.68 | |
| 22 | 3 | | | | | | -0.607807 | 0 | $1.32 \times 10^4$ | -1.61 | |
| 22 | 3 | | | | | | -0.634306 | 0 | $1.30 \times 10^4$ | -1.60 | |
| 18 | 4 | 8.331914 | 0.169468 | 1.022 | | | | | | | |
| 19 | 4 | 3.337283 | 1.046168 | -107.7 | 3.63 | -0.153 | | | | | |
| 19 | 4 | 3.158567 | 1.173177 | -16.86 | 5.31 | -0.036 | 0.857999 | -2.336966 | 33.89 | 1.24 | -1.66 |
| 20 | 4 | | | | | | -1.174250 | -1.388435 | 5.472 | -0.900 | 0.685 |
| 20 | 4 | | | | | | 2.792342 | 0.044332 | 28.92 | -1.60 | 1.80 |
| 21 | 4 | 2.259133 | -0.513454 | -7.811 | 0.470 | 7.87 | | | | | |
| 21 | 4 | -0.712709 | -0.874219 | -130.9 | 0.513 | -1.49 | | | | | |
| 22 | 5 | | | | | | -2.485783 | -6.712073 | 4.038 | 0.528 | 4.25 |
| 23 | 5 | 4.340928 | 1.298681 | -3.002 | 0.738 | -0.849 | 3.772670 | 1.665968 | 22.53 | -0.331 | -6.89 |

strong non linearity ($\mu$ = 0.125), respectively. No attempt was made to compile an exhaustive set of external cycles, because such cycles do not occur in the quadratic and cubic recurrences.

All points $P_m(k_{(r)})$, with r = 1 and 2 of the recurrence (3-80) were drawn in the phase plane together with the main invariant curves (Fig. 3-30 to 3-32). From these figures it is seen that the boundary between internal and external cycles is indeed given by the main invariant curves, and that the points of both internal and external cycles converge separately to the homoclinic points of those curves. The eigenvalues and eigendirections of the external saddles are also found to satisfy asymptotic relations of the form (3-14) and (3-17). Furthermore, the positions of the higher-order cycles ($k/r \gg 1$) appear to guide the higher-order loops of the main invariant curves. This property is expecially striking in Fig. 3-32.

The distribution of invariant curves associated with internal cycles has the same features as that of the quadratic recurrence, i.e. the centre (0,0) is surrounded successively by island structures, a diffusion region, and a region of stochastic instability. There exists, however, a fundamental difference compared to the quadratic recurrence, for which the regions of diffusion and stochastic instability involves only monotonically diverging half-trajectories $\{x_n, y_n\}$, , = 0, 1, ... In the case of the bounded quadratic recurrence the behaviour of the $\{x_n, y_n\}$ is not monotonic. Once the consequents of an initial point $(x_o, y_o)$ of $\{x_n, y_n\}$ have left the

Main invariant curves and distribution of cycles.

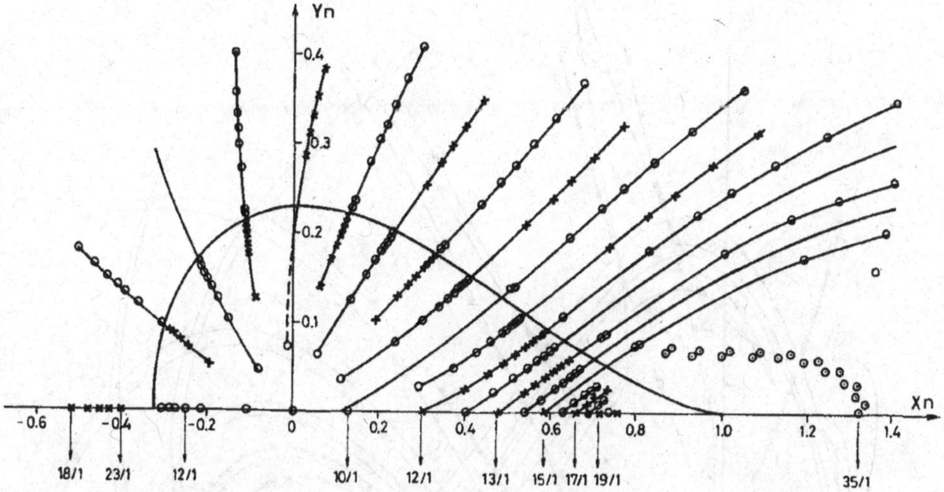


Fig. 3-30. Recurrence (3-80), $\mu = 0.8$, o centres, x saddles

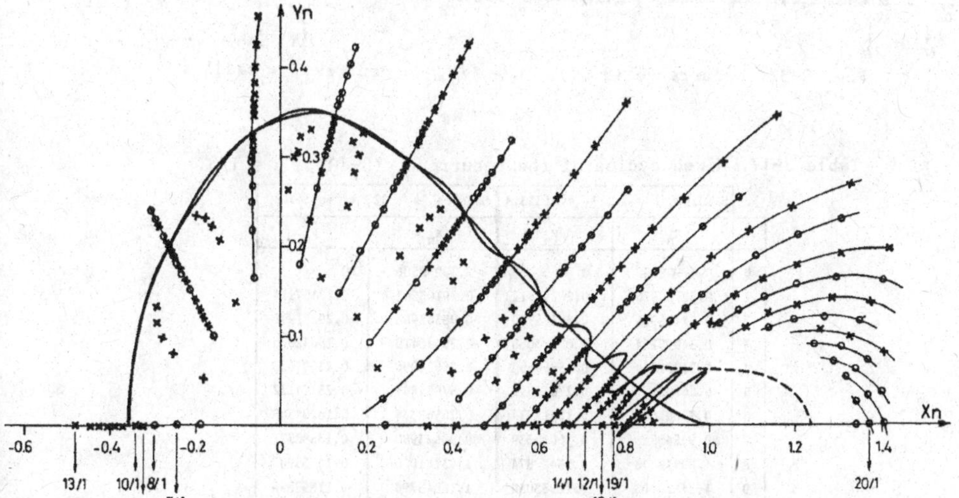


Fig. 3-31. Recurrence (3-80), $\mu = 0.5$, o centres, x saddles

internal part of the region of stochastic instability (at an exponential rate), their rate of divergence slows down, because the total region of stochastic instability is bounded above by the instability rings and island structures of external cycles. The total stochastic region of the bounded quadratic recurrence consists therefore in general of several disjoint parts, as can be seen from the discrete phase portraits of Figs. 3-33 to 3-36. The case $\mu = -\frac{1}{2}$ (exceptional case $\varphi = 2\pi/3$ at (0,0)) is special in the sense that both (0,0) and the "point at infinity" are simultaneously unstable.

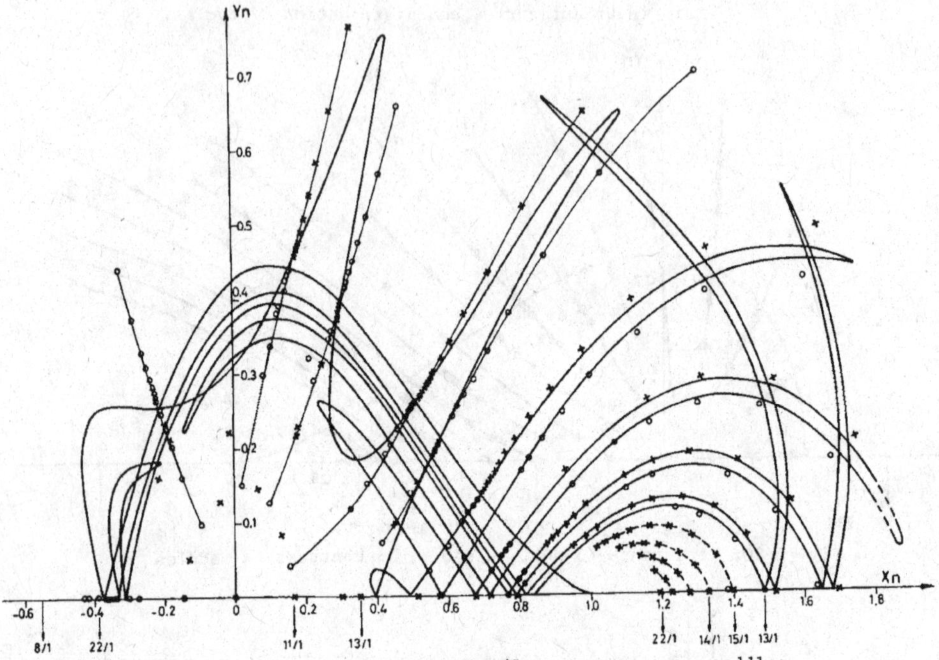


Fig. 3-32. Recurrence (3-80), $\mu = 1/8$, o centres, × saddles

Table 3-17. Mixed cycles of the recurrence (3-80), $\mu = 1/8$.

| | Saddles, $k = 21$, $\lambda_2 = -1333.4$ | | Saddles, $k = 22$, $\lambda_2 = -344.50$ | |
|---|---|---|---|---|
| n | $x_n$ | $y_n$ | $x_n$ | $y_n$ |
| 0 | 0.47988738 | 0 | 0.39970506 | 0 |
| 1 | 0.38755926 | -0.20291412 | 0.29103562 | -0.22667234 |
| 2 | 0.07405915 | -0.36875590 | -0.05363962 | -0.29271991 |
| 3 | -0.34995254 | 0.07313060 | -0.29440419 | 0.15642026 |
| 4 | 0.22032034 | 0.45850690 | 0.25920090 | 0.43697632 |
| 5 | 0.56706127 | 0.27636817 | 0.57954844 | 0.25324122 |
| 6 | 0.77305669 | 0.18418914 | 0.76568334 | 0.16294568 |
| 7 | 0.93543955 | 0.16056339 | 0.90543980 | 0.13586371 |
| 8 | 1.09418348 | 0.15487874 | 1.03741076 | 0.13135907 |
| 9 | 1.24519703 | 0.12533038 | 1.16815794 | 0.11852389 |
| 10 | 1.34484425 | 0.04982361 | 1.27443854 | 0.07429080 |
| 11 | 1.34484425 | -0.04982361 | 1.31673954 | 0 |
| 12 | 1.24519703 | -0.12533038 | 1.27445854 | -0.07429080 |
| 13 | 1.09418348 | -0.15487874 | 1.16815794 | -0.11852389 |
| 14 | 0.93543955 | -0.16056339 | 1.03741076 | -0.13135907 |
| 15 | 0.77305669 | -0.18418914 | 0.90543980 | -0.13586371 |
| 16 | 0.56706127 | -0.27636817 | 0.76568334 | -0.16294568 |
| 17 | 0.22032034 | -0.45850690 | 0.57954844 | -0.25324122 |
| 18 | -0.34995254 | -0.07313060 | 0.25920090 | -0.43697632 |
| 19 | 0.07405915 | 0.36875590 | -0.29440419 | -0.15642026 |
| 20 | 0.38755926 | 0.20291412 | -0.05363962 | 0.29271991 |
| 21 | 0.47988738 | 0 | 0.29103562 | 0.22667234 |
| 22 | 0.38755926 | -0.20291412 | 0.39970506 | 0 |

Table 3-17. Mixed cycles of the recurrence (3-80), $\mu = 1/8$.

| | Saddles, $k = 18$, $\lambda_1 = 20.122$ | | Saddles, $k = 18$, $\lambda_2 = -1507.3$ | |
|---|---|---|---|---|
| n | $x_n$ | $y_n$ | $x_n$ | $y_n$ |
| 0 | 0.43816645 | 0 | 1.92903050 | 0 |
| 1 | 0.33663749 | -0.21795510 | 1.62045710 | -0.45911466 |
| 2 | 0.00225624 | -0.33634655 | 1.01080118 | -0.60970696 |
| 3 | -0.33605561 | 0.13331563 | 0.40104319 | -0.71820548 |
| 4 | 0.26888751 | 0.48766133 | -0.42560977 | -0.18558925 |
| 5 | 0.63926705 | 0.31870756 | 0.02932469 | 0.43077896 |
| 6 | 0.90630263 | 0.26321319 | 0.43594814 | 0.30464334 |
| 7 | 1.16569343 | 0.24751934 | 0.63861138 | 0.15082561 |
| 8 | 1.40134131 | 0.16900756 | 0.73759936 | 0.07020762 |
| 9 | 1.50370856 | 0 | 0.77902661 | 0.02071363 |
| 10 | 1.40134131 | -0.16900756 | 0.77902661 | -0.02071363 |
| 11 | 1.16569343 | -0.24751934 | 0.73759936 | -0.07020762 |
| 12 | 0.90630263 | -0.26321319 | 0.63861138 | -0.15082561 |
| 13 | 0.63926705 | -0.31870756 | 0.43594814 | -0.30464334 |
| 14 | 0.26888751 | -0.48766133 | 0.02932469 | -0.43077896 |
| 15 | -0.33605561 | -0.13331563 | -0.42560977 | 0.18585925 |
| 16 | 0.00225624 | 0.33634655 | 0.40104319 | 0.71820548 |
| 17 | 0.33663749 | 0.21795510 | 1.01080118 | 0.60970696 |
| 18 | 0.43816645 | 0 | 1.62045710 | 0.45911466 |
| 19 | 0.33663749 | -0.21795510 | 1.92903050 | 0 |

All of the bifurcations identified in the case of the quadratic recurrence are found to occur on internal cycles of the recurrence (3-80). From the non-uniqueness of saddles with fixed k and r (several examples can be found in Table 3-15), it can be seen that the bifurcation rate is higher than in the case of the quadratic non-linearity, or in other words, in the recurrence (3-80) the same degree of stochasticity is attained at a lower strength of the non-linearity. The equivalence between the distance from (0,0) and the strength of the non-linearity is also found to hold for internal cycles.

Compared to the quadratic and cubic recurrences, a qualitatively new feature of the bounded quadratic recurrence is the existence of mixed cycles, i.e. of cycles some points of which lie in the internal, others in the external regions (with respect to the first loops of the main invariant curves). Points of mixed cycles were found more or less by lucky accidents, deliberate searches being generally fruitless. Mixed cycles can be characterized alternated by the property that all of their points are located inside the areas bounded by the intersections of the main invariant curves (cf. Table 3-17 and Fig.3-32). The existence of mixed cycles is therefore intimately related to the existence of homoclinic points. An extension of Table 3-17 to other cycles is difficult, because near points of mixed cycles the algebraic equations (0-24) are ill conditioned. The ill conditionning increases in the direction of decreasing max $|\lambda|$, which is the direction of interest for the study of bifurcations. Using a combination of numerical methods the evolution of the mixed saddle 18/2, farthest from the degenerate fixed point (1,0), was determined (Table 3-18), and the companion

Display of some discrete trajectories of the recurrence (3-80).

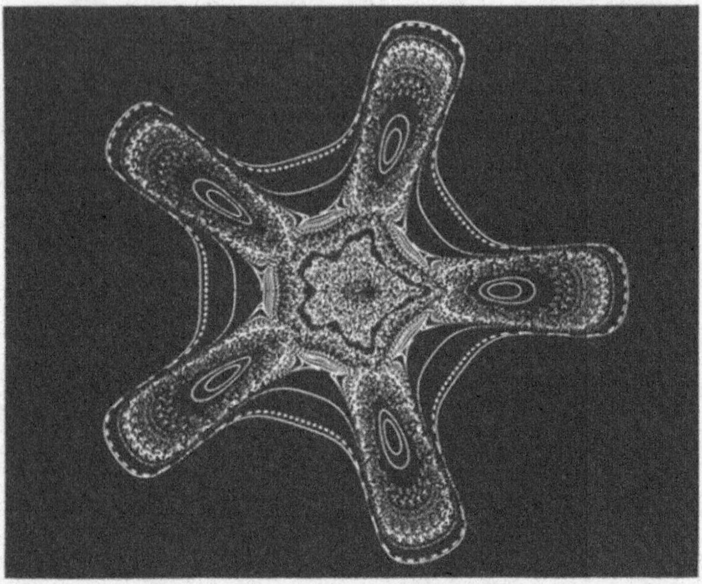

Fig. 3-33. $\mu = 0.3$

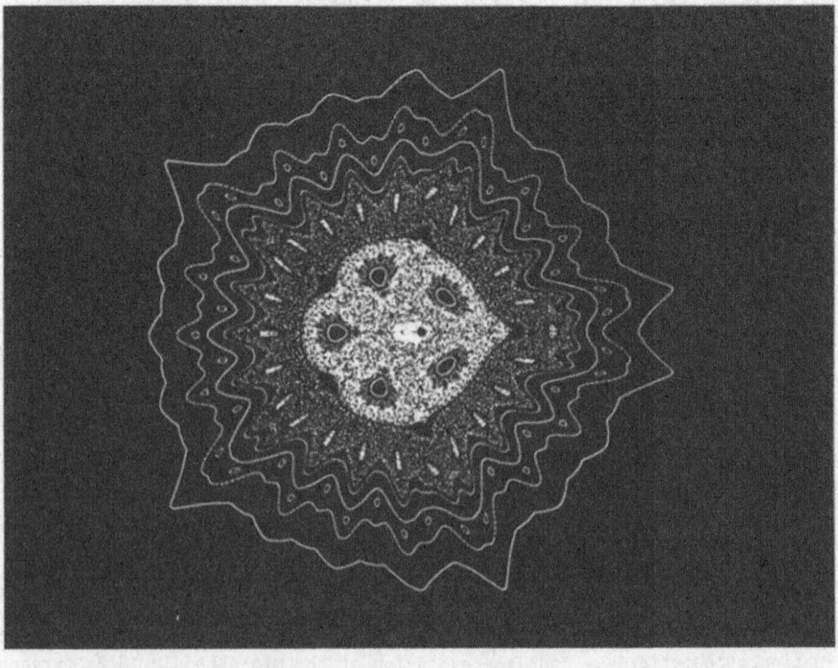

Fig. 3-34. $\mu = 0.25$

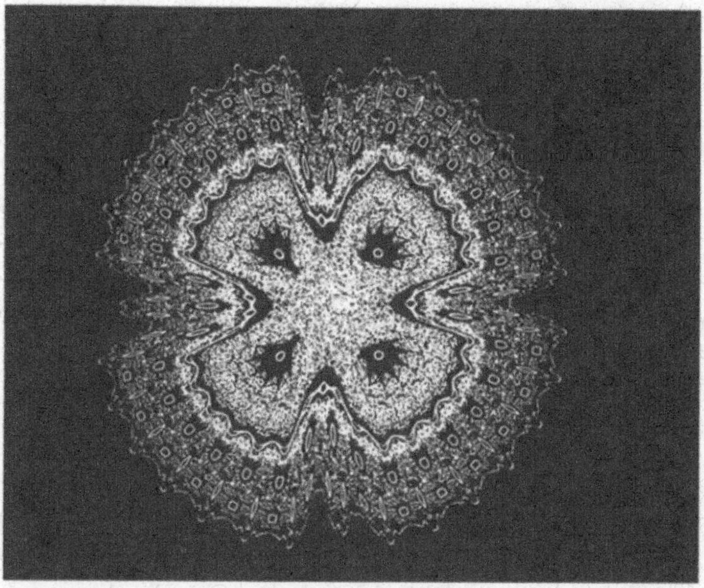

Fig. 3-35. $\mu = 0.05$

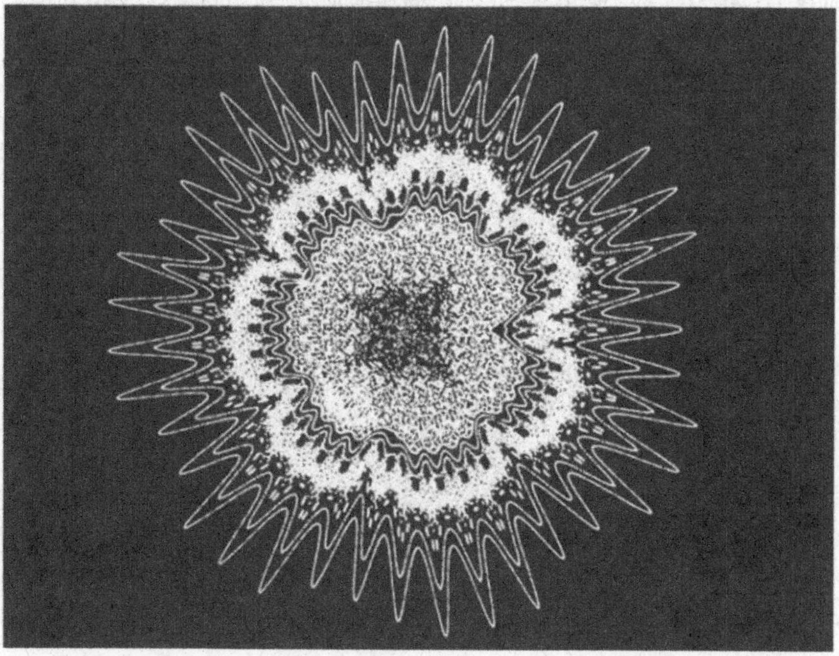

Fig. 3-36. $\mu = -0.2$

Table 3-18. Evolution of the mixed saddle 18/2 of type 1 of the recurrence (3-80).

| $\mu$ | $x$ | $y$ | $\lambda_1$ | $p_1 = -p_2$ |
|---|---|---|---|---|
| 0.125200000 | 0.439462299 | 0 | 16.439 | −0.1792 |
| 0.125400000 | 0.441685098 | 0 | 8.4857 | −0.07315 |
| 0.125438050 | 0.442995315 | 0 | 2.6528 | −0.01765 |
| 0.125438873 | 0.443200515 | 0 | 1.3618 | −0.004779 |
| 0.125438880 | 0.443219079 | 0 | 1.1096 | −0.001585 |

cycle (a saddle of type 3 and then a centre in linear approximation) was found (Table 3-19). The points of the mixed cycle pair 18/2 merge into double roots of the equations (0-24) at $\mu = \mu_0 \simeq 0.1254389$, and cease to exist for $\mu > \mu_0$. One point of the merged cycle is $x \simeq 0.4432194$, $y = 0$. Mixed cycles appear therefore to undergo the same bifurcations as internal and external cycles.

Table 3-19. Evolution of the mixed saddle 18/2 of type 3 of the recurrence (3-80).

| $\mu$ | $x$ | $y$ | $\lambda_2\,(<0)$ or $\varphi\,(>0)$ | $p_1 = -p_2$ |
|---|---|---|---|---|
| 0.125000000 | 0.448569785 | 0 | −28.020 | −0.1407 |
| 0.125205800 | 0.447091455 | 0 | −18.610 | −0.1036 |
| 0.125405800 | 0.444661610 | 0 | −4.8139 | −0.03102 |
| 0.125425800 | 0.444124489 | 0 | −1.7323 | −0.008130 |
| 0.125427400 | 0.444067200 | 0 | −1.1778 | −0.002325 |
| 0.125427600 | 0.444059767 | 0 | 3.043 | − |
| 0.125435800 | 0.443658569 | 0 | 1.594 | − |
| 0.125438880 | 0.443223757 | 0 | 0.1034 | − |

The qualitative difference between a mixed and an inner cycle pair k/r is strikingly illustrated by the shape of the corresponding invariant curves. In the

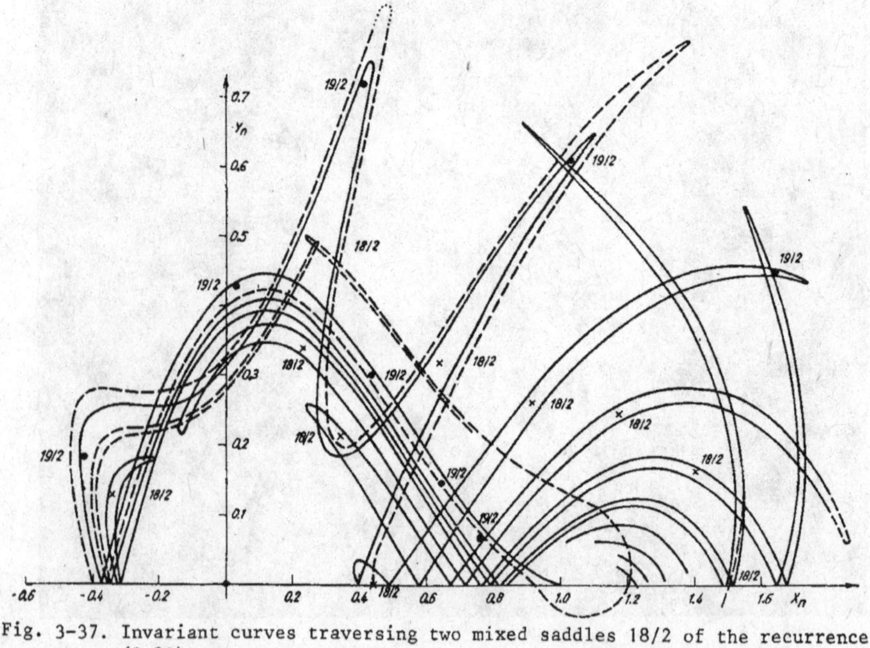


Fig. 3-37. Invariant curves traversing two mixed saddles 18/2 of the recurrence (3-80) : – – –

case of a mixed centre and saddle of type 1 pair, the invariant curves crossing the saddles do not form anything resembling an island structure, even if $\lambda_1 - 1 \ll 1$. These invariant curves appear to ignore the existence of their companion centres, and of other saddles of the same cycle, and follow the outline of the main invariant curves instead (Fig. 3-37). Homoclinic and heteroclinic points are produced even if $\lambda_1 - 1 \ll 1$. The same behaviour is found in the case of saddles k/r of type 3. In Fig. 3-37, the invariant curves of the two 18/2 mixed saddles are too close to be seen separately. Compared to the saddle of type 1, the saddle of type 3 gives rise to longer and narrower secondary invariant curve loops.

Contrarily to internal centres, mixed centres exist only inside essentially stochastic phase plane cells, because, as was already pointed out before, their existence is inherently coupled to the existence of homoclinic points. The stochastic character of the region of existence of mixed centres (in linear approximation) implies their instability in the full non linear formulation. This instability is due to the existence of strongly shape changing invariant curves "originating" far from a mixed centre, and passing arbitrarily close to the latter. Mixed centres appear therefore to meet the validity conditions of a conjecture on the arbitrarily small proximity of centres and homoclinic points [Z 1]. This conjecture (expressed in the form of a "theorem") has been formulated without taking into account the influence of the rotation number on the properties of cycles of an increasing order k, i.e. it does not distinguish between cycles sets satisfying the mutually inconsistent asymptotic properties (3-14) and (3-15). Such an inconsistency implies, for example, that there are no homoclinic points near the main centre (0,0), except possibly for $\mu = -\frac{1}{2}$ in the case of the quadratic and bounded quadratic non-linearities.

## 3.4 Exponential recurrence

The normalized form of the recurrence is

(3-90)
$$x_{n+1} = y_n + F(x_n), \qquad y_{n+1} = -x_n + F(x_{n+1})$$
$$F(x) = \mu x + (1-\mu)x^2 . e^{-(x^2-1)/4}, \qquad -1 < \mu < +1 \quad .$$

As in the case of the bounded quadratic non-linearity, the difference between the recurrence above and the quadratic recurrence consists only of higher order terms in x. The principal similarity is the existence of the main fixed points (0,0) and (1,0), which turn out to be a centre and a saddle of type 1, respectively. A particular feature of the recurrence (3-90) is the existence of a third fixed point $(x_1,0)$ independently of the value of $\mu$. The abscissa $x_1 = 1.8742631665$ is a root of

(3-91)
$$x^2 . e^{-(x^2-1)/4} = 1 \quad .$$

The fixed point is also a centre.

Extrapolating from the results of the bounded quadratic recurrence, it is to be expected that the cycles of the exponential recurrence will subdivide into at least four distinct families :

i) internal ones, surrounding the main centre (0,0),

ii) internal ones, surrounding the secondary centre $(x_1,0)$,

iii) external ones, surrounding both (0,0) and $(x_1,0)$,

iV) mixed ones.

This conjecture is confirmed by numerical computations, since no cycle could be found analytically. A partial list of cycles of types i) to iV) is given in Tables 3-20 to 3-29. These tables have been compiled for a weak ($\mu$ = 0.8), medium ($\mu$ = 0.5), and a strong ($\mu$ = 0.125) non-linearity, similarly to the preceding cases. A precise classification of a cycle is only possible when the main invariant curves, passing through the saddle (1,0) are already known. These curves are shown, together with the positions of cycles, in Fig. 3-38 to 3-40.

No unusual features have been noted in the study of the solution of the recurrence (3-90), except for the already known fact that the addition of higher order

Table 3-20. Internal cycles of the recurrence (3-90), $\mu$ = 0.8.

| | | Centres or Saddles of type 3 | | | | | Saddles of type 1 | | | | |
|---|---|---|---|---|---|---|---|---|---|---|---|
| k | r | x | y | $\phi$ or $\lambda_2$ (> 0) (< 0) | $p_1$ | $p_2$ | x | y | $\lambda_1$ | $p_1$ | $p_2$ |
| 10 | 1 | 0.189432 | 0.028752 | 0.0002 | | | 0.196903 | 0 | 1.0002 | 610 | |
| 11 | 1 | 0.451203 | 0 | 0.0066 | | | 0.441080 | 0.040626 | 1.0066 | -2.063 | -1.951 |
| 12 | 1 | 0.580359 | 0.036561 | 0.0183 | | | 0.589318 | 0 | 1.0185 | 35.55 | |
| 13 | 1 | 0.683756 | 0 | 0.0332 | | | 0.676342 | 0.030490 | 1.0338 | -2.275 | -1.908 |
| 14 | 1 | 0.746870 | 0.024771 | 0.0509 | | | 0.752863 | 0 | 1.0522 | 16.19 | |
| 15 | 1 | 0.805174 | 0 | 0.0717 | | | 0.800365 | 0.019911 | 1.0744 | -2.515 | -1.798 |
| 16 | 1 | 0.841714 | 0.015934 | 0.0965 | | | 0.845555 | 0 | 1.1012 | 9.227 | |
| 17 | 1 | 0.877123 | 0 | 0.1263 | | | 0.874054 | 0.012729 | 1.1345 | -2.928 | -1.647 |
| 18 | 1 | 0.899553 | 0.010164 | 0.1627 | | | 0.902000 | 0 | 1.1762 | 5.638 | |
| 19 | 1 | 0.921717 | 0 | 0.2073 | | | 0.919761 | 0.008117 | 1.2294 | -3.883 | -1.461 |

Table 3-21. Internal cycles of the recurrence (3-90) surrounding the secondary centre, $\mu$ = 0.8.

| | | Centres or Saddles of type 3 | | | | | Saddles of type 1 | | | | |
|---|---|---|---|---|---|---|---|---|---|---|---|
| k | r | x | y | $\phi$ or $\lambda_2$ (> 0) (< 0) | $p_1$ | $p_2$ | x | y | $\lambda_1$ | $p_1$ | $p_2$ |
| 17 | 1 | 2.277303 | 0 | 0.0094 | | | 1.147126 | 0 | 1.0095 | 85.38 | |
| 18 | 1 | 2.281743 | 0 | 0.0126 | | | 1.119276 | 0.011357 | 1.0127 | 2.028 | 2.154 |
| 19 | 1 | 2.284513 | 0 | 0.0164 | | | 1.092604 | 0 | 1.0166 | 52.10 | |
| 34 | 1 | 2.298593 | 0 | 0.1976 | | | | | | | |

Table 3-22. External cycles of the recurrence (3-90), $\mu$ = 0.8.

| | | Centres or Saddles of type 3 | | | | | Saddles of type 1 | | | | |
|---|---|---|---|---|---|---|---|---|---|---|---|
| k | r | x | y | $\phi$ or $\lambda_2$ (> 0) (< 0) | $p_1$ | $p_2$ | x | y | $\lambda_1$ | $p_1$ | $p_2$ |
| 18 | 1 | 2.512161 | 0 | 0.1201 | | | 0.280959 | 0.432038 | 1.127 | -0.121 | -0.137 |
| 19 | 1 | 2.433720 | 0.140754 | 0.1168 | | | 2.474242 | 0 | 1.124 | -72.15 | |
| 20 | 1 | 2.442786 | 0 | 0.1065 | | | 0.241067 | 0.389029 | 1.112 | -0.128 | -0.139 |
| 34 | 1 | 2.298593 | 0 | 0.1976 | | | | | | | |
| 35 | 2 | 2.522336 | 0.081427 | 0.0167 | | | | | | | |

Table 3-23. Internal cycles of the recurrence (3-90), $\mu$ = 0.5.

| | | Centres or Saddles of type 3 | | | | | Saddles of type 1 | | | | |
|---|---|---|---|---|---|---|---|---|---|---|---|
| k | r | x | y | $\phi$ or $\lambda_2$ (> 0) (< 0) | $p_1$ | $p_2$ | x | y | $\lambda_1$ | $p_1$ | $p_2$ |
| 7 | 1 | 0.482854 | 0.100218 | 0.3594 | | | 0.504898 | 0 | 1.4232 | 2.513 | |
| 8 | 1 | 0.677675 | 0 | 0.8712 | | | 0.656810 | 0.079757 | 2.2578 | 6.076 | 1.016 |
| 9 | 1 | 0.766961 | 0.057476 | 1.625 | | | 0.778529 | 0 | 3.8130 | 1.183 | |
| 10 | 1 | 0.847399 | 0 | -2.687 | -0.333 | | 0.836789 | 0.041039 | 6.8371 | 1.298 | -0.851 |
| 11 | 1 | 0.886536 | 0.028693 | -8.977 | -0.670 | 0.468 | 0.892236 | 0 | 12.85 | 0.871 | |
| 12 | 1 | 0.924755 | 0 | -21.21 | -0.684 | | 0.919495 | 0.020361 | 24.88 | 0.880 | -0.785 |
| 13 | 1 | 0.943658 | 0.014226 | -45.67 | -0.739 | 0.689 | 0.932969 | 0.030450 | 48.99 | 0.884 | -0.764 |
| 14 | 1 | 0.962492 | 0 | -94.66 | -0.739 | | 0.939657 | 0.035546 | 97.28 | 0.886 | -0.758 |
| 15 | 1 | 0.957718 | 0.024808 | -192.7 | -0.754 | 0.692 | 0.942987 | 0.037958 | 193.9 | 0.887 | -0.756 |
| 16 | 1 | 0.960049 | 0.026562 | -388.9 | -0.754 | 0.692 | 0.944648 | 0.039203 | 387.4 | 0.887 | -0.756 |
| 17 | 1 | 0.961212 | 0.027438 | -781.3 | -0.754 | 0.692 | 0.971538 | 0.018902 | 774.3 | 0.786 | -0.754 |
| 18 | 1 | 0.980045 | 0.013240 | -1566 | -0.753 | 0.738 | | | | | |
| 19 | 1 | 0.980628 | 0.013679 | -3136 | -0.753 | 0.738 | 0.991641 | 0.003768 | 3096 | 0.754 | -0.752 |
| 13 | 2 | 0.358598 | 0.049089 | 0.2218 | | | 0.365433 | 0 | 1.0224 | 49.35 | |
| 15 | 2 | 0.597912 | 0.045959 | 0.3446 | | | 0.598782 | 0 | 1.4071 | 6.760 | |
| 17 | 2 | 0.729927 | 0.032981 | 1.367 | | | 0.742935 | 0 | 3.3395 | 2.053 | |
| 19 | 2 | 0.813534 | 0.024490 | -8.721 | -0.249 | 1.143 | 0.801565 | 0 | 13.88 | 1.372 | |
| 21 | 2 | 0.820639 | 0.094318 | -54.06 | -0.766 | -0.007 | | | | | |
| 23 | 2 | 0.834833 | 0.104282 | -254.0 | -0.752 | 0.070 | | | | | |

Table 3-24. Internal cycles of the recurrence (3-90) surrounding the secondary centre, $\mu$ = 0.5.

| | | Centres or Saddles of type 3 | | | | | Saddles of type 1 | | | | |
|---|---|---|---|---|---|---|---|---|---|---|---|
| k | r | x | y | $\phi$ or $\lambda_2$ (> 0) (< 0) | $p_1$ | $p_2$ | x | y | $\lambda_1$ | $p_1$ | $p_2$ |
| 8 | 1 | 2.203356 | 0 | 0.0216 | | | 1.440471 | 0.072753 | 1.022 | 1.916 | 2.159 |
| 9 | 1 | 2.262698 | 0 | 0.0522 | | | 1.520086 | 0.260181 | 1.054 | 0.503 | 0.523 |
| 10 | 1 | 2.286700 | 0 | 0.0897 | | | 1.492436 | 0.280200 | 1.094 | 0.476 | 0.490 |
| 11 | 1 | 2.297538 | 0 | 0.1386 | | | 1.128740 | 0 | 1.148 | 0.974 | |
| 18 | 1 | 2.307525 | 0 | 2.052 | | | 1.011686 | 0.002912 | 4.719 | 0.924 | -1.838 |
| 19 | 1 | 1.012381 | 0.007206 | -3.578 | 1.838 | -0.924 | | | | | |
| 17 | 2 | 2.240198 | 0 | 0.0006 | | | | | | | |

Table 3-25. External cycles of the recurrence (3-90), $\mu$ = 0.5.

| | | Centres or Saddles of type 3 | | | | | Saddles of type 1 | | | | |
|---|---|---|---|---|---|---|---|---|---|---|---|
| k | r | x | y | $\phi$ or $\lambda_2$ (> 0) (< 0) | $p_1$ | $p_2$ | x | y | $\lambda_1$ | $p_1$ | $p_2$ |
| 6 | 1 | 4.622389 | 0 | 1.781 | | | | | | | |
| 9 | 1 | 2.774188 | 0 | 1.217 | | | | | | | |
| 15 | 1 | 2.383799 | 0 | 2.620 | | | 0.460968 | 0.455325 | 5.549 | -0.170 | -0.443 |
| 16 | 1 | 0.630101 | 0.355067 | -2.031 | -0.607 | -0.762 | 0.444095 | 0.439549 | 6.420 | -0.204 | -0.446 |
| 17 | 1 | 0.612917 | 0.341959 | -3.357 | -0.597 | -0.828 | 0.431777 | 0.428084 | 7.503 | -0.230 | -0.447 |
| 18 | 1 | 0.600412 | 0.332537 | -5.013 | -0.597 | -0.871 | 0.422794 | 0.419747 | 9.217 | -0.249 | -0.448 |
| 19 | 1 | 1.113146 | 0.164042 | -7.350 | 0.095 | 0.898 | 0.416262 | 0.413698 | 11.32 | -0.264 | -0.448 |
| 20 | 1 | 0.584750 | 0.320879 | -10.40 | -0.601 | -0.921 | 0.411527 | 0.409320 | 14.72 | -0.274 | -0.447 |
| 15 | 2 | 0.594489 | 1.231845 | -6.112 | -0.310 | -0.943 | | | | | |
| 19 | 3 | 2.131051 | 1.648530 | -1.681 | -1.176 | 0.296 | | | | | |

Table 3-26. Internal cycles of the recurrence (3-90), μ = 1/8.

| | | Centres or Saddles of type 3 | | | | | Saddles of type 1 | | | | |
|---|---|---|---|---|---|---|---|---|---|---|---|
| k | r | x | y | ϕ or $\lambda_2$ (> 0) (< 0) | $p_1$ | $p_2$ | x | y | $\lambda_1$ | $p_1$ | $p_2$ |
| 5 | 1 | 0.481414 | 0.175508 | 2.6722 | | | 0.448145 | 0 | 3.9103 | 0.9049 | |
| 6 | 1 | 0.725995 | 0 | -21.97 | -0.807 | | 0.647907 | 0.142270 | 18.16 | 1.093 | -0.880 |
| 7 | 1 | 0.814133 | 0.081397 | -76.60 | -0.986 | 0.898 | 0.803354 | 0 | 58.25 | 1.013 | |
| 8 | 1 | 0.894435 | 0 | -215.1 | -1.016 | | 0.863816 | 0.060160 | 161.2 | 1.040 | -1.031 |
| 9 | 1 | -0.515894 | 0 | -560.1 | 2.223 | | 0.922073 | 0 | 419.2 | 1.042 | |
| 10 | 1 | 0.957581 | 0 | -1417 | -1.038 | | | | | | |
| 11 | 1 | | | | | | 0.968543 | 0 | 2645 | 1.040 | |
| 12 | 1 | 0.982835 | 0 | -8776 | -1.037 | | | | | | |
| 13 | 1 | | | | | | 0.987262 | 0 | 16250 | 1.037 | |
| 14 | 1 | 0.993048 | 0 | -53700 | -1.035 | | | | | | |
| 16 | 1 | | | | | | 0.996376 | 0.001587 | 245000 | 1.034 | |
| 17 | 1 | 0.998022 | 0.000866 | -810000 | -1.033 | -1.034 | 0.996997 | 0.002229 | 606000 | 1.034 | -1.034 |
| 18 | 1 | | | | | | 0.998533 | 0.000642 | $15 \times 10^5$ | 1.033 | |
| 9 | 2 | -0.228948 | 0 | 0.1609 | | | 0.254162 | 0 | 1.1728 | 3.688 | |
| 11 | 2 | -0.449603 | 0 | -78.11 | 3.066 | | 0.514594 | 0 | 64.92 | 1.122 | |
| 12 | 2 | 0.692000 | 0.203607 | -285.4 | -1.089 | 0.897 | 0.537592 | 0 | 181.6 | 1.250 | |
| 13 | 2 | | | | | | 0.546178 | 0 | 463.1 | 1.315 | |
| 13 | 2 | -0.494082 | 0 | -1414 | 2.406 | | 0.798035 | 0 | 1048 | 1.006 | |
| 13 | 2 | | | | | | 0.717918 | 0 | 1565 | 0.785 | |
| 14 | 2 | | | | | | 0.549517 | 0 | 1160 | 1.344 | |
| 14 | 2 | | | | | | 0.715079 | 0 | 4170 | 0.776 | |
| 14 | 2 | | | | | | 0.803354 | 0 | 3393 | 1.013 | |
| 15 | 2 | | | | | | 0.550842 | 0 | 2888 | 1.356 | |
| 15 | 2 | | | | | | 0.805337 | 0 | 9380 | 1.016 | |
| 15 | 2 | | | | | | 0.713985 | 0 | 10600 | 0.772 | |
| 15 | 2 | | | | | | 0.895218 | 0 | 16400 | 1.017 | |
| 15 | 2 | | | | | | 0.919150 | 0 | 7440 | 1.041 | |
| 16 | 2 | | | | | | 0.551372 | 0 | 7165 | 1.361 | |
| 16 | 2 | | | | | | 0.806105 | 0 | 24400 | 1.017 | |
| 16 | 2 | | | | | | 0.713551 | 0 | 26600 | 0.771 | |
| 17 | 2 | | | | | | 0.551586 | 0 | 17700 | 1.363 | |
| 17 | 2 | | | | | | 0.806408 | 0 | 61600 | 1.017 | |
| 17 | 2 | | | | | | 0.894141 | 0 | 120000 | 1.016 | |
| 17 | 2 | 0.963177 | 0.004886 | -70900 | -1.041 | 1.038 | 0.921795 | 0 | 67600 | 1.042 | |
| 17 | 2 | | | | | | 0.967319 | 0 | 46800 | 1.040 | |
| 18 | 2 | | | | | | 0.551672 | 0 | 43900 | 1.364 | |
| 18 | 2 | | | | | | 0.713307 | 0 | 164000 | 0.770 | |
| 18 | 2 | | | | | | 0.806530 | 0 | 154000 | 1.017 | |
| 19 | 2 | | | | | | 0.551707 | 0 | 108000 | 1.364 | |
| 14 | 3 | 0.367959 | 0 | 0.5527 | | | -0.189602 | 0.205930 | 1.7110 | 0.500 | 0.967 |
| 15 | 3 | 0.560544 | 0.164124 | -10.48 | -0.746 | 1.016 | | | | | |
| 16 | 3 | 0.502015 | 0 | -268.6 | -1.054 | | 0.491754 | 0.025244 | 238.2 | 1.173 | -0.965 |
| 17 | 3 | 0.762547 | 0.017296 | -795.3 | -0.980 | 0.857 | 0.777987 | 0 | 560.1 | 0.993 | |
| 17 | 3 | | | | | | 0.517575 | 0 | 1162 | 1.136 | |
| 17 | 3 | | | | | | 0.740168 | 0 | 1246 | 0.834 | |
| 18 | 3 | 0.533941 | 0 | -4453 | -1.225 | | | | | | |
| 18 | 3 | 0.793611 | 0 | -4679 | -1.000 | | 0.730495 | 0.006232 | 6671 | 0.810 | -0.823 |
| 19 | 3 | | | | | | 0.538896 | 0 | 10400 | 1.260 | |
| 19 | 3 | | | | | | 0.722295 | 0 | 30500 | 0.798 | |
| 19 | 3 | | | | | | 0.797746 | 0 | 19000 | 1.006 | |
| 19 | 3 | | | | | | 0.902259 | 0 | 6098 | 1.020 | |
| 19 | 3 | 0.928364 | 0.001710 | -33700 | -1.046 | 1.044 | 0.911336 | 0 | 2739 | 1.041 | |
| 20 | 3 | 0.544877 | 0 | -36800 | -1.304 | | 0.907362 | 0.010234 | 51800 | 1.021 | -1.039 |
| 20 | 3 | | | | | | 0.915786 | 0.003058 | 22000 | 1.040 | -1.040 |
| 20 | 3 | | | | | | 0.800658 | 0.002636 | 61000 | 1.008 | -1.012 |
| 18 | 4 | | | | | | 0.254162 | 0 | 1.375 | 3.688 | |
| 19 | 4 | | | | | | 0.409631 | 0 | 2.958 | 1.568 | |

Table 3-27. Internal cycles of the recurrence (3-90) surrounding the secondary centre, $\mu = 1/8$.

| | | Centres or Saddles of type 3 | | | | | Saddles of type 1 | | | | |
|---|---|---|---|---|---|---|---|---|---|---|---|
| k | r | x | y | $\phi$ or $\lambda_2$ (> 0) (< 0) | $p_1$ | $p_2$ | x | y | $\lambda_1$ | $p_1$ | $p_2$ |
| 6 | 1 | 1.386242 | 0 | 0.9731 | | | | | | | |
| 7 | 1 | | | | | | 1.233888 | 0 | 6.570 | 1.075 | |
| 8 | 1 | 1.136421 | 0 | -15.28 | -0.800 | | | | | | |
| 9 | 1 | | | | | | 1.089503 | 0 | 45.12 | 1.011 | |
| 10 | 1 | 1.053696 | 0 | -121.9 | -0.982 | | | | | | |
| 11 | 1 | | | | | | 1.035728 | 0 | 283.9 | 1.020 | |
| 12 | 1 | 1.021582 | 0 | -776.0 | -1.019 | | | | | | |
| 13 | 1 | | | | | | 1.014408 | 0 | 1744 | 1.027 | |
| 14 | 1 | 1.008719 | 0 | -4770 | -1.028 | | | | | | |
| 15 | 1 | 1.006121 | 0.002674 | -11800 | -1.029 | 1.030 | | | | | |
| 16 | 1 | 1.003528 | 0 | -29100 | -1.031 | | | | | | |
| 17 | 1 | | | | | | 1.007381 | 0.007216 | 65000 | 1.031 | -1.029 |
| 19 | 1 | 1.001003 | 0.000439 | -440000 | -1.032 | 1.032 | 1.002989 | 0.002924 | 397000 | 1.032 | -1.031 |
| 11 | 2 | 2.149745 | 0 | 0.1217 | | | 1.552678 | 0 | 1.1292 | 1.058 | |
| 13 | 2 | | | | | | 1.275921 | 0 | 7.492 | 1.274 | |
| 14 | 2 | 1.208801 | 0.081613 | -3.253 | -0.834 | 0.858 | | | | | |
| 15 | 2 | | | | | | 1.214812 | 0 | 94.86 | 1.136 | |
| 16 | 2 | | | | | | 1.205818 | 0 | 184.5 | 1.204 | |
| 17 | 2 | | | | | | 1.201734 | 0 | 384.1 | 1.253 | |
| 17 | 2 | | | | | | 1.142382 | 0 | 626.3 | 0.777 | |
| 17 | 2 | | | | | | 1.092349 | 0 | 762.2 | 1.009 | |
| 18 | 2 | | | | | | 1.040446 | 0 | 867.4 | 1.020 | |
| 18 | 2 | | | | | | 1.144648 | 0 | 1515 | 0.765 | |
| 19 | 2 | | | | | | 1.199235 | 0 | 2058 | 1.293 | |
| 19 | 2 | | | | | | 1.145557 | 0 | 3685 | 0.760 | |
| 19 | 2 | | | | | | 1.037495 | 0 | 4518 | 1.019 | |
| 19 | 2 | | | | | | 1.088335 | 0 | 5102 | 1.013 | |
| 19 | 2 | | | | | | 1.053002 | 0 | 5531 | 0.983 | |
| 20 | 2 | | | | | | 1.016358 | 0 | 4999 | 1.027 | |
| 20 | 2 | | | | | | 1.145924 | 0 | 9040 | 0.758 | |
| 20 | 2 | | | | | | 1.036369 | 0 | 12600 | 1.020 | |
| 17 | 3 | | | | | | 1.495978 | 0 | 1.1604 | 24.40 | |
| 18 | 3 | 1.386242 | 0 | 2.919 | | | | | | | |
| 19 | 3 | | | | | | 1.295010 | 0 | 8.582 | 1.646 | |
| 20 | 3 | 1.252018 | 0 | -28.01 | -0.981 | | | | | | |

Table 3-28. External cycles of the recurrence (3-90), $\mu = 1/8$.

| | | Centres or Saddles of type 3 | | | | | Saddles of type 1 | | | | |
|---|---|---|---|---|---|---|---|---|---|---|---|
| k | r | x | y | $\phi$ or $\lambda_2$ (> 0) (< 0) | $p_1$ | $p_2$ | x | y | $\lambda_1$ | $p_1$ | $p_2$ |
| 4 | 1 | 3.486895 | 2.397340 | -1.565 | -0.688 | -0.113 | 3.508965 | 0 | 40.0562 | -0.936 | |
| 5 | 1 | -1.074489 | 0 | -35.04 | 1.384 | | | | | | |
| 6 | 1 | | | | | | -0.822654 | 0 | 95.07 | -1.703 | |
| 7 | 1 | -0.750546 | 0 | -42.82 | 1.624 | | | | | | |
| 8 | 1 | | | | | | -0.662404 | 0 | 124.2 | -1.826 | |
| 9 | 1 | -0.625041 | 0 | -58.70 | 1.784 | | | | | | |
| 10 | 1 | | | | | | -0.585845 | 0 | 201.8 | -1.954 | |
| 11 | 1 | -0.566596 | 0 | -137.6 | 1.960 | | | | | | |
| 12 | 1 | | | | | | -0.548559 | 0 | 469.6 | -2.064 | |
| 13 | 1 | -0.539260 | 0 | -500.5 | 2.086 | | | | | | |
| 14 | 1 | | | | | | -0.531152 | 0 | 1555 | -2.302 | |
| 8 | 2 | 3.339427 | 2.150851 | 0.7069 | | | | | | | |
| 9 | 2 | -1.260544 | 0 | -2820 | 1.313 | | | | | | |
| 10 | 2 | | | | | | -1.185240 | 0 | 2060 | -1.404 | |

Table 3-28 continued

| | | Centres or Saddles of type 3 | | | | | Saddles of type 1 | | | | |
|---|---|---|---|---|---|---|---|---|---|---|---|
| k | r | x | y | $\phi$ or $\lambda_2$ (> 0) (< 0) | $p_1$ | $p_2$ | x | y | $\lambda_1$ | $p_1$ | $p_2$ |
| 11 | 2 | -1.111942 | 0 | -431.7 | 1.338 | | | | | | |
| 11 | 2 | -1.159895 | 0 | -624.7 | 1.455 | | | | | | |
| 11 | 2 | -0.861705 | 0 | -4270 | 1.712 | | | | | | |
| 13 | 2 | -0.814759 | 0 | -8290 | 1.703 | | | | | | |
| 14 | 2 | | | | | | -0.793507 | 0 | 4290 | -1.720 | |
| 13 | 3 | 10.57872 | 0 | 1.780 | | | 2.541117 | +4.057222 | 74.58 | -0.859 | 2.586 |
| 14 | 3 | | | | | | -1.193665 | 0 | 104000 | -1.393 | |
| 14 | 3 | | | | | | -1.068469 | 0 | 52100 | -1.384 | |
| 16 | 4 | -1.656422 | 3.854228 | -58.48 | 2.345 | -0.260 | -2.753181 | 3.320456 | 3.260 | 3.075 | |
| 17 | 4 | 2.969207 | 5.012138 | -80.51 | 2.520 | -1.002 | | | | | |
| 18 | 4 | -0.911337 | 0 | -14.92 | 1.549 | | | | | | |
| 20 | 5 | | | | | | -2.070814 | 4.295107 | 22.81 | -0.424 | 1.544 |

Table 3-29. Mixed cycles of the recurrence (3-90), $\mu = 1/8$.

| | | Centres or Saddles of type 3 | | | | | Saddles of type 1 | | | | |
|---|---|---|---|---|---|---|---|---|---|---|---|
| k | r | x | y | $\phi$ or $\lambda_2$ (> 0) (< 0) | $p_1$ | $p_2$ | x | y | $\lambda_1$ | $p_1$ | $p_2$ |
| 16 | | 0.830925 | 0 | -4035 | -1.063 | | | | | | |
| 16 | | 0.678640 | 0 | -1059 | -0.590 | | | | | | |
| 16 | | 0.599748 | 0 | -116.9 | -3.258 | | | | | | |
| 17 | | | | | | | 0.882425 | 0 | 6468 | 1.000 | |
| 17 | | | | | | | 0.823609 | 0 | 2554 | 1.047 | |
| 17 | | | | | | | 0.691068 | 0 | 1284 | 0.673 | |
| 17 | | | | | | | 0.583362 | 0 | 224.7 | 1.960 | |
| 18 | | 0.931045 | 0 | -33560 | -1.047 | | | | | | |
| 18 | | 0.886964 | 0 | -31460 | -1.008 | | | | | | |
| 18 | | 0.817025 | 0 | -12250 | -1.033 | | | | | | |
| 18 | | 0.700203 | 0 | -5980 | -0.717 | | | | | | |
| 18 | | 0.570658 | 0 | -1250 | -1.610 | | | | | | |
| 19 | | 0.953279 | 0.001242 | -79000 | -1.037 | 1.036 | 0.889014 | 0 | 24700 | 1.010 | |
| 19 | | 0.928364 | 0.001710 | -33700 | -1.046 | 1.044 | 0.704408 | 0 | 7401 | 0.736 | |
| 19 | | 0.889043 | 0.001613 | -39600 | -1.012 | 1.009 | 0.564607 | 0 | 1688 | 1.516 | |
| 20 | | 0.959745 | 0.030054 | -215000 | -1.041 | 1.047 | | | | | |
| 20 | | 0.890860 | 0 | -91700 | -1.013 | | | | | | |
| 20 | | 0.707867 | 0 | -25100 | -0.750 | | | | | | |
| 20 | | 0.559594 | 0 | -5910 | -1.448 | | | | | | |

| $\mu$ | x | y | $\lambda_2$ | $p_1$ |
|---|---|---|---|---|
| 0.125 | 0.599 748 093 3 | 0 | -116.9 | 3.258 |
| 0.130 | 0.604 446 178 8 | 0 | -30.62 | 8.330 |
| 0.131 | 0.605 601 325 5 | 0 | -15.37 | 15.04 |
| 0.131 9 | 0.606 723 641 1 | 0 | -1.913 | 149.8 |
| 0.131 931 0 | 0.606 763 842 3 | 0 | -1.016 | -6534 |
| ($\lambda_2=-1$, $\varphi=\pi$) | | | $\varphi$ | |
| 0.131 931 5 | 0.606 764 491 6 | 0 | 3.059 | |
| 0.132 | 0.606 853 712 2 | 0 | 2.114 | |
| 0.132 2 | 0.607 117 299 4 | 0 | 0.515 1 | |
| ($\varphi=0$, $\lambda_1=1$) | | | $\lambda_1$ | |
| 0.132 219 | 0.607 142 583 6 | 0 | 1.048 | -2129 |
| 0.132 220 | 0.607 143 915 6 | 0 | 1.134 6 | -791.1 |

Table 3-30. Recurrence (3-90). Evolution of the mixed cycle 16/2.

| $\mu$ | x | y | $\lambda_1$ | $p_1$ |
|---|---|---|---|---|
| 0.125 | 0.583 | 0 | 224.7 | 1.960 |
| 0.130 | 0.585 | 0 | 150.0 | 2.294 |
| 0.140 | 0.592 | 0 | 35.19 | 5.151 |
| 0.144 | 0.596 035 | 0 | 1.477 | 156.72 |
| 0.144 021 18 | 0.596 059 024 9 | 0 | 1.011 | 5529 |
| ($\lambda=1$, $\varphi=0$) | | | $\varphi$ | |
| 0.144 021 20 | 0.596 059 047 4 | 0 | 0.0042 | |
| 0.144 580 | 0.596 702 804 2 | 0 | 2.972 | |
| 0.144 584 | 0.596 707 513 0 | 0 | 3.114 | |
| ($\varphi=\pi$, $\lambda_2=-1$) | | | $\lambda_2$ | |
| 0.144 585 | 0.596 708 690 4 | 0 | -1.082 | 747.3 |
| 0.145 | 0.597 205 416 6 | 0 | -4.628 | 25.41 |
| 0.151 7 | 0.609 015 543 5 | 0 | -32.57 | 0.924 1 |
| 0.154 3 | 0.622 047 421 4 | 0 | -9.646 | 0.080 0 |
| 0.154 38 | 0.624 095 458 0 | 0 | -1.873 | 0.009 8 |
| 0.154 382 0 | 0.624 200 036 8 | 0 | -1.079 | 0.001 1 |
| ($\lambda_2=-1$, $\varphi=\pi$) | | | $\varphi$ | |
| 0.154 382 1 | 0.624 205 555 1 | 0 | 3.017 1 | |
| 0.154 385 | 0.624 381 626 9 | 0 | 2.280 | |
| 0.154 391 30 | 0.625 183 442 6 | 0 | 0.326 4 | |
| ($\varphi \to 0$) | | | $\lambda_1$ | |
| 0.154 3 | 0.628 444 924 9 | 0 | 16.42 | 0.093 0 |
| 0.154 390 | 0.625 566 946 1 | 0 | 3.161 | 0.019 1 |
| 0.154 391 | 0.625 332 920 5 | 0 | 2.010 | 0.010 3 |
| 0.154 391 125 | 0.625 244 371 8 | 0 | 1.457 | 0.005 3 |
| ($\lambda_1 \to 1$) | | | | |

Table 3-31. Recurrence (3-90).

Evolution of the mixed cycle 17/2.

Main invariant curves and distribution of cycles

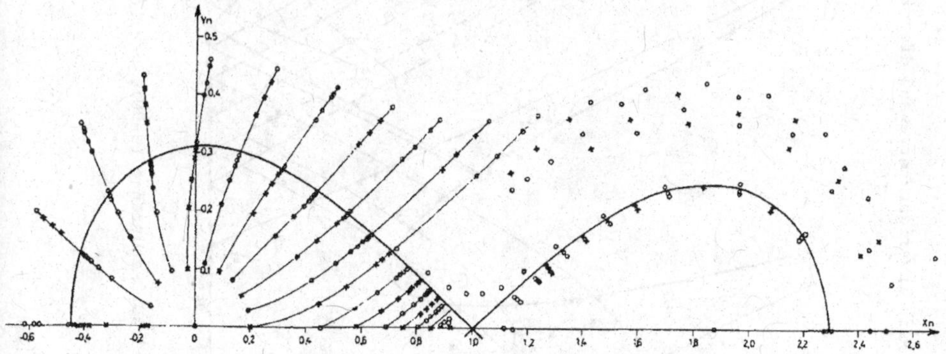


Fig. 3-38. Recurrence (3-90), $\mu$ = 0.8, o centres × saddles

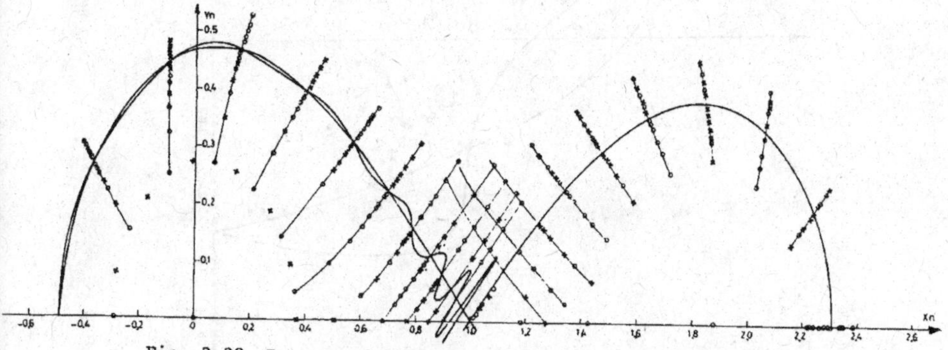


Fig. 3-39. Recurrence (3-90), $\mu$ = 0.5, o centres × saddles

Main invariant curves and distribution of cycles

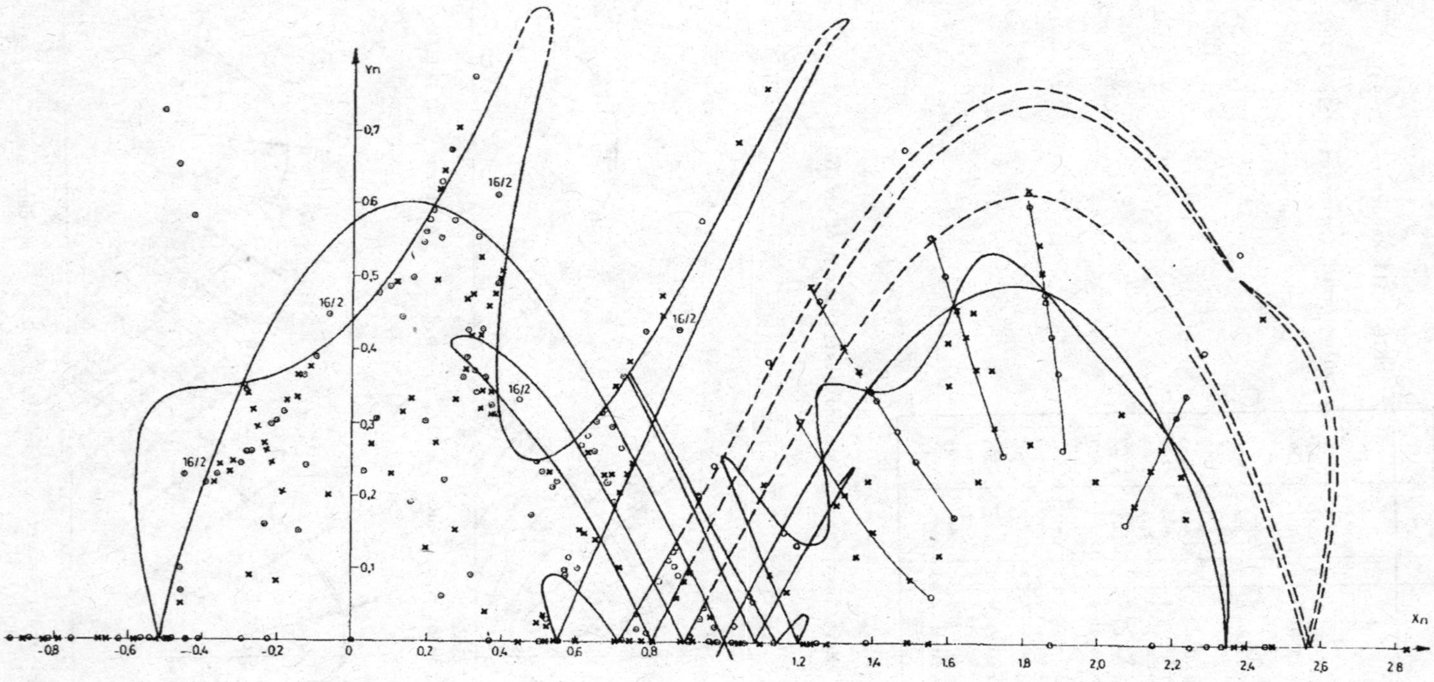


Fig. 3-40. Recurrence (3-90), $\mu = 1/8$, o centres x saddles

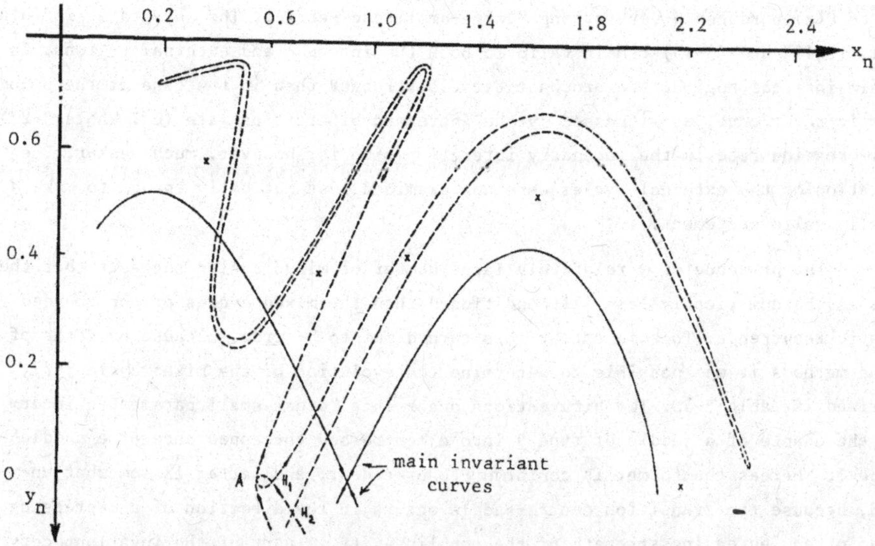


Fig. 3-41. Recurrence (3-90), $\mu = 0.13222$. - - - - : invariant curves crossing the mixed saddle A, $k/r = 16/2$. x other mixed saddles 16/2. $H_1$, $H_2$ two mixed homoclinic points.

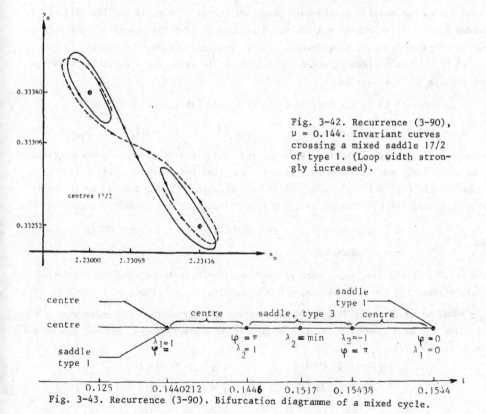

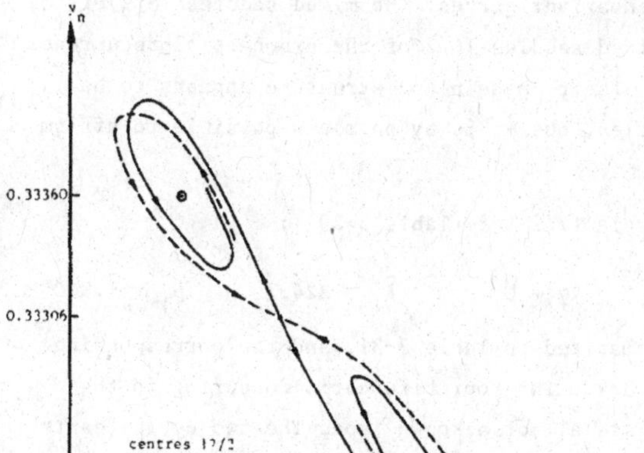


Fig. 3-42. Recurrence (3-90), $\mu = 0.144$. Invariant curves crossing a mixed saddle 17/2 of type 1. (Loop width strongly increased).

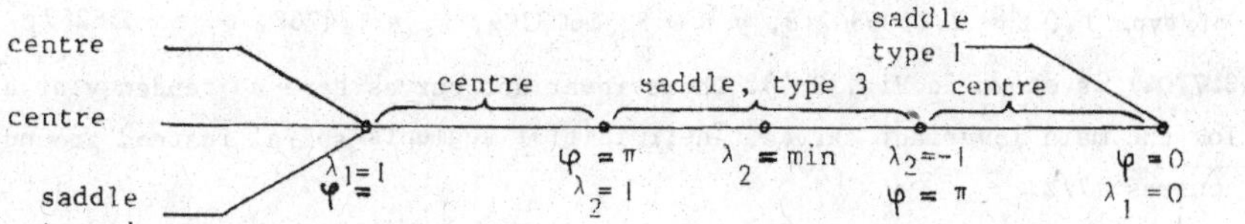


Fig. 3-43. Recurrence (3-90). Bifurcation diagramme of a mixed cycle.

terms to F(x) porduces a very strong "large-amplitude"-effect. The asymptotic formulae (3-14), (3-17) and (3-15) remain valid in both the internal and external regions. In the main internal region, the stochasticity is stronger than in the case of the bounded quadratic recurrence, as witnessed by the increased bifurcation rate (cf. Table 3-27). The bifurcation rate in the secondary internal region is, however, much weaker. The bifurcation of the external cycles were not examined systematically enough to make a generally valid statement.

The presence of a relatively large number of mixed cycles suggests that these cycles may be numerically less ill-conditioned than the mixed cycles of the bounded quadratic recurrence. To some extent this turned out to be true, because by means of routine methods it was possible to determine the evolution of the mixed cycle 16/2, summarized in Table 3-30. Two bifurcations occur in a rather small parametric interval : the change of a saddle of type 3 into a centre and the appearance of a saddle of type 1. Whereas the former is conforming to precedent, the latter is somewhat unexpected, because the transition centre-saddle occurs in the direction of a decreasing instead of an increasing strength of the non linearity. A part of the invariant curves crossing one 16/2 mixed saddle of type 1 are shown in Fig. 3-41. Similarly to the case of the 18/2 mixed saddles of the bounded quadratic recurrence, these invariant curves follow the outline of the main invariant curves. The mixed saddles 18/2 of the bounded quadratic recurrence and the mixed saddles 16/2 of the exponential recurrence are therefore qualitatively equivalent. Their phase plane structure appears to be typical of a certain class of mixed cycles, but it is by no means possible to affirm that this is so in all cases.

Consider in fact the mixed cycle 17/2 (cf. Table 3-29) :

$$\mu = 0.125 \qquad x = 0.583361 \qquad y = 0 \qquad \lambda_1 = 224.7 \qquad p_1 = 1.960 \ .$$

Its evolution as a function of $\mu$ is summarized in Table 3-31, and the corresponding bifurcation diagramme is shown in Fig. 3-43. The four bifurcations occuring in the parametric interval $0.125 < \mu < 0.1544$ are all of a known type. The two cycles centre existing before the bifurcation at $\mu \simeq 0.1440212$ have for $\mu = 0.1440212$ a point at

$$x = 0.5960590502 \qquad y = -0.0000124776 \qquad (\varphi = 0.0071)$$

and

$$x = 0.5960590473 \qquad y = 0 \qquad (\varphi = 0.0042)$$

respectively. For $\mu = 0.144$, a part of the invariant curves crossing the mixed 17/2 saddle of type 1 ( $x = 2.230680218$, $y = 0.3330608306$, $\lambda_1 = 1.4768$, $p_1 = -0.8287$, $p_2 = -0.7704$) is shown in Fig. 3-43. These invariant curves have no tendency at all to follow the main invariant curves. Their initial segments spiral instead around two mixed centres 17/2

$$x = 2.230006518 \qquad y = 0.3336009942$$

and

$$x = 2.231355383 \qquad y = 0.3325230390$$

of rotation angle $\varphi = 0.56213$, with the formation of homoclinic points. The computation of the 17/2 invariant curves becomes extremely ill-conditioned after more than about one turn around the 17/2 centres, and for this reason the part close to the latter is not shown. It appears, hower, that the configuration of Fig. 3-42 represents an unstable centre (in the sense of Birkhoff, see p. 576, vol. 2 of [B.4]) approached simultaneously by intersecting consequent and antecedent invariant curve spirals. Such a centre admits homoclinic points in an arbitrary small neighbourhood.

If one imagines a continuous variation of F(x) in (3-90), such that the secondary centre $(x_1,0)$ is made to coalesce with the main saddle (1,0), then the solution structure of (3-90) will go over continuously into that of a recurrence like (3-80). The presence of a cusp at (1,0) thus becomes self-evident in the latter case.

### 3.5 Sinusoïdal recurrence

The normalized one-parameter form of the recurrence is

$$(3\text{-}92)\qquad \begin{aligned} x_{n+1} &= y_n + F(x_n), & y_{n+1} &= -x_n + F(x_{n+1}) \\ F(x) &= x - \frac{1}{\pi}(1-\mu)\,\sin\pi x, & -1 &< \mu < +1 \end{aligned}$$

Compared to the cubic recurrence the difference consists in higher order terms in x only. A very pronounced similarity of the solution structure is found to exist. Because of the extra symmetry produced by $F(x) = -F(-x)$, there are three fixed points (0,0), (±1,0), the former a centre and the latter two, saddles of type 1. No positions of cycles are known analytically. The main invariant curves together with the position of numerically computed cycles are shown in Figs. 3-44 to 3-47. They differ only in a minor way from Figs. 3-24 to 3-27 of the cubic recurrence. A partial list of internal cycles, corresponding to the interval $|x_n| < 1$ is given in Tables 3-32 to 3-35, for a weak ($\mu = 0.8$), medium ($\mu = 0.6$, $\mu = 0.5$) and strong non-linearity ($\mu = 0.125$), respectively. Because of the peiodicity of sin $\pi x$ there exist also "image" cycles in all image intervals $n < |x| < n+1$, $n = 1, 2, \ldots$ Uniqueness of representation is restored if the cycles and invariant curves of the recurrence (3-92)

Table 3-32. Internal cycles of the recurrence (3-92), $\mu = 0.8$

| | | Centres or Saddles of type 3 | | | | | Saddles of type 1 | | | | |
|---|---|---|---|---|---|---|---|---|---|---|---|
| k | r | x | y | $\phi$ or $\lambda_2$ (> 0) (< 0) | $p_1$ | $p_2$ | x | y | $\lambda_1$ | $p_1$ | $p_2$ |
| 10 | 1 | 0.192375 | 0 | 0.0006 | | | 0.183142 | 0.034641 | 1.0006 | -1.8455 | -1.8191 |
| 12 | 1 | 0.528528 | 0.063406 | 0.0558 | | | 0.544200 | 0 | 1.0574 | 22.70 | |
| 14 | 1 | 0.694362 | 0 | 0.1552 | | | 0.681487 | 0.053592 | 1.1678 | -2.534 | -1.809 |
| 16 | 1 | 0.776297 | 0.041148 | 0.2819 | | | 0.785994 | 0 | 1.3250 | 7.855 | |
| 18 | 1 | 0.847170 | 0 | 0.4418 | | | 0.839956 | 0.030677 | 1.5527 | -3.498 | -1.606 |
| 11 | 1 | 0 | 0.236398 | 0.0001 | | | 0.421529 | 0 | 1.0001 | 750.7 | |
| 13 | 1 | 0.626152 | 0.029660 | 0.0026 | | | 0.629726 | 0 | 1.0026 | 622.0 | |
| 15 | 1 | 0.742328 | 0.023428 | 0.0115 | | | 0.745203 | 0 | 1.0116 | 178.5 | |
| 17 | 1 | 0.817383 | 0.017633 | 0.0318 | | | 0.819313 | 0 | 1.0323 | 73.10 | |
| 19 | 1 | 0.868735 | 0.013033 | 0.0720 | | | 0.870509 | 0 | 1.0746 | 34.01 | |

Table 3-33. Internal cycles of the recurrence (3-92), $\mu = 0.6$

| | | Centres or Saddles of type 3 | | | | | Saddles of type 1 | | | | |
|---|---|---|---|---|---|---|---|---|---|---|---|
| k | r | x | y | $\phi$ or $\lambda_2$ (> 0) (< 0) | $p_1$ | $p_2$ | x | y | $\lambda_1$ | $p_1$ | $p_2$ |
| 8 | 1 | 0.453033 | 0.125940 | 0.3067 | | | 0.483433 | 0 | 1.3563 | 4.911 | |
| 10 | 1 | 0.707779 | 0 | 1.0843 | | | 0.681794 | 0.107117 | 2.8662 | -11.46 | -1.314 |
| 12 | 1 | 0.803887 | 0.073576 | 2.1970 | | | 0.819270 | 0 | 7.3406 | 1.853 | |
| 14 | 1 | 0.886224 | 0 | 1.9597 | | | 0.874874 | 0.048771 | 23.78 | 2.486 | -1.110 |
| 16 | 1 | 0.937687 | 0.020427 | -18.77 | -0.766 | 1.311 | 0.926662 | 0 | 96.66 | 1.169 | |
| 18 | 1 | 0.952461 | 0.028047 | -340.2 | -0.954 | 1.229 | 0.966782 | 0 | 340.7 | 1.026 | |
| 7 | 1 | 0.173157 | 0.107260 | 0.0002 | | | 0.219843 | 0 | 1.0002 | 1811 | |
| 9 | 1 | 0.610616 | 0.061379 | 0.1091 | | | 0.616059 | 0 | 1.1152 | 21.55 | |
| 11 | 1 | 0.766536 | 0.043664 | 0.5972 | | | 0.776399 | 0 | 1.7898 | 5.158 | |
| 13 | 1 | 0.853557 | 0.029785 | 2.411 | | | 0.847633 | 0 | 5.4974 | 2.490 | |
| 15 | 1 | 0.906677 | 0.018970 | -23.04 | -1.131 | 0.323 | 0.918686 | 0 | 30.72 | 1.259 | |
| 17 | 1 | 0.939894 | 0.012642 | -155.7 | -0.853 | 1.183 | 0.930622 | 0 | 183.5 | 1.177 | |
| 19 | 1 | 0.957288 | 0.034207 | -720.2 | -0.972 | 1.264 | 0.933603 | 0 | 688.2 | 1.215 | |
| 15 | 2 | 0.365446 | 0.087740 | 0.00004 | | | 0.383555 | 0 | 1.00004 | 5065 | |
| 17 | 2 | 0.555704 | 0.031531 | 0.0046 | | | 0.557121 | 0 | 1.0046 | 832.5 | |
| 19 | 2 | 0.661868 | 0.030820 | 0.0394 | | | 0.662609 | 0 | 1.0402 | 158.2 | |
| 21 | 2 | 0.744079 | 0.026456 | 0.2464 | | | 0.748085 | 0 | 1.2780 | 30.19 | |
| 23 | 2 | 0.802143 | 0.014487 | 0.7927 | | | 0.804504 | 0 | 2.1318 | 5.382 | |
| 25 | 2 | 0.831790 | 0.010761 | -6.079 | -0.748 | 1.741 | | | | | |
| 19 | 3 | 0.661868 | 0.030820 | 0.0394 | | | | | | | |
| 25 | 3 | 0.533419 | 0.020885 | 0.0003 | | | 0.533998 | 0 | 1.0003 | 19100 | |

Table 3-34. Internal cycles of the recurrence (3-92), $\mu = 0.5$

| | | Centres or Saddles of type 3 | | | | | Saddles of type 1 | | | | |
|---|---|---|---|---|---|---|---|---|---|---|---|
| k | r | x | y | $\phi$ or $\lambda_2$ (> 0) (< 0) | $p_1$ | $p_2$ | x | y | $\lambda_1$ | $p_1$ | $p_2$ |
| 8 | 1 | 0.582482 | 0.153842 | 1.330 | | | 0.614404 | 0 | 2.772 | 2.594 | |
| 10 | 1 | 0.783714 | 0 | 2.696 | | | 0.756743 | 0.110130 | 10.27 | 5.134 | -1.207 |
| 12 | 1 | 0.826017 | 0.082724 | -2.436 | 1.439 | -1.078 | 0.869073 | 0 | 48.63 | 1.381 | |
| 14 | 1 | | | | | | 0.921351 | 0 | 170.6 | 0.929 | |
| 14 | 1 | | | | | | 0.910939 | 0.043952 | 279.4 | 1.449 | -1.139 |
| 16 | 1 | | | | | | 0.951040 | 0 | 1784 | 1.157 | |
| 16 | 1 | | | | | | 0.964557 | 0.017685 | 1649 | 1.166 | -1.121 |
| 7 | 1 | 0.453251 | 0.077538 | 0.0507 | | | 0.464549 | 0 | 1.052 | 298.2 | |
| 9 | 1 | 0.706256 | 0.067193 | 0.7390 | | | 0.704403 | 0 | 2.035 | 5.167 | |
| 11 | 1 | 0.827897 | 0.041883 | -6.293 | -1.248 | -0.182 | 0.847201 | 0 | 11.12 | 1.652 | |
| 13 | 1 | 0.895605 | 0.027717 | -77.14 | -0.895 | 1.414 | 0.879485 | 0 | 92.84 | 1.415 | |
| 15 | 1 | 0.936293 | 0.016235 | -612.5 | -1.151 | 0.948 | 0.887071 | 0 | 326.8 | 1.549 | |
| 17 | 1 | 0.897807 | 0.014480 | -2188 | -0.793 | 1.511 | 0.888515 | 0 | 1829 | 1.604 | |
| 13 | 2 | 0.336146 | 0.034562 | 0.0001 | | | 0.338709 | 0 | 1.0001 | $\approx 10^4$ | |
| 15 | 2 | 0.550870 | 0.037106 | 0.0208 | | | 0.554182 | 0 | 1.021 | 175.7 | |
| 17 | 2 | 0.661074 | 0.032258 | 0.3025 | | | 0.660400 | 0 | 1.350 | 16.34 | |
| 19 | 2 | 0.742438 | 0.097574 | 1.484 | | | 0.732013 | 0 | 3.580 | 13.52 | |
| 19 | 3 | 0 | 0.235774 | 0.00001 | | | 0.279538 | 0 | 1.00001 | $> 10^4$ | |

Table 3-35. Internal cycles of the recurrence (3-92), $\mu = 1/8$

| | | Centres or Saddles of type 3 | | | | | Saddles of type 1 | | | | |
|---|---|---|---|---|---|---|---|---|---|---|---|
| k | r | x | y | $\phi$ or $\lambda_2$ (> 0) (< 0) | $p_1$ | $p_2$ | x | y | $\lambda_1$ | $p_1$ | $p_2$ |
| 6 | 1 | 0.639539 | 0 | 2.938 | | | 0 | 0.550145 | 10.19 | -0.345 | |
| 8 | 1 | | | | | | 0.445982 | 0.617745 | 85.02 | 0.120 | -0.514 |
| 8 | 1 | | | | | | 0.809131 | 0 | 140.8 | 1.617 | |
| 10 | 1 | | | | | | 0.823337 | 0 | 1631 | 1.658 | |
| 10 | 1 | | | | | | 0.907343 | 0 | 1930 | 1.452 | |

Table 3-35 continued

| | | Centres or Saddles of type 3 | | | | | Saddles of type 1 | | | | |
|---|---|---|---|---|---|---|---|---|---|---|---|
| k | r | x | y | $\phi$ or $\lambda_2$ (> 0) (< 0) | $p_1$ | $p_2$ | x | y | $\lambda_1$ | $p_1$ | $p_2$ |
| 12 | 1 | | | | | | 0.824597 | 0 | 19400 | 1.663 | |
| 12 | 1 | | | | | | 0.946971 | 0 | 23000 | 1.58 | |
| 12 | 1 | | | | | | 0.974101 | 0 | 25300 | 1.57 | |
| 14 | 1 | 0.996109 | 0.003718 | -39900 | -1.58 | 1.58 | | | | | |
| 16 | 1 | 0.998966 | 0.000814 | -478000 | -1.59 | 1.59 | 0.997122 | 0.004350 | 280000 | 1.59 | -1.59 |
| 5 | 1 | 0 | 0.404297 | 0.2602 | | | 0.409829 | 0 | 1.2944 | 8.412 | |
| 7 | 1 | 0 | 0.623699 | -19.45 | 0.343 | | 0.775672 | 0 | 24.42 | 1.654 | |
| 9 | 1 | 0.454494 | 0.633308 | -479.6 | -0.560 | 0.178 | 0.820012 | 0 | 481.4 | 1.646 | |
| 9 | 1 | | | | | | 0.914325 | 0 | 304.9 | 1.462 | |
| 9 | 1 | | | | | | 0.932892 | 0 | 74.35 | 1.627 | |
| 11 | 1 | | | | | | 0.824314 | 0 | 5611 | 1.662 | |
| 11 | 1 | | | | | | 0.946118 | 0 | 6478 | 1.582 | |
| 11 | 1 | | | | | | 0.976091 | 0 | 3287 | 1.575 | |
| 11 | 1 | | | | | | 0.980547 | 0 | 592.7 | 1.591 | |
| 13 | 1 | 0.988404 | 0.006802 | -75500 | -1.58 | 1.59 | 0.824678 | 0 | 67100 | 1.66 | |
| 13 | 1 | 0.994205 | 0.000700 | -116000 | -1.59 | 1.58 | 0.947219 | 0 | 80300 | 1.58 | |
| 13 | 1 | | | | | | 0.993114 | 0 | 39000 | 1.58 | |
| 15 | 1 | 0.996659 | 0.004590 | -138000 | -1.59 | 1.58 | | | | | |
| 17 | 1 | 0.999035 | 0.001326 | -1650000 | -1.59 | 1.59 | 0.997170 | 0.004426 | 972000 | 1.59 | -1.59 |
| 11 | 2 | 0 | 0.512730 | 0.6820 | | | 0.500990 | 0 | 1.931 | 25.30 | |
| 13 | 2 | | | | | | 0.760273 | 0 | 153.6 | 1.744 | |
| 15 | 2 | | | | | | 0.783281 | 0 | 469.8 | 1.674 | |
| 15 | 2 | | | | | | 0.791988 | 0 | 781.0 | 1.561 | |
| 15 | 2 | | | | | | 0.806352 | 0 | 3078 | 1.613 | |
| 17 | 2 | 0.818761 | 0.001288 | -25200 | -1.64 | 1.64 | 0.819183 | 0 | 17300 | 1.64 | |
| 17 | 2 | | | | | | 0.923787 | 0 | 2848 | 1.414 | |
| 17 | 2 | | | | | | 0.930958 | 0 | 1684 | 1.663 | |
| 17 | 2 | | | | | | 0.937924 | 0 | 13100 | 1.58 | |
| 18 | 2 | | | | | | 0.821175 | 0.003092 | 34000 | 1.65 | -1.65 |
| 18 | 2 | | | | | | 0.922369 | 0 | 16000 | 1.44 | |
| 18 | 3 | 0.533747 | 0 | -19.13 | 5.356 | | | | | | |
| 20 | 3 | | | | | | 0.764264 | 0 | 3210 | 1.710 | |

Main invariant curves and distribution of cycles

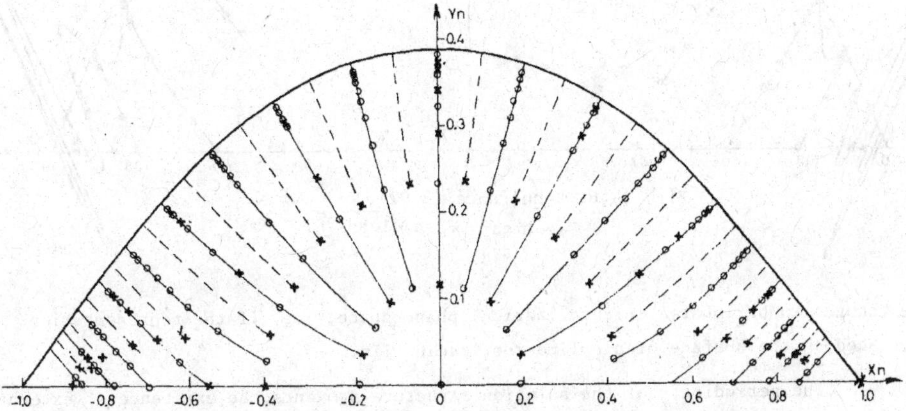


Fig. 3-44. Recurrence (3-92), $\mu$ = 0.8. o centres × saddles

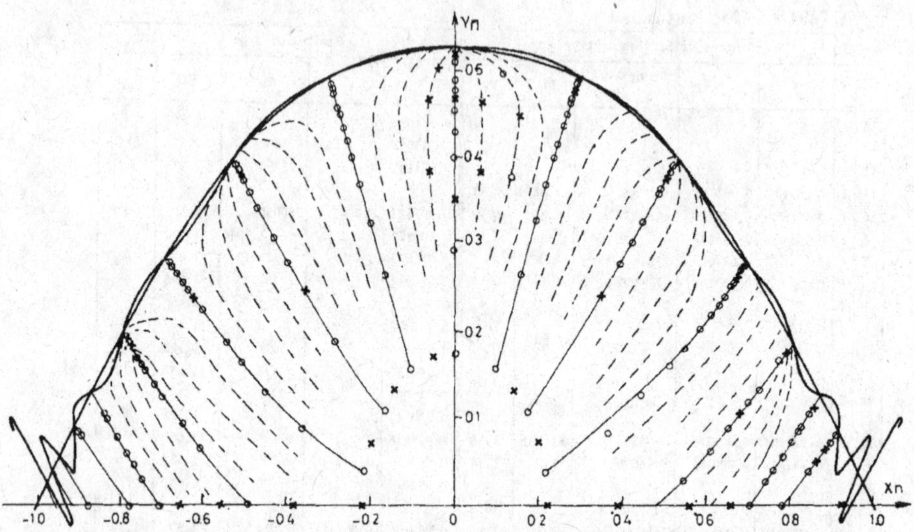


Figure 3-45. Recurrence (3-92), μ = 0.6. o centres x saddles

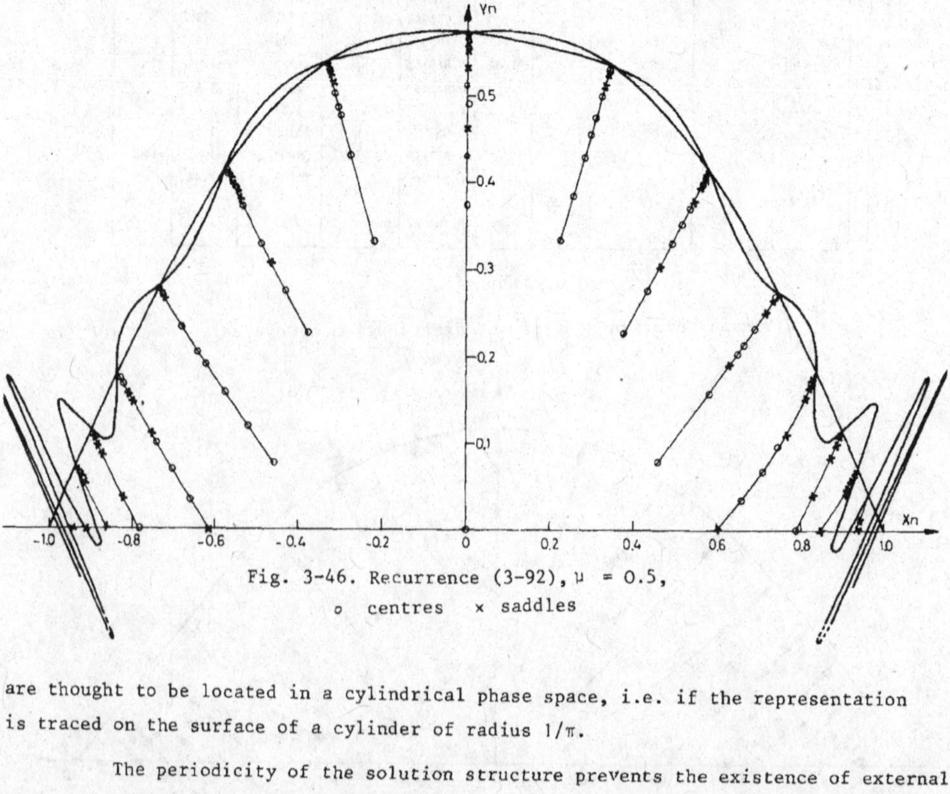


Fig. 3-46. Recurrence (3-92), μ = 0.5,
o centres x saddles

are thought to be located in a cylindrical phase space, i.e. if the representation is traced on the surface of a cylinder of radius $1/\pi$.

The periodicity of the solution structure prevents the existence of external cycles, in spite of the fact that the non-linearity in F(x) is bounded. There exist,

Main invariant curves and distribution of cycles

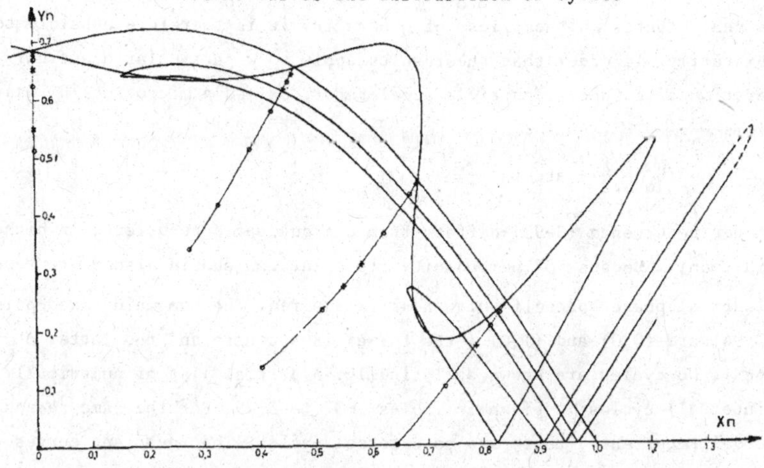


Fig. 3-47. Recurrence (3-92), $\mu = 1/8$. o centres × saddles

Table 3-36. Mixed cycles of the recurrence (3-92), $\mu = 1/8$.

| | | Centres or Saddles of type 3 | | | | | Saddles of type 1 | | | | |
|---|---|---|---|---|---|---|---|---|---|---|---|
| k | r | x | y | $\phi$ or $\lambda_2$ (> 0) (< 0) | $p_1$ | $p_2$ | x | y | $\lambda_1$ | $p_1$ | $p_2$ |
| 16 | | | | | | | 0.844072 | 0 | 4730 | 1.827 | |
| 16 | | | | | | | 0.883652 | 0 | 6614 | 1.287 | |
| 17 | | | | | | | 0.831163 | 0 | 11200 | 1.70 | |
| 17 | | | | | | | 0.851425 | 0 | 24200 | 1.98 | |
| 18 | | | | | | | 0.842549 | 0 | 12500 | 1.80 | |
| 19 | | 0.903668 | 0.000082 | -49400 | -1.44 | 1.44 | | | | | |

however, mixed cycles, the points of which are located in more than one image interval (Table 3-36). The recurrence (3-92) possesses the same bifurcations as those of the cubic recurrence. The nature of bifurcations producing mixed cycles is not yet known, the numerical problem being quite ill conditioned. The asymptotic formulae (3-14), (3-17) and (3-15) are found to hold like in the case of the cubic recurrence.

## 3.6 Unsymmetric sinusoidal recurrence

The normalized recurrence is a generalization of the form (3-92) providing a continuous indirect link between the quadratic and cubic recurrences :

(3-93)
$$x_{n+1} = y_n + F(x_n),\; y_{n+1} = -x_n + F(x_{n+1}), \qquad -1 < \mu < +1 ,$$
$$F(x) = x - \frac{1-\mu}{b\cos b_o}\left[\sin(bx+b_o) - \sin b_o\right],\; 0 < b_o < \frac{\pi}{2},\; b = \pi - 2b_o$$

Since it involves two independent parameters $\mu$ and $b_o$, the numerical and graphical representation of the solution is more complicated than in the preceding cases. A preliminary study has shown, however, that the qualitative properties of the recurrence (3-93) do not change as long as $b_o \neq 0$. The value of $b_o$ appears to play merely the role of a scale factor. Without loss of generality it is therefore possible to choose $b_o$ arbitrarily. In order that the results apply to a particular practical case (the operation of a special particle accelerator, called a microtron) $b_o$ was fixed as follows

$$b_o = \frac{\pi}{2} - \text{arc tg}\,\frac{1-\mu}{\pi} \quad . \tag{3-94}$$

The one-parameter recurrence (3-93) differs from the quadratic recurrence in higher-order terms in x only. Because of periodicity its solutions should also be represented in a cylindrical phase space if uniqueness is desired. The two main fixed points of (3-93), (3-94) are (0,0) and (1,0) ; the former is a centre and the latter a saddle of type 1. No cycles are known analytically. A partial list of numerically determined (internal) cycles is given in Tables 3-37 to 3-39. For the same reason as in the sinusoidal recurrence, there are no external cycles. The invariant curves passing through the main saddle (1,0) are shown in Figs. 3-48 to 3-50. A comparison of the corresponding cycles and invariant curves of the unsymmetric sinusoidal and quadratic recurrences discloses a remarkable degree of both qualitative and quantita-

Table 3-37. Cycles of the recurrence (3-93), (3-94), $\mu = 0.8$.

| | | Centres or Saddles of type 3 | | | | | Saddles of type 1 | | | | |
|---|---|---|---|---|---|---|---|---|---|---|---|
| k | r | x | y | $\phi$ or $\lambda_2$ (> 0) (< 0) | $p_1$ | $p_2$ | x | y | $\lambda_1$ | $p_1$ | $p_2$ |
| 10 | 1 | 0.250529 | 0 | 0.0001 | | | 0.241025 | 0.036592 | 1.0001 | -1.900 | -1.893 |
| 11 | 1 | 0.541454 | 0.049673 | 0.0037 | | | 0.553773 | 0 | 1.0037 | 138.1 | |
| 12 | 1 | 0.701253 | 0 | 0.0103 | | | 0.690883 | 0.042725 | 1.0103 | -2.135 | -2.021 |
| 13 | 1 | 0.784250 | 0.033848 | 0.0188 | | | 0.792358 | 0 | 1.0190 | 48.78 | |
| 14 | 1 | 0.852864 | 0 | 0.0296 | | | 0.846697 | 0.025965 | 1.0300 | -2.269 | -2.023 |
| 15 | 1 | 0.889912 | 0.019596 | 0.0434 | | | 0.894539 | 0 | 1.0444 | 25.71 | |
| 16 | 1 | 0.923864 | 0 | 0.0616 | | | 0.920418 | 0.014651 | 1.0635 | -2.425 | -1.964 |
| 17 | 1 | 0.942223 | 0.010888 | 0.0859 | | | 0.944776 | 0 | 1.0897 | 14.19 | |
| 18 | 1 | 0.959820 | 0 | 0.1187 | | | 0.957934 | 0.008060 | 1.1258 | -2.706 | -1.840 |
| 19 | 1 | 0.969314 | 0.005949 | 0.1631 | | | 0.970704 | 0 | 1.1767 | 7.818 | |

Table 3-38. Cycles of the recurrence (3-93), (3-94), $\mu = 0.5$.

| | | Centres or Saddles of type 3 | | | | | Saddles of type 1 | | | | |
|---|---|---|---|---|---|---|---|---|---|---|---|
| k | r | x | y | $\phi$ or $\lambda_2$ (> 0) (< 0) | $p_1$ | $p_2$ | x | y | $\lambda_1$ | $p_1$ | $p_2$ |
| 7 | 1 | -0.453247 | 0 | 0.4374 | | | 0.611988 | 0 | 1.531 | 2.349 | |
| 8 | 1 | -0.519721 | 0 | 1.1355 | | | 0.763198 | 0.090499 | 2.766 | 1.084 | -4.284 |
| 9 | 1 | -0.541178 | 0 | 2.9517 | | | 0.872783 | 0 | 5.685 | 1.277 | |
| 10 | 1 | -0.548847 | 0 | -9.366 | 30.47 | | 0.915006 | 0.038910 | 13.10 | 1.057 | -1.375 |
| 11 | 1 | -0.551694 | 0 | -29.21 | 25.90 | | 0.931111 | 0.054332 | 32.41 | 1.040 | -1.419 |
| 12 | 1 | 0.971773 | 0 | -81.08 | -1.043 | | 0.968120 | 0.015436 | 82.94 | 1.120 | -1.082 |
| 13 | 1 | 0.961612 | 0.037471 | -216.9 | -1.072 | 0.881 | 0.973774 | 0.021412 | 215.3 | 1.127 | -1.086 |
| 13 | 2 | -0.364999 | 0 | 0.0319 | | | 0.455375 | 0 | 1.032 | 36.39 | |
| 15 | 2 | -0.495642 | 0 | 0.5326 | | | 0.707595 | 0 | 1.685 | 5.359 | |
| 17 | 2 | -0.532975 | 0 | -2.357 | 128.0 | | 0.848055 | 0 | 6.938 | 1.665 | |
| 19 | 2 | 0.899388 | 0.024261 | -43.33 | -0.732 | 1.318 | 0.885209 | 0 | 53.98 | 1.345 | |
| 19 | 3 | -0.313851 | 0 | 0.0013 | | | 0.375926 | 0 | 1.0013 | 986.2 | |

Table 3-39. Cycles of the recurrence (3-93), (3-94), $\mu = 1/8$

| | | Centres or Saddles of type 3 | | | | | Saddles of type 1 | | | | |
|---|---|---|---|---|---|---|---|---|---|---|---|
| k | r | x | y | $\phi$ or $\lambda_2$ (> 0) (< 0) | $p_1$ | $p_2$ | x | y | $\lambda_1$ | $p_1$ | $p_2$ |
| 5 | 1 | -0.502858 | 0 | 2.525 | | | 0.551511 | 0 | 3.849 | 1.011 | |
| 6 | 1 | -0.577941 | 0 | -25.97 | 2.731 | | 0.754058 | 0.163018 | 20.18 | 1.297 | -1.084 |
| [illegible] | 1 | -0.594668 | 0 | -114.5 | 2.511 | | 0.892309 | 0 | 80.35 | 1.347 | |
| 8 | 1 | 0.953152 | 0 | -423.3 | -1.458 | | 0.932867 | 0.054883 | 291.3 | 1.423 | -1.446 |
| 9 | 1 | -0.600442 | 0 | -1494 | 2.452 | | 0.970033 | 0 | 1024 | 1.514 | |
| 9 | 2 | -0.288774 | 0 | 0.1467 | | | 0.318515 | 0 | 1.157 | 4.245 | |
| 11 | 2 | -0.545848 | 0 | -85.46 | 3.311 | | 0.621978 | 0 | 70.28 | 1.259 | |
| 12 | 2 | 0.861806 | 0.029972 | -395.1 | -1.282 | 1.141 | 0.641711 | 0 | 251.8 | 1.377 | |
| 13 | 2 | | | | | | 0.647291 | 0 | 874.8 | 1.418 | |
| 13 | 2 | | | | | | 0.802214 | 0.226028 | 1611 | 1.202 | |
| 14 | 3 | 0.455590 | 0 | 0.4915 | | | 0.440250 | 0.052857 | 1.617 | -4.440 | -1.292 |
| 15 | 3 | -0.460141 | 0 | -8.748 | 1.772 | | -0.511499 | 0 | 5.072 | -4.768 | |
| 16 | 3 | 0.608929 | 0 | -286.8 | -1.183 | | 0.597869 | 0.029811 | 253.1 | 1.089 | -1.300 |
| 17 | 3 | | | | | | 0.613179 | 0.040647 | 802.0 | 1.117 | -1.527 |
| 19 | 4 | -0.387920 | 0 | 1.048 | | | 0.505232 | 0 | 2.615 | 1.917 | |
| 20 | 4 | | | | | | -0.525779 | 0 | 40.96 | -0.413 | |
| 25 | 5 | -0.504186 | 0 | 0.0378 | | | | | | | |

Main invariant curves and distribution of cycles

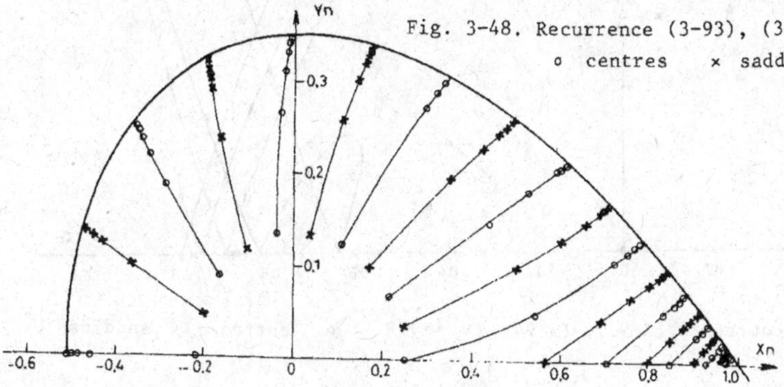


Fig. 3-48. Recurrence (3-93), (3-94), $\mu = 0.8$
o centres × saddles

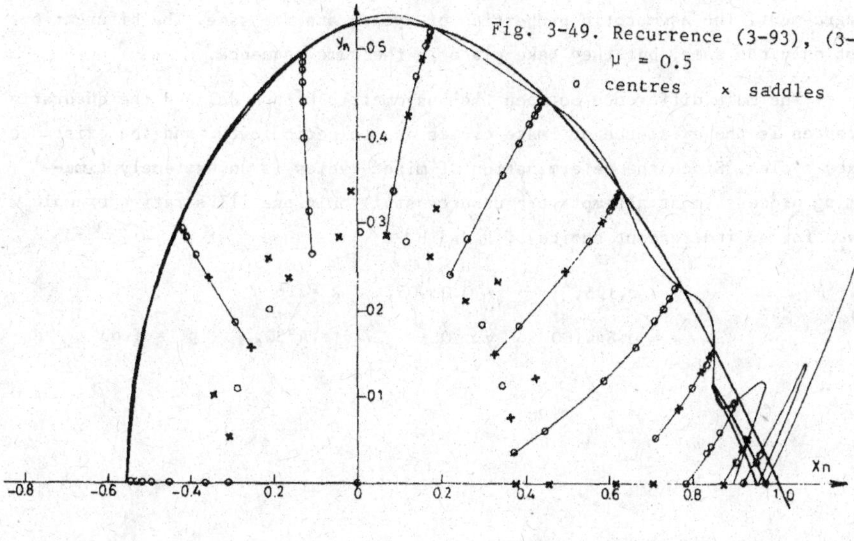


Fig. 3-49. Recurrence (3-93), (3-94), $\mu = 0.5$
o centres × saddles

Main invariant curves and distribution of cycles.

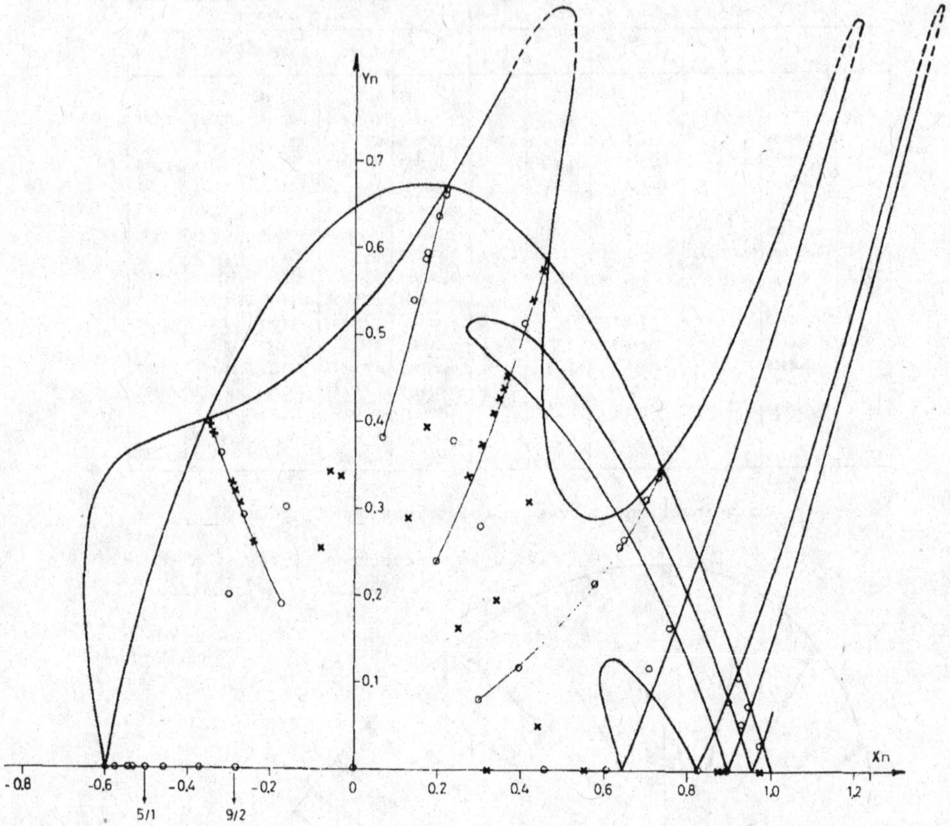


Fig. 3-50. Recurrence (3-93), (3-94), $\mu = 1/8$ o centres x saddles

tive agreement. The asymptotic properties of cycles are the same. The bifurcations are not only the same, but they take place in the same sequence.

The main difference between the unsymmetric sinusoidal and the quadratic recurrences is the existence of image cycles due to periodicity, and the existence of mixed cycles. Since the determination of mixed cycles is an extremely time-consuming process (most attempts are unsuccessful) only one illustrative example will be given for an independent choice of $\mu$ and $b_o$ :

$$\mu = 0.125, \quad b_o = 0.00005, \quad k = 16 \tag{3-95}$$
$$x \simeq 0.844100 \quad y = 0 \quad \lambda_1 = 4730, \quad p_1 \simeq 1.83 .$$

## 3.7 Box-within-a-box structure in the phase plane

It is worthwhile to summarize the various "microscopic" data collected so far and to see to what kind of global phase portrait of the recurrence (3-1) they add up. The recurrence (3-1) has a centre $x = y = 0$, $k = 1$ by construction. Since the rotation angle of this centre is given by $\varphi = \arccos \mu$, an infinity of exceptional cases exist in the interval $-1 < \mu < 1$. Only the case $\varphi = 2\pi/3$ turns out to be unstable in the presence of an even non-linearity, all others being stable in the sense that they do not admit any real invariant angles. It has been found that in the stable cases the traversal of every exceptional value

(3-96) $$\mu = \cos(2\pi r/k), \qquad k > r \text{ and relatively prime}$$

in the sense of decreasing $\mu$, releases from (0,0) one or two island structure of order $k/r$, depending on whether $F(x)$ is an even or odd function. The first enumerable set of cycles consists therefore of cycles bifurcated from (0,0) at the exceptional parameter values (3-96), according to one of the schematic rules

(3-97) $$\begin{aligned} &\text{centre } (0,0) \rightarrow \text{centre } (0,0) + \text{centre } k_{(r)} + \text{saddle } k_{(r)} \text{ of type } 1 \quad , \\ &\text{centre } (0,0) \rightarrow \text{centre } (0,0) + 2 \text{ centres } k_{(r)} + 2 \text{ saddles } k_{(r)} \text{ of type } 1 \quad . \end{aligned}$$

The bifurcated centres start their existence with the rotation angle $\varphi = 0$ and the bifurcated saddles with the eigenvalue $\lambda_1 = 1$. The mean geometrical distance $s(x, y)$ between (0,0) and the points $P_m(k)$ of bifurcated cycles increases with the parametric distance $s(\mu) = \mu_o - \mu$ from the bifurcation value $\mu_o$, and so do $\varphi(k_{(r)})$, $\lambda_1(k_{(r)})$ and the angles between the eigenslopes $p_1$, $p_2$ of the saddle of type 1. For any given $\mu_o$, except $\mu_o = \cos(2\pi/3)$ when $F(x)$ is an even function, the centre (0,0) is thus surrounded by an infinite but enumerable set of "concentric" cycle pairs $k_{(r)}$, with $k$, $r$ relatively prime and verifying $\mu_o < \cos(2\pi r/k)$.

When the difference $\lambda_1(k_{(r)}) - 1$ is small, the numerically computed (singular) invariant curves traversing the saddles $k_{(r)}$ are for all practical purposes indistinguishible from closed regular invariant curves. An invariant curve is said to be regular if it does not contain any singular points (such as cycles). The assumption that invariant curves which traverse saddles are not "analytically" closed for arbitrarily small values of $s(x, y)$ implies that homoclinic points can exist arbitrarily close to (0,0). It has been found, however, that for $\mu$ constant, homoclinic points are accumulation points of cycles with an increasing ratio $k/r$, satisfying the asymptotic relations (3-14), (3-17), whereas cycles approaching (0,0) have a decreasing ratio $k/r$, and satisfy the asymptotic relations (3-15). The assumption that sufficiently close to (0,0) the invariant curves are not truly closed leads to a contradiction between the two sets of qualitatively distinct asymptotic properties. A small neighbourhood of (0,0) is therefore filled with an infinite but enumerable set of island structures, defined by singular invariant curves which traverse the saddles $k_{(r)}$ of type 1.

Because two close island structures have been bifurcated from (0,0) at two distinct values of $\mu$, the area between them contains, in addition to other island structures, quasi-regular closed invariant curves, resulting from degenerate island structures, i.e. closed curves passing through an infinity of singular points (points of saddles $k_{(r)}$ of type 1 for which $k \to \infty$, $r \to \infty$, $k/r \to$ const., $\lambda_1(k_{(r)}) \to 1$, but $\cos(2\pi r/k) > \mu_0$). The distribution of closed invariant curves near (0,0) is thus extremely irregular in theory, but quite smooth in practice.

When $r/k = 1/3$, $\mu_0' = -1/2$ and $F(x)$ is an even function the scheme (3-97) is replaced by

(3-98) $\left\{\begin{matrix}\text{saddle } 3_{(1)} \text{ of type 1} \\ \text{centre } (0,0)\end{matrix}\right\} \to \left\{\begin{matrix}\text{three branch} \\ \text{saddle } (0,0)\end{matrix}\right\} \to \left\{\begin{matrix}\text{saddle } 3_{(1)} \text{ of type 1} \\ \text{centre } (0,0)\end{matrix}\right\}$

The saddles $3_{(1)}$, and the accompanying centres $3_{(1)}$ come into existence at a larger $\mu$ via the bifurcation

(3-99) $\left\{\begin{matrix}\text{double cycle } k_{(r)} & \to & \text{centre } k_{(r)} & + \text{ saddle } k_{(r)} \text{ of type 1} \\ \lambda = +1 & & \varphi(k_r) = 0 & \lambda_1(k_{(r)}) = 1\end{matrix}\right\}$

The double cycle $k_{(r)}$ is traversed by two invariant curves forming a stochastic cusp (cf. Fig. 3-20).

Because the construction of an iterated recurrence does not produce any intrinsically new data, the cycle and invariant curve structure near any centre $k_{(r)}$, $k > 1$, is the same as near the main centre (0,0), i.e. near every centre of the recurrence (3-1) there exists a "box-within-a-box" solution structure. The growth of the rotation angle $\varphi(\bar{k}_{(\bar{r})})$ as a function of the parametric distance $s(\mu)$ of the centre $\bar{k}_{(\bar{r})}$, bifurcated from the centre $k_{(r)}$ and characterized by the property that $\bar{k}/\bar{r}$ contains the factor $k/r$, is much faster than the growth of $\varphi(k_{(r)})$. In other words, the bifurcation speed is higher inside the "inner" boxes. The same property holds for a fixed $\mu$ and an increase of the Euclidian distance $s(x, y)$. The higher bifurcation speed of cycles is the primary source of stochasticity inside Birkhoff's instability rings, and of the global phenomenon of diffusion and stochastic instability of the recurrence (3-1). In the case of an even function $F(x)$ there is a secondary source of stochasticity produced by the bifurcations (3-98) and (3-99).

Since the rotation angle of a centre may reach the value $\varphi = \pi$, which produces the stochasticity-generating bifurcation (cf. Fig. 3-17)

(3-100) $\qquad$ centre $k_{(r)} \to$ saddle $_{k(r)}$ of type 3 + centre $(2k)_{(2r)}$ ,

there is a destruction of the regular region close to the centres $k_{(r)}$. This region is replaced by a considerably smaller region near the centres $(2k)_{(2r)}$. The chain of bifurcation $\varphi \to \pi$ increases thus the number of local regularity regions and reduces simultaneously their individual size. In this splitting up and dispersal process the total centre-induced regularity area diminishes after each bifurcation.

As the distance s(x, y) from (0,0) increases, the phase plane cells filled with island structures and quasi-regular invariant curves (corresponding to degenerate or practically to very narrow island structures) is further reduced by the appearance of additional bifurcations, of type (3-99) when F(x) is even, and of the three types

(3-101) centre $k_{(r)} \rightarrow$ centre $k_{(r)}$ + 2 centres $(2k)_{(2r)}$ + 2 saddles $(2k)_{(2r)}$ of type

(3-102) saddle $k_{(r)}$ of type 3 $\rightarrow$ centre $k_{(r)}$ + saddle $(2k)_{(2r)}$ of type 1

(3-103) centre $k_{(r)} \rightarrow$ saddle $k_{(r)}$ of type 1 + 2 centres $k_{(r)}$ with same $\varphi(k_{(r)})$ but

complementary symmetry, when F(x) is odd. When the non-linearity in F(x) is bounded, the stochasticity is enhanced some more by the appearance of mixed cycles. As a result, the total phase plane area is subdivided into four parts :

i) a dispersed centre-dominated regular region
ii) a bounded stochastic regions inside Birkhoff's instability rings
iii) a diffusion region characterized by a slight overlapping of separatrices of neighbouring island structures ("broken up", as illustrated in Figs. 3-8 to 3-14 and 3-14 to 3-28), and
iV) a region of stochastic instability characterized by an overlapping of separatrices of former island structures and of the main invariant curves.

## 3.8 The method of sections

It is well known that many continuous dynamic systems, described by ordinary differential equations, can be studied by projecting the trajectories on a suitable surface of section. The result is a recurrence at least one unit lower in dimensionality. As an illustration consider the Hamiltonian differential equation

(3-104) $$\ddot{z}(t) + w^2 z(t) + k^2 f(t).z^3(t) = 0 \quad ,$$

where w, k are real constants and f(t) is a periodic function of period T. For simplicity, let

(3-105) $$f(t) = \left\{ \begin{array}{ll} 1 & mT < t < (m + \frac{1}{2}) T \\ -1 & (m + \frac{1}{2}) T < t < (m + 1) T \end{array} \right\}, \quad m = 0, \pm 1, \ldots \; .$$

The solution of eq. (3-104), (3-105) passing through a phase plane point $(z_o, z'_o)$ can be expressed explicitly by means of elliptic functions. Such a solution is, however, practically inconvenient because the arguments, moduli and kinds of the elliptic functions involved depend all both on the values of w, k, T and on the point $(z_o, z'_o)$. Before proceding further, it is convenient to reduce first the number of parameters by means of the scaling

(3-106) $$x = \frac{k}{w} z, \quad y = \frac{k}{w^2} z, \qquad \tau = t/w$$

which leads to

$$\ddot{x}(\tau) + x(\tau) + f(\tau).x^3(\tau) = 0$$

(3-107)
$$f(\tau) = \begin{cases} 1, & mT < w\tau < (m + \frac{1}{2})\, T \\ -1, & (m + \frac{1}{2})T < w\tau < (m + 1)\, T \end{cases}, \qquad m = 0, \pm 1, \ldots .$$

If one is only interested in the solutions for points not too far from (0,0), then a locally equivalent recurrence can be found by chosing the "intersection planes"

(3-108)
$$x_n = x(t_o + nT), \quad y_n = \dot{x}_n = \dot{x}(t_o + nT), \qquad n = 0, \pm 1, \ldots ,$$

and computing the relations between, say, $(x_o, y_o)$ and $(x_1, y_1)$ by expanding the elliptic functions in power series. These power series can of course also be sought directly. A short exploration shows that the coefficients of these series depend only on $x_o$, $x'_o$, $x''_o$, ... The values $x_o$, $x'_o$ are known by definition, $x''_o$ is defined by equation (3-107), and the values of the higher derivatives can be obtained by (one-sided) differentiation. By a tedious but straightforward procedure, the locally equivalent recurrence is found to be [P 5]

(3-109)
$$\begin{aligned} x_{n+1} &= ax_n + by_n + c_{30}x^3_n + c_{21}x^2_n y_n + c_{12}x_n y^2_n + c_{03}y^3_n + \ldots \\ y_{n+1} &= -bx_n + ay_n + d_{30}x^3_n + d_{21}x^2_n y_n + d_{12}x_n y^2_n + d_{03}y^3_n + \ldots , \end{aligned}$$

where the first neglected terms are of fifth order in $x_n$, $y_n$. The constants in the above equation are

$$a = \cos wT, \; b = \sin wT, \qquad \theta = \frac{1}{2} wT,$$

$$c_{30} = \left[R_{30} + \frac{1}{2}(S_{30} - S_{12})\sin 2\theta + S_{21}\cos^2\theta + S_{03}\sin^2\theta\right]\sin\theta$$

$$c_{21} = \left[R_{21} - S_{12} + 3(S_{12} - S_{30})\cos^2\theta + \frac{3}{2}(S_{21} - S_{03})\sin 2\theta\right]\sin\theta$$

$$c_{12} = \left[R_{12} + S_{21} + \frac{3}{2}(S_{12} - S_{30})\sin 2\theta + 3(S_{03} - S_{21})\cos^2\theta\right]\sin\theta$$

$$c_{03} = \left[R_{03} + \frac{1}{2}(S_{03} - S_{21})\sin 2\theta - S_{30}\sin^2\theta - S_{12}\cos^2\theta\right]\sin\theta$$

$$d_{30} = \left[-S_{30} + \frac{1}{2}(R_{30} - R_{12})\sin 2\theta + R_{21}\cos^2\theta + R_{03}\sin^2\theta\right]\sin\theta$$

$$d_{21} = \left[3(R_{12} - R_{30})\cos^2\theta + \frac{3}{2}(R_{21} - R_{03})\sin 2\theta - R_{12} - S_{21}\right]\sin\theta$$

$$d_{12} = \left[3(R_{03} - R_{21})\cos^2\theta + \frac{3}{2}(R_{12} - R_{30})\sin 2\theta - S_{12} + R_{21}\right]\sin\theta$$

$$d_{03} = \left[-S_{03} + \frac{1}{2}(R_{03} - R_{21})\sin 2\theta - R_{12}\cos^2\theta - R_{30}\sin^2\theta\right]\sin\theta$$

$$S_{30} = -\frac{\theta^2}{2!} + 4\frac{\theta^4}{4!} - 25\frac{\theta^6}{6!} + 208\frac{\theta^8}{8!} + \ldots$$

$$S_{21} = -3\frac{\theta^3}{3!} + 24\frac{\theta^5}{5!} - 207\frac{\theta^7}{7!} + \ldots$$

$$S_{12} = -6\frac{\theta^4}{4!} + 66\frac{\theta^6}{6!} - 612\frac{\theta^8}{8!} + \ldots$$

$$S_{03} = -6\frac{\theta^5}{5!} + 66\frac{\theta^7}{7!} + \ldots$$

$$R_{30} = -\theta + 4\frac{\theta^3}{3!} - 25\frac{\theta^5}{5!} + 208\frac{\theta^7}{7!} + \dots$$

$$R_{21} = -3\frac{\theta^2}{2!} + 24\frac{\theta^4}{4!} - 207\frac{\theta^6}{6!} + 1848\frac{\theta^8}{8!} + \dots$$

$$R_{12} = -6\frac{\theta^3}{3!} + 66\frac{\theta^5}{5!} - 612\frac{\theta^7}{7!} + \dots$$

$$R_{03} = -6\frac{\theta^4}{4!} + 66\frac{\theta^6}{6!} - 612\frac{\theta^8}{8!} + \dots$$

Contrary to the cubic recurrence (3-59), the recurrence (3-109) contains all terms in the third-order forms. The coefficients $c_{ij}$ and $d_{ij}$ are not independent because the Jacobian determinant of (3-109) is unity by construction. The fixed point (0,0) of (3-109) is a centre with the rotation angle $\varphi = WT$. The exceptional cases of the recurrence (3-109) can be analysed in the same way as the exceptional cases of (3-6) and (3-59). Since eqs. (3-109) contain no quadratic terms, the case $\varphi = 2\pi/3$ is not essentially different from a centre. Applying the method of canonical forms to the exceptional case $\varphi = 2\pi/4$, for which $4WT = 2\pi$, it is found that (0,0) is traversed by four invariant curves, i.e. it is an eight-segment saddle. This instability does not occur in the case of the cubic recurrence (3-59) because the special choice of F(x) excludes the existence of the required dominant cubic terms.

At least in principle it is of course possible to choose the non-linearities in the recurrence (3-1) in such a manner that any exceptional case $\varphi = 2\pi r/k$ of a centre leads to an unstable fixed point traversed by several invariant curves. For fixed non-linearities, however, the probability of a destabilization of a centre decreases rapidly as k (the order of the resonance) decreases.

### 3.9 Non-vanishing Jacobian and uniqueness of antecedents

The recurrences discussed in the preceding sections have a Jacobian determinant equal to unity and they admit unique antecedents. Contrary to widespread belief, this property is not universal, because the validity of the implicit-function theorem is only local and not global. Instead of going into the details of the proof of this theorem (see for example [A7]), it is sufficient to examine some particularly simple recurrences. Let

(3-111) $$x_{n+1} = \frac{a}{2} y_n^{-2}, \quad y_{n+1} = bx_n y_n^3 + f(y_n),$$

where f(y) is an even but otherwise arbitrary differentiable function. It is easy to see by inspection that all pairs of points $(x_n, y_n)$ and $(-x_n, -y_n)$ of (3-111) have the same consequent $(x_{n+1}, y_{n+1})$. The recurrence (3-111) has therefore non-unique antecedents. Since

$$\frac{\partial x_{n+1}}{\partial x_n} = 0, \quad \frac{\partial x_{n+1}}{\partial y_n} = -ay_n^{-3}, \quad \frac{\partial y_{n+1}}{\partial x_n} = by_n^3, \quad \frac{\partial y_{n+1}}{\partial y_n} = 3bx_n y_n^2 + f'(y_n),$$

its Jacobian determinant is,

$$J(x_n,\ y_n) = \frac{\partial x_{n+1}}{\partial x_n}\frac{\partial y_{n+1}}{\partial y_n} - \frac{\partial x_{n+1}}{\partial y_n}\frac{\partial y_{n+1}}{\partial x_n} \equiv ab = \text{const.}$$

The recurrence (3-111) is conservative, i.e. $J(x_n,\ y_n) \equiv 1$ when $a = 1/b$ .

In the special case $a = b = 1$ and $f(y_n) \equiv 0$, an explicit solution of the recurrence (3-111) is known :

$$\text{(3-112)}\qquad \begin{aligned} x_n &= 2^{2n-2n-1} / (x_o^{\,2^n-2} \cdot y_o^{\,2^{n+1}-2}) \\ y_n &= x_o^{\,2^n-1} \cdot y_o^{\,2^{n+1}-1} / 2^{2^n-n-1} . \end{aligned}$$

A still simpler example of a conservative recurrence with non-unique antecedents is

$$\text{(3-113)}\qquad x_{n+1} = \frac{1}{2}x_n^2, \qquad y_{n+1} = y_n/x_n, \qquad J(x_n,\ y_n) \equiv 1 \quad ,$$

in which $x_n$ is not coupled to $y_n$. The general solution of (3-113) is

$$\text{(3-114)}\qquad y_n x_n = 2^{-n}\, x_o y_o \quad .$$

Other examples of recurrences with non-unique antecedents and constant or variable Jacobians of constant sign are readily constructed. The problem of uniqueness of antecedents of a given nominally conservative recurrence has therefore to be examined on its own merits, without undue reliance on the implicit function theorem. In particular, it is necessary to pay particular attention to the case when $J(x, y) = 1$ almost everywhere (but not everywhere), i.e. when there exist points (x, y) at which at least one partial derivative of f, g in $J(x_n, y_n)$ is either unbounded or undefined (cf. chapter VI).

### 3.10 The variational entropy

The study of zero and one-dimensional singularities of the recurrence (3-1) with the non-linearities (3-3) suggests that its phase portrait separates into an orderly and into a stochastic part, with an extremely complicated boundary between them. The notion of the variational entropy (1-64) allows to obtain a confirmation of this deduction as well as an estimate of the size of the orderly region. The required numerical computations are carried out conveniently using the equivalent formulation (1-65). The starting point is a discrete consequent half-trajectory $\{x_n,\ y_n\}$, n = 0, 1, 2, ... At each point $(x_n,\ y_n)$ partial derivatives (3-4) are evaluated and the variational equation

$$\text{(3-115)}\qquad \lambda^2 - (a+d)\lambda + J = 0$$

formed. The difference J-1 is used as an accuracy test. The root of (3-115) corresponding to consequents is used at each step in the formula (1-65).

Choosing the initial conditions $(x_o, y_0)$ of the generating trajectories on the x-axis it was found that $h_v = 0$ when $0 < x_o < \bar{x}_o$, $y_0 = 0$. For example, when $\mu = 0.5$, then $\bar{x}_o \simeq 0.5$ for the quadratic and sinusoidal recurrences, and $\bar{x}_o \simeq 0.6$ for the cubic one. As one would expect from the relative sizes of the main invariant curves, the orderly region of the quadratic exponential and the bounded quadratic recurrences are somewhat smaller : $\bar{x}_o \simeq 0.4$ for the former, and $\bar{x}_o \simeq 0.35$ for the latter. Similar numbers were found for $\mu = 1/8$.

When $x_o > \bar{x}_o$, i.e. $(x_o, y_0)$ is inside a stochastic region, the sequence of the $h_v$ does not converge. It is oscillatory without possessing a visible pattern. Except in the orderly region, conservative recurrences of form (3-1) possess therefore no variational entropy. Since the $h_v$ happen to be bounded, it was attempted to determine their mean arithmetical value

(3-116) $$H_n = \frac{1}{n} \sum_{m=0}^{n-1} h_{vm} \quad .$$

This attempt turned out to be successful. In the case of the quadratic recurrence $H = \lim_{n\to\infty} H_n$ appears to increase monotonically as $x_o - \bar{x}_o$ increases. For $\mu = 0.5$ and $x_o = 0.55$ one finds, for example $H = 0.014$, and for $x_o = 0.70$, $H = 0.163$. A further increase of $x_o$ leads to very large $x_n$ and $y_n$, and a numerical determination of H becomes inefficient.

In the case of the bounded quadratic and quadratic-exponential recurrences, however, the half-trajectories $\{x_n, y_n\}$, $n > 0$ remain bounded for all positive $x_o$. Values of the mean variational entropy H as a function of $x_o$ are given in Tables 3-40a and b. For $\mu = 0.5$, the increase of H with $x_o$ is monotonic, whereas for $\mu = 1/8$ it is not. The disorder in the stochastic region of a conservative recurrence increases therefore simultaneously with the effective strength of the non-linearity, but it is to a certain extent location - dependent.

Table 3-40a. Mean variational entropy of the recurrence (3-90).

$\mu = 0.5$

| $x_o$ | 0.35 | 0.40 | 0.45 | 0.50 | 0.55 | 0.60 | 0.65 | 0.70 | 0.75 | 0.80 | 0.85 | 0.90 | 0.95 |
|---|---|---|---|---|---|---|---|---|---|---|---|---|---|
| H | 0.049 | 0.101 | 0.159 | 0.182 | 0.215 | 0.265 | 0.274 | 0.279 | 0.259 | 0.255 | 0.313 | 0.284 | 0.184 |

| $x_o$ | 0.30 | 0.35 | 0.40 | 0.45 | 0.50 | 0.55 | 0.60 | 0.65 | 0.70 | 0.75 | 0.80 | 0.85 | 0.90 | 0.95 |
|---|---|---|---|---|---|---|---|---|---|---|---|---|---|---|
| H | 0.011 | 0.013 | 0.023 | 0.013 | 0.019 | 0.012 | 0.010 | 0.009 | 0.022 | 0.016 | 0.012 | 0.012 | 0.021 | 0.011 |

$\mu = 1/8$

Table 3-40b. Mean variational entropy of the recurrence (3-92).

$\mu = 0.5$

| $x_o$ | 0.50 | 0.55 | 0.60 | 0.65 | 0.70 | 0.75 | 0.80 | 0.85 | 0.90 | 0.95 |
|---|---|---|---|---|---|---|---|---|---|---|
| H | 0.042 | 0.095 | 0.139 | 0.191 | 0.192 | 0.239 | 0.338 | 0.326 | 0.318 | 0.261 |

$\mu = 1/8$

| $x_o$ | 0.45 | 0.50 | 0.55 | 0.60 | 0.65 | 0.70 | 0.75 | 0.80 | 0.85 | 0.90 | 0.95 |
|---|---|---|---|---|---|---|---|---|---|---|---|
| H | 0.013 | 0.011 | 0.013 | 0.022 | 0.011 | 0.017 | 0.011 | 0.011 | 0.009 | 0.018 | 0.014 |

## Chapter IV

## STOCHASTICITY IN ALMOST CONSERVATIVE RECURRENCES

### 4.0 Introduction

Discrete dynamic systems describing macroscopic evolution processes are as a rule non conservative, in the sense that their Jacobian determinant $J(x_n, y_n) \not\equiv 1$. This property has significant consequences as far as qualitative properties of the phase plane portrait are concerned. Compared to a conservative recurrence of form (3-1), the first consequence is the disappearance of the integral invariant

$$D_o = \iint_{\mathcal{D}_o} dx_n \; dy_n \quad , \tag{4-1}$$

where $D_o$ is the size of the region $\mathcal{D}_o$ enclosed by a continuous contour $\mathcal{C}_o$ in the phase plane. Let $\mathcal{C}_o$ be a generating element of a sequence of contours $\{\mathcal{C}_n\}$, determined by means of the recurrence

$$x_{n+1} = g(x_n, y_n), \quad y_{n+1} = f(x_n, y_n), \qquad n = 0, \pm 1, \pm 2, \ldots \; . \tag{0-22a}$$

The consequent sequence $\{\mathcal{C}_n\}$, $n > 0$ defines indirectly the sequences $\{\mathcal{D}_n\}$ and $\{D_n\}$. If (0-22a) is interpreted as a transformation of integration variables in (4-1), then

$$D_1 = \iint_{\mathcal{D}_1} J(x_n, y_n) \; dx_n \; dy_n \tag{4-2}$$

A similar expression holds for $\mathcal{D}_n$ and $D_n$, $n > 1$. When the recurrence (0-22a) is conservative (in a narrow sense), then $J(x_n, y_n) \equiv 1$ and $D_1 = D_o$, and by induction $D_n = D_o$, $n > 1$. In spite of area conservation (Liouville's theorem) the shapes of the contours $\mathcal{C}_n$, $n > 0$, may differ considerably from that of $\mathcal{C}_o$. Area conservation is not affected by the presence of stochasticity, but it deforms the shapes of $\mathcal{C}_n$ strongly and increases their length $S_n$. When $n \to \infty$, the length $S_n$ becomes usually unbounded, or the curve $\mathcal{C}_n$ ceases to be rectifiable in some other way. When $J(x_n,y_n) \not\equiv 1$ the relation (4-2) loses most of its usefullness. In fact, let $|J(x_n, y_n)| < 1$ inside a region $\mathcal{D}_o$ of finite area $D_o$. It would be tempting to conclude that $D_1 < D_o$ implies that $\mathcal{D}_1$ is contained inside $\mathcal{D}_o$. Unfortunately, this property does not hold in general, because $|J(x_n, y_n)| < 1$ does not exclude the existence of unstable singularities inside $\mathcal{D}_o$ and it is quite possible that the latter "push" a part of $\mathcal{C}_1$ outside of $\mathcal{D}_o$. Moreover, even if $\mathcal{C}_1$ is inside $\mathcal{D}_o$, and $S_o$, $S_1$ are both finite, it may happen that $S_1 > S_o$ and that $S_n$ diverges as $n \to \infty$.

If no consequent $\mathcal{D}_n$ of $\mathcal{D}_o$ falls outside of $\mathcal{D}_o$, then $\mathcal{D}_o$ is a region of no escape for the points $(x_n, y_n)$ of the discrete trajectories $\{x_n, y_n\}$, $n > 0$, $(x_o,y_o) \in \mathcal{D}_o$. The regions $\mathcal{D}_n$, $n \geqslant 0$ lie all inside the composite influence domain $\mathcal{D}$ of some (not necessarily fully known) set of attractors. It happens frequently that $\mathcal{D}$ is larger than $\mathcal{D}_o$. No attractor can exist of course in the finite part of the phase plane when the recurrence (0-22a) is conservative. The absence of attractors has, however, no influence on the existence of a region of no escape $\mathcal{D}_o$. In other words, $\mathcal{D}_o$ may be considered as a composite influence domain of the stable point-singularities contained therein. The existence of a region of no-escape is usually hard to establish analytically, but relatively easy to observe numerically. The simplest method consists in the computation of a certain number of suitably chosen samples of trajectories $\{x_n, y_n\}$, $0 < n < N$, N sufficiently large. As in the case of conservative recurrences (cf. Fig. 3-33 to 36), the point sets $\{x_n, y_n\}$ possess often rather unexpected macroscopic regularities (Fig. 4-1). When a no-escape region is known to exist, numerical convergence tests can be applied to some sequences derived from $\{x_n, y_n\}$, $0 < n < N$ ; for example to $\{x_{n+k}\}$ or $\{\sqrt{x^2_{n+k}+y^2_{n+k}}\}$, $m < n < N$, k and m fixed, subject to the assumption that for $n > m$ the sequence $\{x_n, y_n\}$ has reached its characteristic asymptotic form. If the convergence tests yield positive results, an estimate of the location of some attractors inside $\mathcal{D}_o$ becomes possible.

The characteristic properties of the boundary points of a "natural" no-escape region $\mathcal{D}_o$ are unknown in general. But if the boundary $\mathcal{C}_o$ of $\mathcal{D}_o$ coincides with the boundary $\mathcal{C}$ of a composite influence domain $\mathcal{D}$, then $\mathcal{C}_o$ is composed of : a) invariant curves, b) isolated point-singularities, c) continuous loci of point-singularities, d) combinations of a, b, c, e) excess antecedents of a, b, c, and f) possible accumulations of some of the preceding elements. There exist many possibilities for a non-natural $\mathcal{C}_o$, and in particular $V(x_n, y_n) = C$, where V is a Liapunov function and C a constant. When the recurrence (0-22a) admits critical curves (in the sense of Julia-Fatou), then a natural $\mathcal{C}_o$ may consist entirely of critical curve segments.

Stochasticity of a non-conservative recurrence like (0-22) or (0-22a) is known to possess three distinct sources : i) survival from a generating first-order recurrence, ii) survival from a second-order conservative recurrence, and iii) two-dimensional non-uniqueness of antecedents. These sources can act separately or in combination. As an illustration of the rather pure case ii) consider the recurrence (3-1) to which a dissipative term has been added :

$$\text{(4-3)} \quad \begin{aligned} &x_{n+1} = G(y_n) + F(x_n), \quad y_{n+1} = -x_n + F(x_{n+1}), \qquad n = 0, \pm 1, \pm 2, \ldots \\ &G(y) = (1+\alpha)y + \beta y^3, \qquad \alpha\ ,\ \beta = \text{real constants}\ . \end{aligned}$$

The Jacobian determinant of (4-3) is

$$\text{(4-4)} \quad J(x_n, y_n) = 1 + \alpha + 3\beta\, y_n^{\,2}$$

When $|\alpha| \ll 1$ and $|\beta| \ll 1$, and $y_n$ is not too large, the deviation from $J(x_n, y_n) \equiv 1$ is small, and (4-3) can be said to be almost conservative. The recurrence (4-3) admits one inverse for $\beta > 0$, and either one or three inverses for $\beta < 0$ (depending on the values of $x_n$ and $y_n$).

Display of some discrete trajectories of the recurrence (4-5).

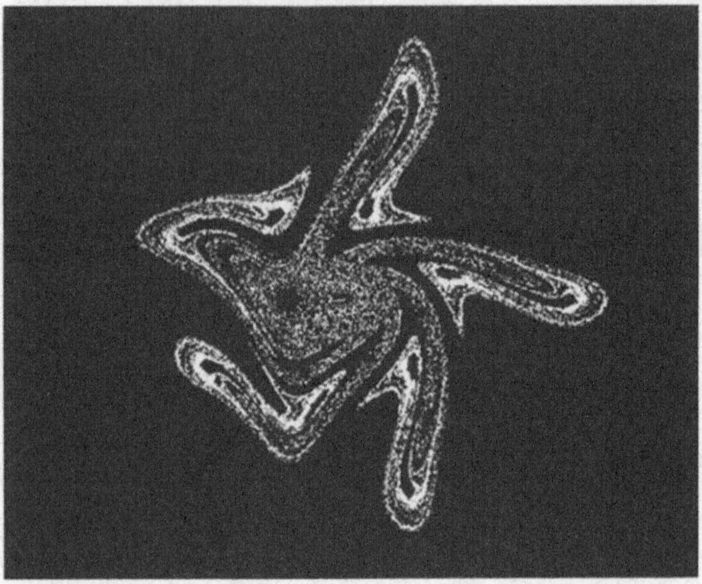

Fig. 4-1a. $\mu = 0.5$ $\alpha = -0.001$

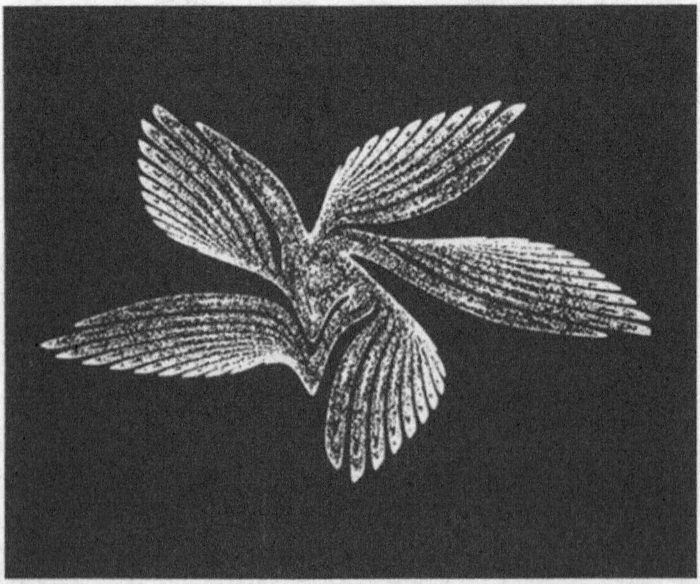

Fig. 4-1b. $\mu = 1/8$ $\alpha = -0.001$

4.1 Quadratic recurrence

The simplest form of (4-3), corresponding to a conservative case already studied, is :

$$x_{n+1} = (1+\alpha)y_n + F(x_n), \quad y_{n+1} = -x_n + F(x_{n+1}) \tag{4-5}$$
$$F(x) = \mu x + (1-\mu)x^2, \quad -1 < \mu < 1, \quad -1 << \alpha < 0, \quad J = 1+\alpha$$

The existence of the main fixed points (0,0) and (1,0) is unaffected by the perturbation term $\alpha y_n$. Whereas (1,0) remains a saddle of type 1, (0,0) becomes a stable focus with eigenvalues

$$\lambda_{1,2} = (1+\frac{\alpha}{2})\mu \pm i\sqrt{(1+\alpha)(1-\mu^2) - \frac{1}{4}\alpha^2\mu^2}, \qquad |\lambda_{1,2}| = \sqrt{1+\alpha} \tag{4-6}$$

No cycles of (4-5) are known explicitly in terms of elementary functions but many can be found numerically. For example for $\alpha = -0.001$ a partial list of cycles is given in Table 4-1a for $\mu = 0.5$ and in Table 4-1b for $\mu = 0.125$. The first observation to be made from such numerical data is that, at least for $|\alpha|$ sufficiently small, cycles of all orders k continue to possess a rotation number r, and that the locations of cycles having a large and increasing ratio k/r are little changed by the presence of dissipation (cf. Tables 3-1 to 3-4). The situation is, however, radically different for cycles having a small or a decreasing ratio k/r. For example, when $\alpha = -0.001$, the counterparts of the cycle family k/r = (4m+1)/m, m = 1, 2, ..., $\mu = 0$ (cf. Table 3-5) exist only for m < 6. One saddle and one focus 21/5 are given by

(4-7)

| | | | |
|---|---|---|---|
| x = 0.459303 | y = 0.012505 | $\lambda_1$ = 2.055 | $\lambda_2$ = 0.4765 |
| x = -0.387356 | y = 0.028498 | $\|\lambda\|$ = 0.9896 | $\varphi$ = 0.7648 |

The two cycles 25/6 exist when $|\alpha| \leq 0.0002$ and cycles like 29/7, 33/8, ... exist only when $|\alpha|$ is still smaller. A small damping affects therefore catastrophically the existence of cycles having a small ratio k/r. Only a finite number of these cycles survive a small deviation of $\alpha$ from $\alpha = 0$, the others disappear simply into the domain of complex numbers.

The bifurcation of type (3-97) can no longer occur at (0,0), because (0,0) is an isolated focus. The exceptional cases (weak resonances), $\varphi = \arg\lambda$ are not of the same type as those of centres. As can be seen from Tables 4-1 and 4-2, all cycle pairs centre-saddle of type 1 have been replaced by the cycle pairs focus-saddle of type 1. The cycle pairs "saddle of type 3-saddle of type 1" are still there and only the eigenvalue product $\lambda_1.\lambda_2$ in somewhat changed ($\lambda_1.\lambda_2 < 1$ when $\alpha < 0$).

The positions of cycles of the set k/1 are shown in Fig. 4-2 and 4-3 together with the main invariant curves. As expected from continuity considerations, all cycle points k/1 converge to homoclinic points on the main invariant curves, in essentially the same manner as in the conservative case. The points of the cycles k/r, r > 1 approach the homoclinic points in a more irregular manner than for $\alpha = 0$.

Table 4-1. Cycles of the recurrence (4-5), $\mu = 0.5$, $\alpha = -0.001$

| | | | | |
|---|---|---|---|---|
| focus 7/1 | x = -0.451464 | y = 0.028942 | $\|\lambda\|$ = 0.996504 | $\varphi$ = 0.448365 |
| saddle 7/1 | x = 0.611834 | y = 0.013867 | $\lambda_2$ = 0.644448 | $p_2$ = -2.10300 |
| | | | $\lambda_1$ = 1.54089 | $p_1$ = 2.51952 |
| focus 8/1 | x = 0.786483 | y = -0.006615 | $\|\lambda\|$ = 0.996006 | $\varphi$ = 1.19421 |
| saddle 8/1 | x = 0.760313 | y = 0.097121 | $\lambda_2$ = 0.345556 | $p_2$ = -1.05241 |
| | | | $\lambda_1$ = 2.87082 | $p_1$ = 4.16114 |
| saddle 9/1 | x = 0.863286 | y = 0.055439 | $\lambda_1$ = -0.551520 | $p_1$ = -0.537706 |
| | | | $\lambda_2$ = -1.79692 | $p_2$ = -0.089269 |
| saddle 9/1 | x = 0.872834 | y = 0.003681 | $\lambda_2$ = 0.164002 | $p_2$ = -1.23516 |
| | | | $\lambda_1$ = 6.04283 | $p_1$ = 1.26069 |
| saddle 10/1 | x = 0.924527 | y = -0.002461 | $\lambda_1$ = -0.094398 | $p_1$ = -0.802000 |
| | | | $\lambda_2$ = -10.4880 | $p_2$ = 0.825471 |
| saddle 10/1 | x = 0.914039 | y = 0.041357 | $\lambda_2$ = 0.070047 | $p_2$ = -1.04977 |
| | | | $\lambda_1$ = 14.1339 | $p_1$ = 1.34754 |
| saddle 11/1 | x = 0.949533 | y = 0.022580 | $\lambda_1$ = -0.030699 | $p_1$ = -1.01674 |
| | | | $\lambda_2$ = -32.2176 | $p_2$ = 0.879188 |
| saddle 11/1 | x = 0.953382 | y = 0.001386 | $\lambda_2$ = 0.028078 | $p_2$ = -1.09622 |
| | | | $\lambda_1$ = 35.2241 | $p_1$ = 1.09806 |
| saddle 12/1 | x = 0.971823 | y = $-9.73 \cdot 10^{-4}$ | $\lambda_1$ = -0.0110981 | $p_1$ = -1.04427 |
| | | | $\lambda_2$ = -89.0303 | $p_2$ = 1.04739 |
| saddle 12/1 | x = 0.967758 | y = 0.016383 | $\lambda_2$ = 0.010927 | $p_2$ = -1.08164 |
| | | | $\lambda_1$ = 90.4209 | $p_1$ = 1.11516 |
| saddle 13/1 | x = 0.980929 | y = 0.008812 | $\lambda_1$ = -0.004151 | $p_1$ = -1.08756 |
| | | | $\lambda_2$ = -237.740 | $p_2$ = 1.05840 |
| saddle 13/1 | x = 0.982425 | y = $5.27 \cdot 10^{-4}$ | $\lambda_2$ = 0.004201 | $p_2$ = -1.09821 |
| | | | $\lambda_1$ = 234.912 | $p_1$ = 1.09690 |
| saddle 14/1 | x = 0.964161 | y = 0.037569 | $\lambda_1$ = $-1.57 \cdot 10^{-3}$ | $p_1$ = -1.07814 |
| | | | $\lambda_2$ = -626.945 | $p_2$ = 0.907456 |
| saddle 14/1 | x = 0.987758 | y = 0.006341 | $\lambda_2$ = $1.60 \cdot 10^{-3}$ | $p_2$ = -1.10315 |
| | | | $\lambda_1$ = 613.080 | $p_1$ = 1.10292 |
| saddle 15/1 | x = 0.985749 | y = 0.014030 | $\lambda_1$ = $-5.98 \cdot 10^{-4}$ | $p_1$ = -1.10238 |
| | | | $\lambda_2$ = -1645.47 | $p_2$ = 1.06421 |
| saddle 15/1 | x = 0.993317 | y = $1.99 \cdot 10^{-4}$ | $\lambda_2$ = $6.14 \cdot 10^{-4}$ | $p_2$ = -1.10939 |
| | | | $\lambda_1$ = 1602.73 | $p_1$ = 1.10780 |
| saddle 16/1 | x = 0.995930 | y = $-1.48 \cdot 10^{-4}$ | $\lambda_1$ = $-2.28 \cdot 10^{-4}$ | $p_1$ = -1.11182 |
| | | | $\lambda_2$ = -4310.74 | $p_2$ = 1.11055 |
| focus 17/2 | x = 0.840215 | y = 0.025152 | $\|\lambda\|$ = 0.991532 | $\varphi$ = 2.75364 |
| saddle 17/2 | x = 0.840215 | y = 0.025152 | $\lambda_2$ = 0.991532 | $p_2$ = -1.36264 |
| | | | $\lambda_1$ = 5.80318 | $p_1$ = 2.92603 |
| saddle 19/2 | x = 0.902304 | y = -0.025360 | $\lambda_1$ = -0.019296 | $p_1$ = -1.27626 |
| | | | $\lambda_2$ = -50.8471 | $p_2$ = 0.773605 |
| saddle 19/2 | x = 0.884614 | y = 0.004117 | $\lambda_2$ = 0.015813 | $p_2$ = -1.35142 |
| | | | $\lambda_1$ = 62.0459 | $p_1$ = 1.27572 |
| focus 18/8 | x = 0.849170 | y = 0.059168 | $\|\lambda\|$ = 0.991036 | $\varphi$ = 1.84721 |
| saddle 20/2 | x = 0.891276 | y = 0.005683 | $\lambda_2$ = 0.013570 | $p_2$ = -1.82207 |
| | | | $\lambda_1$ = 72.2312 | $p_1$ = 1.32817 |

Table 4-2. Cycles of the recurrence (4-5), $\mu = 1/8$, $\alpha = -0.001$

| | | | | |
|---|---|---|---|---|
| saddle 5/1 | x = -0.500193 | y = +0.001229 | $\lambda_1$ = -0.614942 | $p_1$ = 11.2728 |
| | | | $\lambda_2$ = -1.61805 | $p_2$ = -12.1934 |
| saddle 5/1 | x = 0.550083 | y = $7.36 \cdot 10^{-4}$ | $\lambda_2$ = 0.244686 | $p_2$ = -0.974598 |
| | | | $\lambda_1$ = 4.06647 | $p_1$ = 0.969667 |
| saddle 6/1 | x = 0.827881 | y = $-5.54 \cdot 10^{-4}$ | $\lambda_1$ = -0.033019 | $p_1$ = -1.05619 |
| | | | $\lambda_2$ = -30.1035 | $p_2$ = 1.05665 |
| saddle 6/1 | x = 0.753465 | y = 0.162888 | $\lambda_2$ = 0.045075 | $p_2$ = -1.07506 |
| | | | $\lambda_1$ = 22.0523 | $p_1$ = 1.25762 |

Table 4-2 continued

| | | | | |
|---|---|---|---|---|
| saddle 7/1 | $x = -0.585830$ | $y = 0.001210$ | $\lambda_1 = -0.007574$ | $p_1 = 2.44727$ |
| | | | $\lambda_2 = -131.107$ | $p_2 = -2.46858$ |
| saddle 7/1 | $x = 0.892102$ | $y = 8.80 \cdot 10^{-5}$ | $\lambda_2 = 0.011224$ | $p_2 = -1.34063$ |
| | | | $\lambda_1 = 88.4700$ | $p_1 = 1.33859$ |
| saddle 8/1 | $x = 0.953810$ | $y = -2.11 \cdot 10^{-4}$ | $\lambda_1 = -0.002052$ | $p_1 = -1.45992$ |
| | | | $\lambda_2 = -483.372$ | $p_2 = 1.45849$ |
| saddle 8/1 | $x = 0.932775$ | $y = 0.054932$ | $\lambda_2 = 0.003088$ | $p_2 = -1.44386$ |
| | | | $\lambda_1 = 321.241$ | $p_1 = 1.41569$ |
| saddle 9/1 | $x = 0.970704$ | $y = 0.024797$ | $\lambda_1 = -5.81 \cdot 10^{-4}$ | $p_1 = -1.52362$ |
| | | | $\lambda_2 = -1704.54$ | $p_2 = 1.49049$ |
| saddle 9/1 | $x = 0.969996$ | $y = 1.06 \cdot 10^{-6}$ | $\lambda_2 = 8.77 \cdot 10^{-4}$ | $p_2 = -1.51325$ |
| | | | $\lambda_1 = 1129.07$ | $p_1 = 1.51139$ |
| focus 10/2 | $x = 0.611292$ | $y = 0.207303$ | $\lvert\lambda\rvert = 0.995010$ | $\varphi = 1.47947$ |
| saddle 11/2 | $x = -0.539507$ | $y = 0.001766$ | $\lambda_1 = -0.009073$ | $p_1 = 3.08432$ |
| | | | $\lambda_2 = -109.010$ | $p_2 = -3.15803$ |
| saddle 11/2 | $x = 0.617277$ | $y = 8.48 \cdot 10^{-4}$ | $\lambda_2 = 0.011766$ | $p_2 = -1.20611$ |
| | | | $\lambda_1 = 84.0605$ | $p_1 = 1.19559$ |
| saddle 12/2 | $x = 0.635477$ | $y = 9.69 \cdot 10^{-4}$ | $\lambda_2 = 0.003140$ | $p_2 = -1.30487$ |
| | | | $\lambda_1 = 314.656$ | $p_1 = 1.28953$ |
| saddle 15/3 | $x = -0.511289$ | $y = 1.13 \cdot 10^{-3}$ | $\lambda_2 = 0.077035$ | $p_2 = 4.25157$ |
| | | | $\lambda_1 = 12.7877$ | $p_1 = -4.33360$ |
| saddle 15/3 | $x = -0.448096$ | $y = 3.30 \cdot 10^{-3}$ | $\lambda_1 = -0.059425$ | $p_1 = 2.49834$ |
| | | | $\lambda_2 = -16.5771$ | $p_2 = -2.51500$ |
| saddle 16/3 | $x = 0.593684$ | $y = -0.026894$ | $\lambda_2 = 3.00 \cdot 10^{-3}$, | $p_2 = -1.04531$ |
| | | | $\lambda_1 = 327.136$ | $p_1 = 1.22235$ |

For a given $\alpha$, the displacement of cycle points k/r is considerable and it increases as k/r decreases. Since the presence of small damping affects the main invariant curves very little, producing merely some asymmetry with respect to $y_n$, Fig. 4-2 and 4-3 do not differ very much from Fig. 3-4 and 3-5, respectively.

Substantial qualitative differences become, however, apparent when a local analysis of the properties of cycles is carried out, because $\alpha < 0$ has converted all "surviving centres" into asymptotically stable foci, i.e. into attractors with finite immediate influence domains. Except for the saddle (1,0), the invariant curves are qualitatively changed. As an illustration, consider the cycle 3/1 and let its points be designated by $(x, y)$, $(x_1, y_1)$ and $(x_2, y_2)$. For simplicity, assume that $\mu$ and $\alpha$ are fixed. Inserting (4-5) into (0-24) and rearranging leads to two simultaneous equations for x and $x_1$ :

$$G_1(x, x_1, \mu, \alpha) = \gamma F(x_1) + \frac{1}{c} x_1 - cx - \frac{\gamma}{c} F(x) = 0 \qquad (4\text{-}8)$$

$$G_2(x, x\ , \mu, \alpha) = \gamma F(g) - c\, x_1 - x = 0 \quad ,$$

where

$$c = 1+\alpha, \qquad \gamma = 2+\alpha \qquad \text{and } g = g(x, x_1) = \frac{\gamma}{c} F(x) - \frac{1}{c} x_1 \ .$$

As mentioned before, these equations cannot be solved explicitly in terms of $\mu$ and $\alpha$ except for $\alpha = 0$ (cf. eq. (3-10)). From the known generating solution for $\alpha = 0$, it

Main invariant invariant curves and distribution of cycles

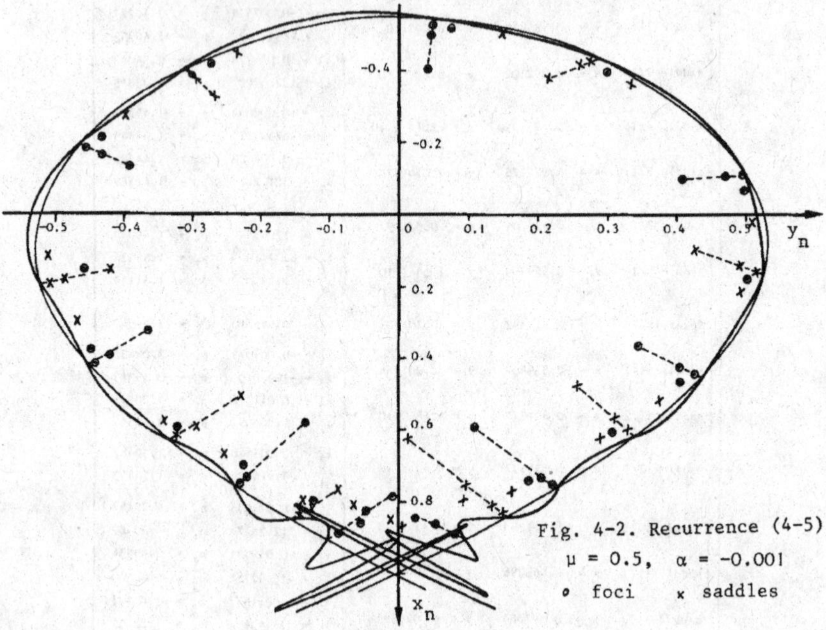


Fig. 4-2. Recurrence (4-5), μ = 0.5, α = -0.001
∘ foci × saddles

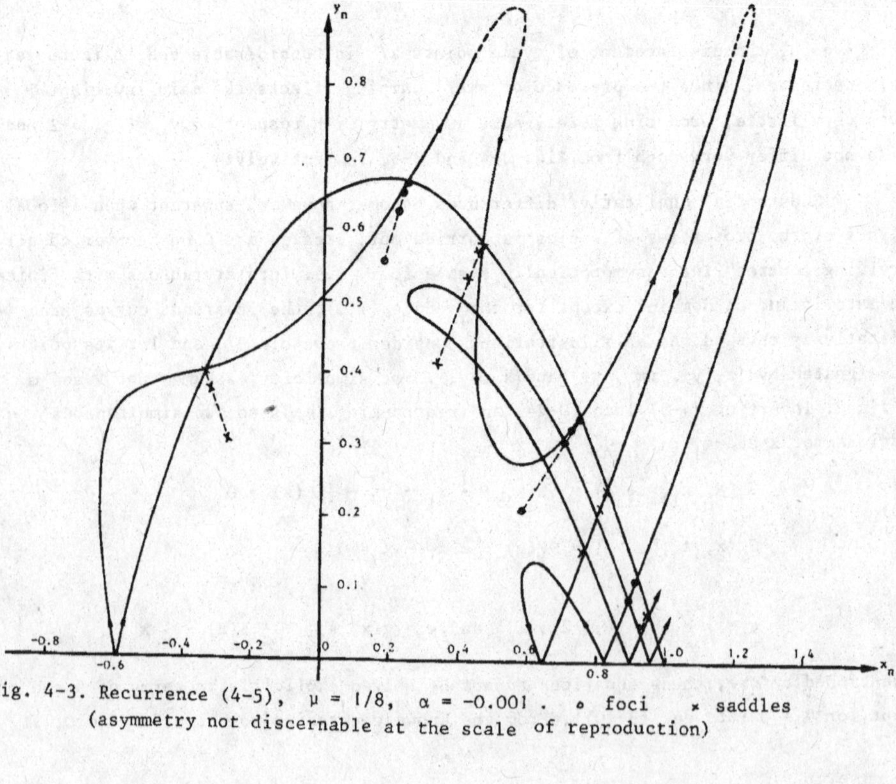


Fig. 4-3. Recurrence (4-5), μ = 1/8, α = -0.001 . ∘ foci × saddles
(asymmetry not discernable at the scale of reproduction)

is in principle possible to construct a series expansion in powers of $\alpha$ for $|\alpha|$ sufficiently small. Consider, for example, the case of a double root of (4-8). Adding to (4-8) the third equation

$$G_3(x, x_1, \mu, \alpha) = \frac{\partial G_1}{\partial x}\cdot\frac{\partial G_2}{\partial x_1} - \frac{\partial G_1}{\partial x_1}\cdot\frac{\partial G_2}{\partial x} = 0 \qquad (4\text{-}9)$$

and considering $\mu$ as a third independent variable, after some manipulations there results

$$\begin{aligned} x &= -\frac{1}{2}(\sqrt{2}-1) + a_0\,\alpha^2 + \dots \quad , \qquad x_1 = \frac{1}{4}(2-\sqrt{2}) + \frac{\sqrt{2}}{b}\,\alpha + \dots \quad , \\ y &= -\frac{\sqrt{2}}{8}\,\alpha + \dots \quad , \qquad \mu = 1 - \sqrt{2} - a_1\,\alpha^2 + \dots \quad , \end{aligned} \qquad (4\text{-}10)$$

where $a_o \approx -0.385$ and $a_1 \approx -0.5166$. The omitted terms are of higher order in $\alpha$. The existence of the double cycle 3/1 is therefore qualitatively unaffected by the presence of damping. For $|\alpha|$ small, there is merely a small change of the bifurcation value $\mu$ and a small shift of the positions of the cycle points. The double cycle turns out to be a node-saddle with one eigenvalue equal to unity.

The third iterate of (4-5), established with the help of an algebraic computer programme, possesses quadratic dominating terms. For $\alpha = -0.001$, the bifurcation value is $\mu \approx -0.414214$ and the explicit form becomes (with (x, y) transfered to (0,0) and all coefficients rounded off to six digits) :

$$\begin{aligned} x_{n+3} &= x_n - 0.998002\,y_n - 0.002813\,x_n^2 - 0.000023\,x_n y_n + 1.41140\,y_n^2 + \dots \\ y_{n+3} &= 0.997003\,y_n - 1.41998\,x_n^2 + 2.82563\,x_n y_n - 2.82563\,y_n^2 + \dots \ . \end{aligned} \qquad (4\text{-}11)$$

An elementary analysis of (4-11) shows that the points of the double cycle 3/1 are crossed by two priviledged invariant curves. The shape of some branches of these curves is shown in Fig. 4-4. The consequent branches spiral towards the main focus

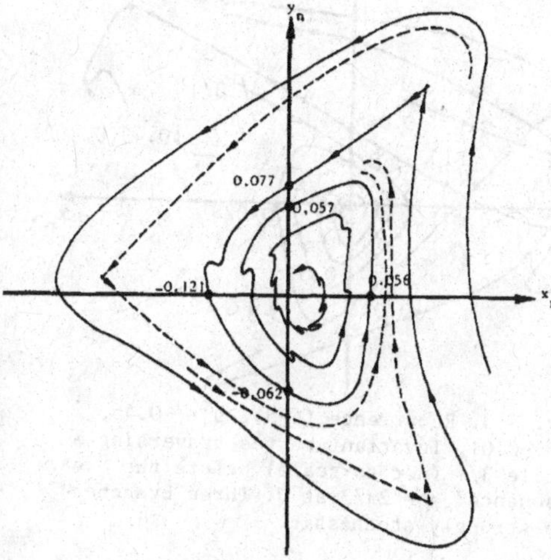


Fig. 4-4. Recurrence (4-5), $\alpha = -0.01$, $\mu = -0.4142671287$. Critical case $\lambda = +1$. Invariant curves traversing a node-saddle 3/1 (not to scale). This node-saddle is a stochastic cusp when $\alpha = 0$.

(0,0), forming many S-loops of the form : , which become longer and flatter as they approach (0,0). Near (0,0) the S-loops just described become so narrow that they cannot be resolved at a given reproduction scale. The consequent branches look therefore more and more like ordinary spirals. The antecedent branches behave differently as they recede from the 3/1 cycle points. Instead of becoming gradually smoother, they develop increasing oscillations and have thus a more and more stochastic appearance.

When $\mu$ decreases below its bifurcation value, the double cycle 3/1 splits into two simple ones : a saddle of type 1 and a node (stable when $\alpha < 0$, and unstable when $\alpha > 0$). The parametric existence interval of the node is very small, it changes soon into a stable focus. Similarly to the case $\alpha = 0$, as $\mu$ decreases, the saddles 3/1 move towards the main fixed point (0,0), but since the latter is a non degenerate focus with a finite influence domain for all admissible $\mu$, the saddles 3/1 cannot merge with (0,0). The distance between (0,0) and the 3/1 saddles passes through a minimum and then increases. The successive positions of one saddle 3/1 are given in Table 4-3 as a function of $\mu$ for $\alpha = -0.01$ and $-0.55 < \mu < -0.45$. When $|\alpha|$ is smaller, the saddles 3/1 come simply closer to (0,0) before moving away again. For example, for $\alpha = -0.0004$, the successive positions are :

$$\begin{array}{llll} & \mu = -0.4995 & x = -0.000702 & y = 0.000068 \\ (4\text{-}12) & \mu = -0.5000 & x = -0.000200 & y = 0.000068 \\ & \mu = -0.5005 & x = -0.000700 & y = 0.000068 \end{array}$$

The invariant curves crossing the saddles 3/1 are shown in Fig. 4-5 and 4-6 for

Table 4-3. Recurrence (4-5). Positions of one 3/1 saddle of type 1 as $\mu$ crosses the exceptional case $\mu = -1/2$, $\alpha = -0.01$.

| $\mu$ | x | y |
|---|---|---|
| -0.450 | 0.046686 | 0.061683 |
| -0.460 | 0.035219 | 0.046629 |
| -0.470 | 0.025378 | 0.033264 |
| -0.480 | 0.016778 | 0.021211 |
| -0.490 | 0.009302 | 0.010381 |
| -0.496 | 0.005738 | 0.004867 |
| -0.498 | 0.005056 | 0.003480 |
| -0.500 | 0.005031 | 0.002542 |
| -0.502 | 0.005903 | 0.002068 |
| -0.504 | 0.007507 | 0.001869 |
| -0.510 | 0.014012 | 0.001704 |
| -0.518 | 0.023533 | 0.001679 |
| -0.530 | 0.037561 | 0.001656 |
| -0.540 | 0.048826 | 0.001643 |
| -0.550 | 0.059697 | 0.001631 |

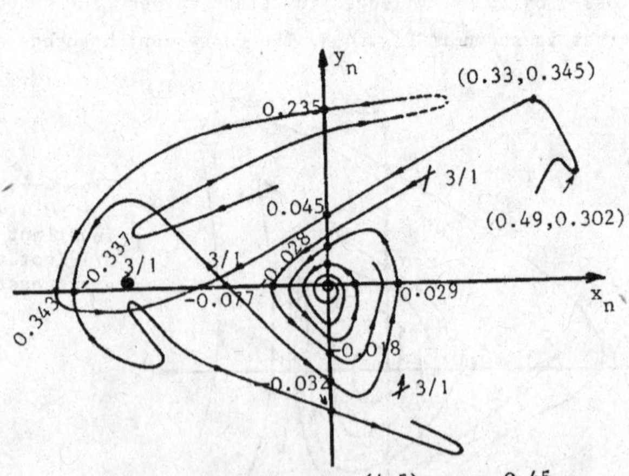


Fig. 4-5. Recurrence (4-5), $\mu = -0.45$, $\alpha = -0.01$. Invariant curves traversing a saddle 3/1 (not to scale) before the "weak resonance" $\varphi = 2\pi/3$ at 0. Three branches are strongly stochastic.

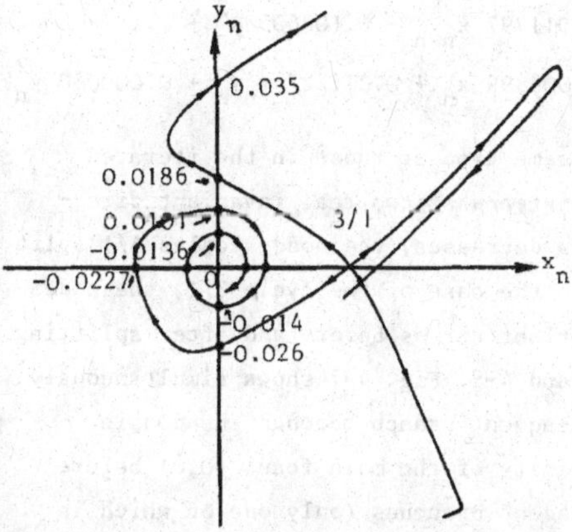


Fig. 4-6. Recurrence (4-5), $\mu = 0.55$, $\alpha = -0.01$. Invariant curves traversing a saddle 3/1 (not to scale) after the "weak resonance" $\varphi = 2\pi/3$ at O. Three branches are strongly stochastic.

$\mu = -0.45$ and $\mu = -0.55$, respectively. One consequent branch emerging from a saddle spirals smoothly towards the main focus (0,0), whereas the other consequent branch, as well as the two antecedent branches, recede stochastically with the formation of numerous homoclinic points. The exceptional case $\varphi = 2\pi/3$ of the main focus (0,0) manifests itself therefore not by a singularity change at (0,0) but by a very pronounced reduction of the size of its influence domain. The latter is surrounded by a strongly stochastic region.

For the cycle 4/1, the two equations analoguous to (4-8) are

$$\begin{aligned} G_1(x,x_2,\mu,\alpha) &= (1-\mu)\left[F^2(x)+F^2(x_2)\right]+\mu\left[F(x)+F(x_2)\right]+\frac{2}{\alpha^2}(1-\mu)(1+\alpha)\left[F(x_2)-F(x)\right]^2+ \\ &\qquad -x_2-x = 0\,, \\ G_2(x,x_2,\mu,\alpha) &= (1-\mu)F^2(x)+\left[\mu+(1-\mu)\frac{1+\alpha}{\alpha}\big(F(x_2)-F(x)\big)\,F(x) + \frac{1+\alpha}{\alpha}\big(F(x_2)-F(x)\big)\right]\cdot \\ &\qquad \cdot\left[\mu+\frac{1}{\alpha}(1-\mu)(1+\alpha)\big(F(x_2)-F(x)\big)\right] - \frac{x_2+(1+\alpha)x}{2+\alpha} = 0\,. \end{aligned} \tag{4-13}$$

When $\alpha \neq 0$, the cycles 4/1 do not appear from the domain of complex numbers by splitting off from the multiple fixed point (0,0), like it did when $\alpha = 0$, but by producing four distinct double roots of (4-13) near (0,0). For $\alpha = -0.001$, for example, the bifurcation value of $\mu$ and the coordinates of one double root are

$$\mu = -0.0000949280 \qquad x = 0.0635315579 \qquad y = -0.0039871388\ . \tag{4-14}$$

Equation (4-13) and (4-9) can also be solved by means of series expansion in terms of $\alpha$, but these series are much more complex than in the case of the cycle 3/1. The first terms are :

$$\begin{aligned} x &= +\left(\frac{\alpha}{4}\right)^{\frac{1}{3}}.\ \text{sign}\ \alpha + \ldots\ , \qquad y = -\left(\frac{\alpha}{4}\right)^{\frac{2}{3}}.\ \text{sign}\ \alpha + \ldots\ , \\ \mu &= -\frac{\alpha}{2}\ .\ \left(\frac{\alpha}{4}\right)^{\frac{1}{3}} + \ldots\ . \end{aligned} \tag{4-15}$$

Like the cycle 3/1, the cycle 4/1 is (at its appearance) a node-saddle with one eigenvalue equal to unity. The iterated recurrence with (x, y) transfered to (0,0), $\alpha = -0.0001$ and the bifurcation value $\mu = -0.0000002$ (determined by means of an algebraic programme) is

$$
(4\text{-}16)\quad
\begin{aligned}
x_{n+4} &= 0.999980\,x_n - 0.000340\,x_n^2 + 0.011797\,x_n y_n - 0.108605\,y_n^2 + \dots \\
y_{n+4} &= 0.001474\,x_n + 0.999980\,y_n - 0.005899\,x_n^2 + 0.217227\,x_n y_n + 0.000080\,y_n^2 .
\end{aligned}
$$

It possesses quadratic dominating terms of the same type as those in the iterated recurrence (4-11). This property implies the existence of two real invariant directions at each double root of the cycle 4/1. As $\mu$ decreases, the node-saddles 4/1 split into stable nodes and saddles of type 1. Like in the case of the cycle 3/1, the nodes 4/1 turn rapidly into stable foci. Typical invariant curves before and after splitting of the node-saddles 4/1 are shown in Figs. 4-7 and 4-8. Fig. 4-7 shows simultaneously the positions of the cycle points 21/5. The consequent branch through the double cycle 4/1 proceeds monotonically to a small vicinity of the main focus (0,0) before starting to spiral around the latter. The antecedent branches (only one of which is shown in Fig. 4-7) form at first rather smooth outgoing spirals, but after about seven turns around the group of four node-saddles, they become "irregular" by developing gradually increasing S-loops. These irregualr parts appear precisely when the antecedent-branches 4/1 traverse the region is which the points of the two cycles 21/5 are located. Since one set of consequent branches passing through the 21/5 saddles of type 1 spirals inwards, it traverses the 4/1 structure and then approaches the main focus (0,0). Continuity implies therefore that the invariant curves originating at the 4/1 and 21/5 saddles intersect, forming heteroclinic points. The exis-

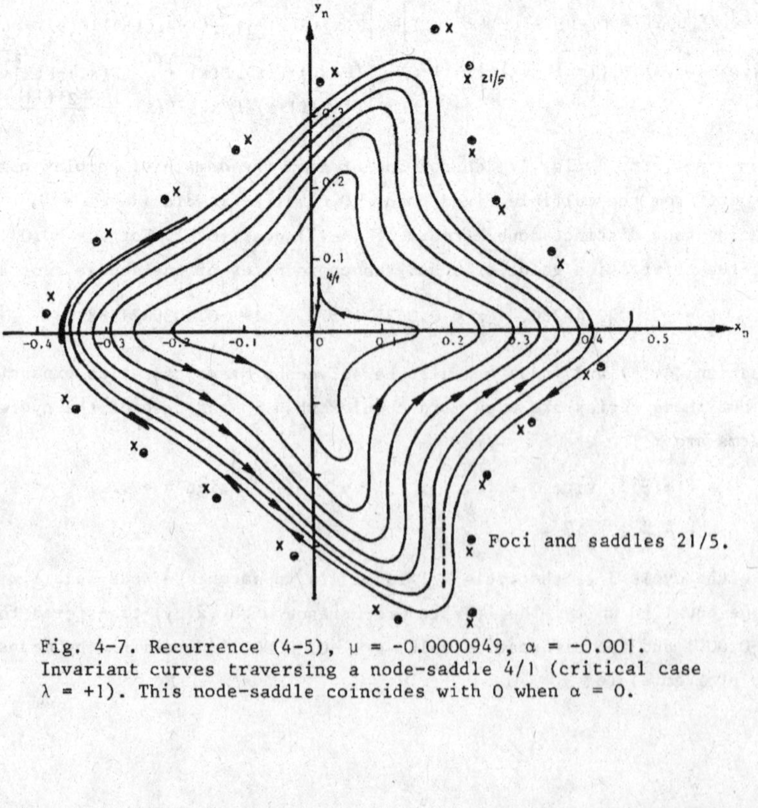


Fig. 4-7. Recurrence (4-5), $\mu = -0.0000949$, $\alpha = -0.001$.
Invariant curves traversing a node-saddle 4/1 (critical case $\lambda = +1$). This node-saddle coincides with 0 when $\alpha = 0$.

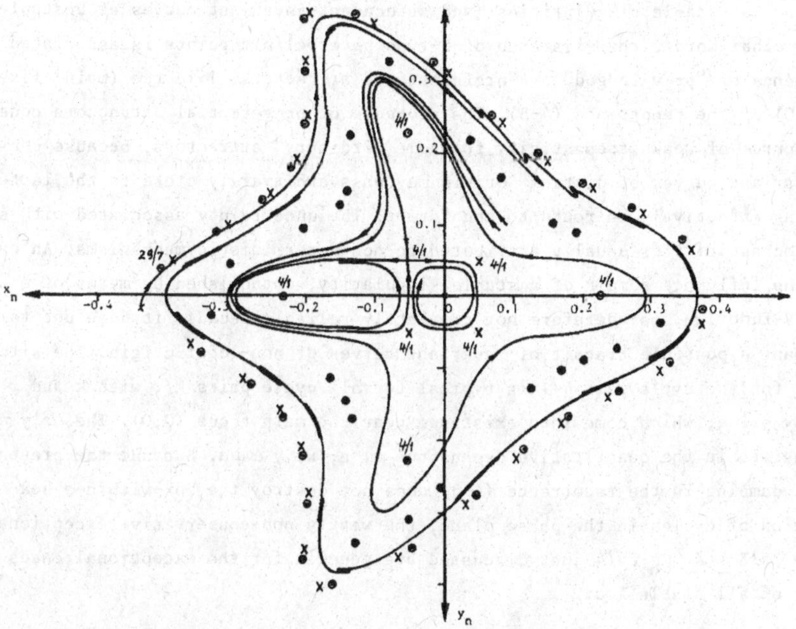


Fig. 4-8. Recurrence (4-5) $\mu = -0.005$, $\alpha = -0.0002$. Invariant curves traversing a saddle 4/1 and a saddle 29/7. • Points on the 29/7 consequent branch after many iterations.

tence of these heteroclinic points does not generate any stochasticity beyond the presence of S-loops on invariant curves.

Fig. 4-8 illustrates the effect of a small damping on the former island structure 4/1. When $\alpha < 0$, one consequent branch emanating from the saddles 4/1 of type 1 approaches the stable focus 4/1, and the other spirals towards the main focus (0,0). The antecedent-branches spiral outwards and enclose both the four 4/1 saddle structure and the four 4/1 foci. Since the cycle pair 29/7 is closer to the 4/1 structure than the cycle pair 21/5 mentioned before, the positions of its points are also shown in Fig. 4-8. One consequent branch emanating from the 29/7 saddles (of type 1) approaches the closest stable 29/7 focus, and the other spirals inwards with the formation of rather pronounced S-loops. As these loops become longer and flatter (cf. Fig. 4-8), the consequent-invariant curve becomes practically indistinguishable from a smooth spiral. It is this apparently smooth part of the 29/7 invariant curve which traverses the 4/1 structure and approaches the main focus (0,0). Again heteroclinic points are formed without any evidence of an accompanying stochasticity. Non-conservative recurrences admit therefore heteroclinic points of an additional type (N-type, for conciseness) which is qualitatively different from that of conservative recurrences. In conservative recurrences, heteroclinic points characterize the overlapping (and thus the destruction) of neighbouring island structures, whereas the N-type heteroclinic points characterize the tendency of consequent invariant curves to terminate at stable singularities (and antecedent-invariant curves at unstable ones). In other words, the existence of N-type heteroclinic points is associated with the existence of "priviledged" or "preferential" attractors, like the (main) fixed point (0,0) of the recurrence (4-5). The presence of preferential attractors constitutes a source of weak stochasticity for some "ordinary" attractors, because consequent invariant curves of distant origin may pass arbitrarily close to the latter, while being effectively en route to the former. The uncertainty associated with such a weak stochasticity is usually attributed to non-deterministic mechanisms. An estimate of the influence domain of a stable singularity, established by means of a Liapunov V-function, is therefore not completely reliable because it does not take into account a possible transit of invariant curves of non-local origin. The situation described for the cycle pair 4/1 is typical for all cycle pairs k/r with k and r relatively prime, which come into existence near the main focus (0,0). The only difference consists in the quantitative dependence on $\mu$, x, y and $\alpha$. Because the presence of a weak damping in the recurrence (4-5) does not destroy the box-within-a-box distribution of cycles in the phase plane, the weakly non-conservative exceptional cases $\varphi = 2\pi/3$ and $\varphi = 2\pi/4$ just discussed are generic for the exceptional cases $\varphi = 2\pi r/k$ of all stable foci.

An inspection of Tables (4-1) and (4-2) suggests that the stochasticity generating bifurcation centre → saddle of type 3 is not markedly affected by the presence of a weak damping. The role of the centre is taken over by a stable focus, which changes quickly (i.e. after traversal of a very small parametric distance) into a stable node, before changing definitely into a saddle of type 3. The bifurcated cycle of doubled order and rotation number is at first a stable node and then a stable focus. As an illustration, let $\alpha = -0.001$ and consider the cycle 4/1 :

| | | | | | |
|---|---|---|---|---|---|
| | $\mu = -0.1034$ | $x = 0.532709$ | $y = -0.005265$ | $\lvert\lambda\rvert = 0.9801$ | $\varphi = 3.062$ |
| (4-17) | $\mu = -0.1036$ | $x = 0.532997$ | $y = -0.005262$ | $\lambda_1 = -0.9035$ | $\lambda_2 = -1.063$ |
| | $\mu = -0.1040$ | $x = 0.533570$ | $y = -0.005255$ | $\lambda_1 = -0.8184$ | $\lambda_2 = -1.174$, |

which gives rise to a cycle 8/2, one point of which is given by

| | | | | | |
|---|---|---|---|---|---|
| | $\mu = -0.1034$ | x = none | y = none | | |
| (4-18) | $\mu = -0.1036$ | $x = 0.538170$ | $y = -0.005194$ | $\lvert\lambda\rvert = 0.9606$ | $\varphi = 0.2198$ |
| | $\mu = -0.1040$ | $x = 0.520847$ | $y = -0.005434$ | $\lvert\lambda\rvert = 0.9606$ | $\varphi = 0.5107$. |

The parametric existence interval of the nodes 4/1 and 8/2 is smaller than $\Delta\mu = 0.0002$. As expected from the conservative case, the cycle 8/2 undergoes the same bifurcation with an increase of the strength of the non-linearity :

| | | | | | |
|---|---|---|---|---|---|
| (4-19) | $\mu = -0.111$ | $x = 0.486902$ | $y = -0.006062$ | $\lvert\lambda\rvert = 0.9606$ | $\varphi = 2.694$ |
| | $\mu = -0.112$ | $x = 0.484016$ | $y = -0.006128$ | $\lambda_1 = -0.5600$ | $\lambda_2 = -1.648$ |

generating a node and then a focus 16/4, etc.

A weak damping has also no marked effect on the increase of the spreading of invariant curves, traversing points of a cycle saddle k/r, as the strength of non linearity $\delta = 1-\mu$ (or the distance from (0,0)) increases. This spreading leads to the appearance of homoclinic and (ordinary) heteroclinic points, and an accompanying stochasticity, like in the case of a conservative recurrence. For an unbounded non linearity, the resulting region of diffusion and stochastic instability form a part of the influence domain of the attractor at infinity.

In the case of bounded quadratic recurrence with a linear damping term

(4-20)
$$x_{n+1} = (1+\alpha)y_n + F(x_n), \quad y_{n+1} = -x_n + F(x_{n+1})$$
$$F(x) = \mu x + 2(1-\mu)x^2 / (1+x^2), \quad -1 < \mu < 1, \quad -1 \ll \alpha < 0,$$

the rough features of the phase portrait happen to be also little changed. As expected, cycles with an increasing ratio k/r are only slightly shifted in position, and the majority of cycles with an decreasing ratio k/r cease to exist. The presence of a weak damping affects qualitatively the nature of the degenerate main fixed point (1,0) : it is no longer a cusp but a node-saddle. The two main invariant curves passing through (1,0) are described locally by the two series

(4-21)
$$y_n = \frac{1-\mu}{\alpha}(x_n - 1)^2 + \ldots$$

and

(4-22)
$$y_n = \frac{\alpha}{1+\alpha}(x_n - 1) - (1-\mu)\frac{2(1+\alpha)^2 + \alpha^2(1+\alpha) - \alpha}{2\alpha(2+\alpha)}(x_n - 1)^2 + \ldots .$$

The global shape of these curves is almost the same as for $\alpha = 0$, except for the presence of a slight asymmetry. The influence of the damping term on mixed cycles is the same as on other cycles : the centre is converted into a stable focus and the parametric existence interval of the cycle pair is shortened. The evolution of the mixed saddle 18/2 is illustrated in Table 4-4. As $\alpha$ decreases, the positions of the

| $+\alpha.10^4$ | x | y | $\lambda_1$ |
|---|---|---|---|
| 0 | 0.266573 | 0.488625 | 5.631 |
| -1 | 0.266715 | 0.488683 | 5.544 |
| -2 | 0.266835 | 0.488750 | 5.282 |
| -3 | 0.266928 | 0.488829 | 4.791 |
| -4 | 0.266975 | 0.488927 | 3.899 |
| -4.5 | 0.266959 | 0.488994 | 3.066 |
| -4.81 | 0.266871 | 0.489069 | 1.618 |
| -4.8188 | 0.266847 | 0.489080 | 1.117 |
| -4.82 | none | none | |

Table 4-4. Recurrence (4-20). Evolution of the mixed saddle 18/2 with an increase of damping, $\mu = 0.1254268$

cycle points change slowly, but the consequent eigenvalue $\lambda_1$ changes rapidly, till the bifurcation $\lambda_1 = +1$ is reached ($\mu \simeq 0.1254268$, $\alpha \simeq -0.000481$). The invariant curve structure is almost the same as in the absence of damping. The invariant curves crossing mixed saddles of type 1 and type 3 follow the outline of the main invariant curves (Fig. 4-9). In the case of saddles of type 3 on a rough scale, the invariant curves are the same as in the case of saddles of type 1, but on a finer scale (not visible in Fig. 4-9), they exhibit very long and very narrow S-loops. Heteroclinic points associated with preferential attractors are also produced.

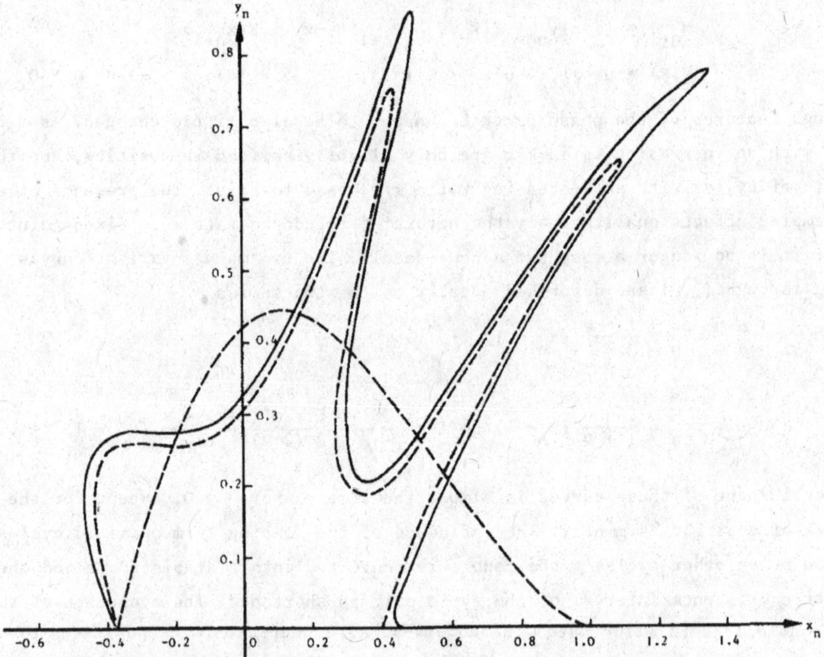


Fig. 4-9. Recurrence (4-20), $\mu = 0.1254268$, $\alpha = -0.0001$. —— Main invariant curve, – – – consequent invariant curve traversing a mixed saddle 18/2 of type 1. The consequent invariant curve crossing the saddle 18/2 of type 3 (too close to – – – to be resolved) has the same shape, but it is considerably more stochastic, i.e. it consists essentially of long and flat S-loops.

Consider now the quadratic recurrence with a small non linear damping term

$$\text{(4-23)}\qquad \begin{aligned} &x_{n+1} = y_n + \beta y_n^3 + F(x_n), \qquad && y_{n+1} = -x_n + F(x_{n+1}) \quad , \\ &F(x) = \mu x + (1-\mu)x^2, && -1 < \mu < 1, \qquad |\beta| \ll 1 \end{aligned}$$

The main fixed point (1,0) of (4-23) is still a saddle of type 1. A degeneration occurs at the main fixed point (0,0) : in linear approximation it is a centre, whereas for $\beta \neq 0$ it is a stable focus when $\beta < 0$ and un unstable one when $\beta > 0$, except for $\mu = -\frac{1}{2}$. In the latter case (0,0) is a six-branch saddle. The inverse recurrence of (4-23) is single-valued when $\beta > 0$, and single or triple-valued when $\beta < 0$. In both cases explicit expressions for antecedents can be obtained in terms of elementary functions, involving the well known roots of a cubic polynomial. A comparison with the recurrences (3-6) and (4-5) is still possible because in the phase plane regions depicted in Figs. 3-4 to 5 and 4-2 to 3, the recurrence (4-23) involves only one inverse branch.

Like in the case of the linearly damped recurrence (4-5) no cycles are known analytically. Numerically it is of course possible to compile a list like that given in Tables 4-1 and 4-2, and this has been done. The properties of all cycles found so far are similar to those of the recurrence (4-5) : i) each cycle possesses an unambiguously defined rotation number r, ii) there exists an infinity of cycles with an increasing ratio k/r, iii) cycle sets with a decreasing ratio k/r terminate suddenly. The shapes of the invariant curves passing through the main saddle (1,0) are about the same as those shown in Fig. 4-2 and 4-3 , with roughly the same amount of asymmetry. Points of cycles whose k/r increase converge to the main homoclinic points, as expected. There appears therefore little visual difference between the phase portraits of the recurrences (4-5) and (4-23). This conclusion is confirmed by a local analysis of some cycles. Consider in fact the cycle 3/1. In the case of a double root it is possible to reduce the seven simultaneous equations, relating the unknowns $\mu$ and the coordinated $(x, y), (x_1, y_1), (x_2, y_2)$ to only three, but their structure is much more complicated than that of the equations (4-8). After some rather lenghthy manipulations, the following series representation has been obtained :

$$x = -\frac{1}{2}(\sqrt{2}-1) + a_o\beta^2 + \ldots \quad , \qquad y = a_1\beta + \ldots \quad ,$$
(4-24)
$$x_1 = \frac{1}{4}(2-\sqrt{2}) + a_1\beta + \ldots \quad , \qquad \mu = 1-\sqrt{2} + a_2\beta^2 + \ldots \quad ,$$

where $a_o \simeq 0.002254$, $a_1 \simeq -0.005524$, $a_2 \simeq 0.017457$ and $|\beta| \ll 1$. The bifurcation value of $\mu$ and the positions of the cycle points happen to be only slightly shifted with respect to the conservative case. For $\beta = 0.001$, the bifurcation occurs at $\mu \simeq -0.414214$ and the iterated recurrence at (x, y), shifted to (0,0), is

(4-25)
$$x_{n+3} = x_n - y_n - 0.000442\, y_n^2 + 0.001060\, x_n y_n + 1.41412\, x_n^2 + \ldots$$
$$y_{n+3} = 1.00012 y_n - 1.41364 y_n^2 + 2.82719 x_n y_n - 2.82878 x_n^2 + \ldots$$

As before, the point (x, y) is a node-saddle with one eigenvalue equal to unity, splitting into a node and a saddle of type 1. For $\beta = 0.001$ and $\mu = -0.415$, the invariant curves traversing a saddle 3/1 are shown in Fig. 4-10. The invariant curves are quite regular ; one antecedent-branch spirals towards the main focus (0,0), another antecedent-branch towards the focus 3/1, and the two consequent branches surround the total 3/1 cycle-structure. Outside of the 3/1 cycle structure, there exists a set of cycle-pairs k/r = (3m+1)/m, m = 2, 3, ... (Fig. 4-11). The lowest k/r cycle of this set which could be determined numerically for $\beta = 0.001$ is 40/13. The location of the 40/13 cycle points are also shown in Fig. 4-10, together with one consequent-invariant curve traversing a 40/13 saddle. N-type heteroclinic points occur and

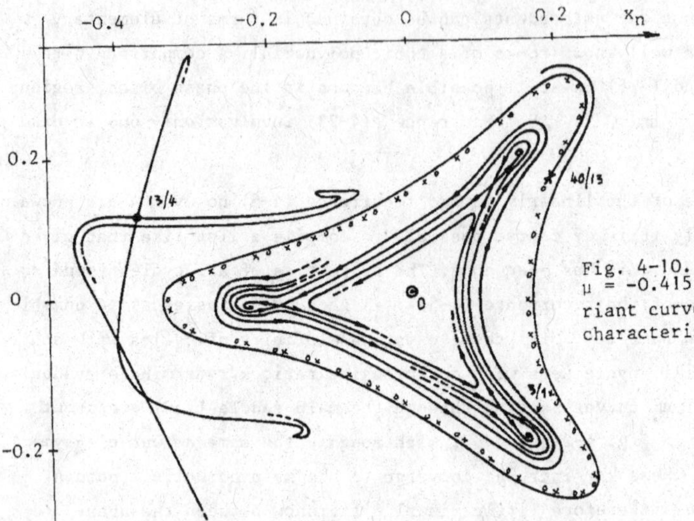


Fig. 4-10. Recurrence (4-23), $\mu = -0.415$, $\beta = 0.001$. Some invariant curves (almost to scale) characterizing the 3/1 cycle pair.

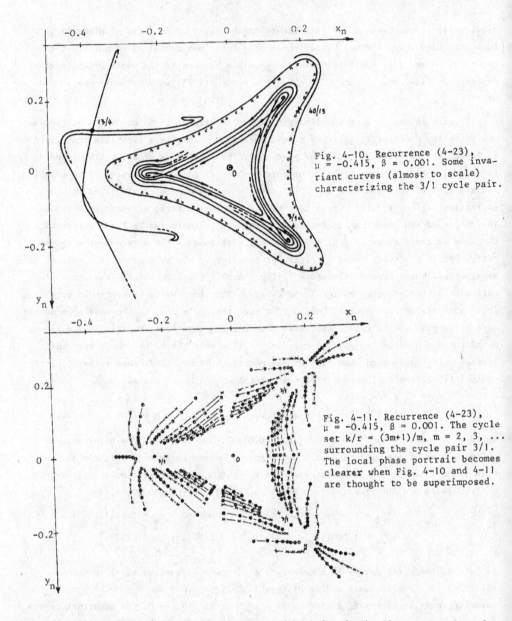

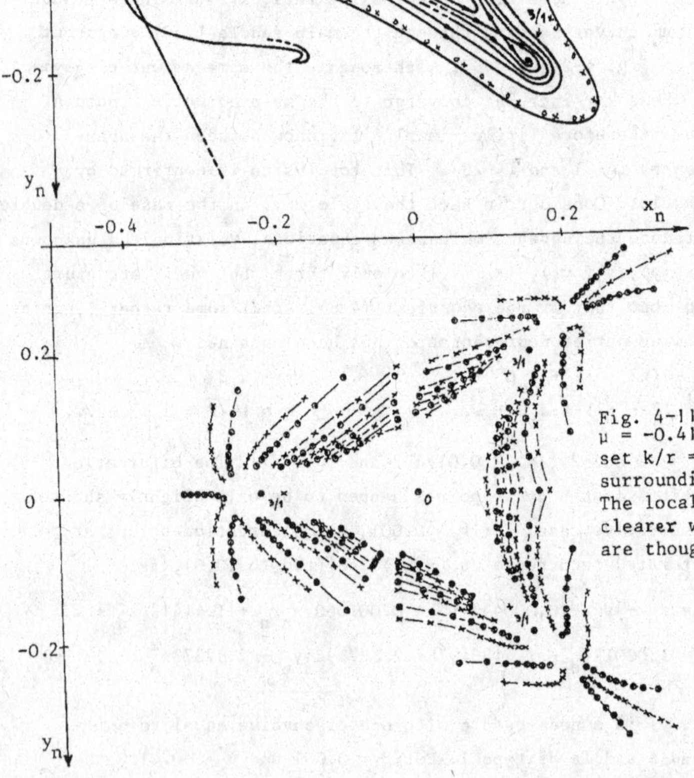


Fig. 4-11. Recurrence (4-23), $\mu = -0.415$, $\beta = 0.001$. The cycle set $k/r = (3m+1)/m$, $m = 2, 3, \ldots$ surrounding the cycle pair 3/1. The local phase portrait becomes clearer when Fig. 4-10 and 4-11 are thought to be superimposed.

confirm the preferential role played by the fixed point (0,0). Since the cycles of the set $k/r = (3m+1)/m$, $m > 2$, recede from the 3/1 structure as m decreases, homoclinic points and stochasticity are generated by the increased spreading of invariant curves. Moreover, at some critical m, the antecedent curves cease to approach the main focus (0,0), and diverge to infinity instead. For $\beta = 0.001$, this change of preferential fixed point takes place in the interval $4 < m < 13$ ; for the cycle 40/13 it is already quite pronounced.

The exceptional case $\varphi = 2\pi/3$ at (0,0) of the recurrence (4-23) is different from that of the recurrence (4-5). The corresponding iterated recurrence for $\mu = -1/2$ is

$$\text{(4-26)}\quad \begin{aligned} x_{n+3} &= x_n + 6x_n y_n + \frac{27}{4}(1+\beta/8)x_n^3 + \frac{9}{2}x_n^2 y_n + 9(1+\beta/8)x_n y_n^2 - \frac{1}{2}(36+7\beta)y_n^3 + \dots , \\ y_{n+3} &= y_n + \frac{9}{4}x_n^2 - 3y_n^2 + \frac{9}{8}x_n^3 + \frac{27}{4}(1+\beta/8)x_n^2 y_n - 9(1+\beta/16)x_n y_n^2 + \frac{1}{8}(72+23\beta)y_n^3 + \dots \end{aligned}$$

The main fixed point (0,0) is traversed by three invariant curves ; it is thus a six-branch saddle, like in the conservative case.

The appearance of the cycles 4/1, i.e. the weakly exceptional case $\varphi = 2\pi/4$ at (0,0) can be described by two simultaneous algebraic equations, but no analytic exact or approximate solution of these equations could be found so far. Numerical studies suggest that except for $\varphi = 2\pi/3$, all exceptional cases $\varphi = 2\pi r/k$ at (0,0) are similar to those of the linearly damped recurrence (4-5). Because the box-within-a-box phase plane structure persists, the same type of exceptional cases occurs at all foci.

## 4.2 Effect of non-unique antecedents

Consider the particular cubic recurrence

$$\text{(4-27)}\quad \begin{aligned} & x_{n+1} = y_n + \beta y_n^3 + F(x_n), \qquad y_{n+1} = -x_n + F(x_{n+1}) \\ & F(x) = \mu x + 2(1-\mu)\, x^3/(1+x^2), \qquad -1 < \mu < 1, \quad -1 \ll \beta < 0 \end{aligned}$$

In the phase plane region $|x_n| < 1.5$, $|y_n| < 1$, its properties are very similar to that of the recurrence (3-59), except for the absence of some cycles and the slight asymmetry of invariant curves produced by the non-conservative term $\beta y_n^3$. The main fixed points (1,0), (-1,0) are saddles of type 1, and (0,0) is a stable focus. Due to the absence of quadratic terms there are no invariant directions at (0,0) when $\mu = -1/2$, i.e. the exceptional case $\varphi = 2\pi/3$ is not different from other exceptional cases. For $\mu = 0.5$, and $\beta = -0.001$, the main invariant curves, and the positions of some cycles, are shown in Fig. 4-12. Compared to the case $\beta = 0$, the main feature of Fig. 4-12 is the absence of cycles like 7/1, 13/2, 15/2, ..., verifying the inequality $6 < k/r \leq 8$. Since these cycles have a low ratio k/r, their absence is a normal consequence of the presence of damping.

A study of the invariant curves traversing the saddles 8/1, 9/1 and 10/1 shows that the main focus (0,0) is a preferential attractor (Fig. 4-13). Whereas one consequent branch originating at a saddle 8/1 or 9/1 goes to its own focus, the other consequent branches go to (0,0). One consequent branch emanating from each of the 9/1 saddles traverses therefore the 8/1 invariant curve structure. As expected the resulting N-type heteroclinic points do not generate any stochasticity. The invariant curve structure of the 10/1 saddles is slightly different. One consequent branch goes to (0,0), traversing both the 9/1 and 8/1 structures, but the other consequent branch

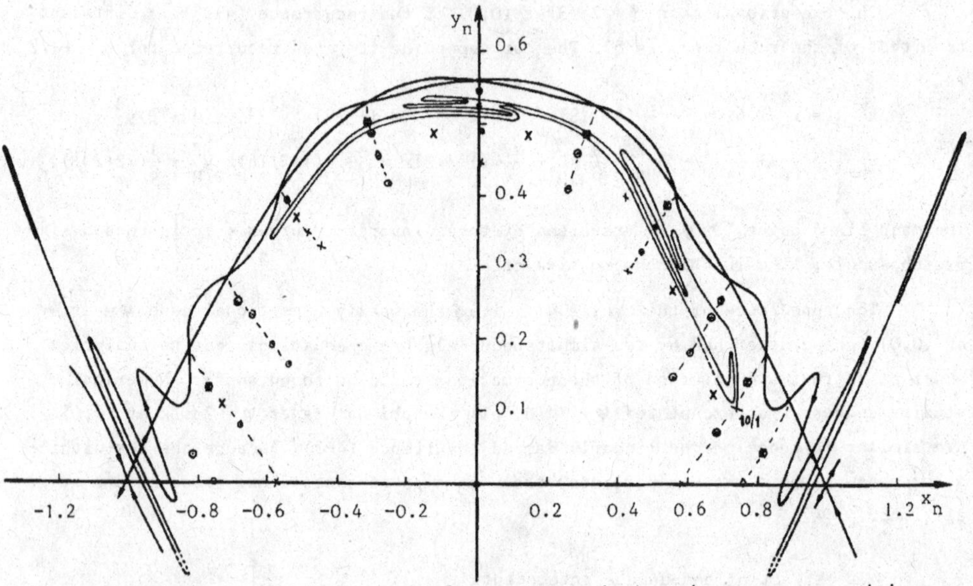


Fig. 4-12. Recurrence (4-27), μ = 0.5, β = -0.001. Main invariant curves, positions of some cycles and one stochastic consequent invariant curve traversing a 10/1 saddle of type 1 (S-loops somewhat simplified).

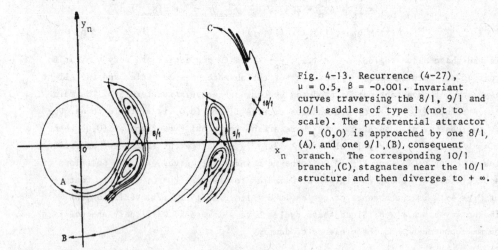

Fig. 4-13. Recurrence (4-27), μ = 0.5, β = -0.001. Invariant curves traversing the 8/1, 9/1 and 10/1 saddles of type 1 (not to scale). The preferential attractor O = (0,0) is approached by one 8/1, (A), and one 9/1, (B), consequent branch. The corresponding 10/1 branch, (C), stagnates near the 10/1 structure and then diverges to + ∞.

overshoots its own focus, circles around the 10/1 structure creating homoclinic and ordinary heteroclinic points, and diverges finally to infinity by following a main invariant curve. The situation just described is generic, i.e. it occurs for other cycles of a high ratio k/r. For example, when μ = 0.464 and β = -0.001, one set of consequent branches of the saddles 6/1, 7/1 and 8/1 spirals towards (0,0), whereas the curves of the other set spiral towards their own foci. A change of "final destination" occurs whenever the larger eigenvalue of a saddle of type 1 exceeds a certain threshold. This threshold is reached while the companion cycle is still a stable focus, i.e. before the bifurcation of the focus into a node and a stochasticity generating saddle of type 3 has taken place.

When the study of the phase portrait of the recurrence (4-27) is not restricted to a region just slightly larger than the rectangle $|x_n| < 1$, $|y_n| < 1$, the damping term $\beta y^3$ becomes dominant, and there appear singular features which are not found in two-dimensional recurrences with unique antecedents. The simplest new possibility is the passage of an invariant curve through the excess antecedent of one of its points. The result is a point of self-intersection. A study of invariant curves of the recurrence (4-27) shows that such a situation does indeed occur [G16]. An exterior main consequent branch $\mathcal{C}$ of (4-27) is shown in Fig. 4-14. $\mathcal{C}$ is self-intersecting at the point M = (68.7, 64.4). The point M possesses three distinct antecedents $M_1 \approx (38.8, 23.4)$, $M_2 \approx (38.8, -36.0)$ and $M_3 \approx (38.8, 12.6)$, the first two being also points of $\mathcal{C}$. The loop of $\mathcal{C}$ responsible for M increases in size as $\mu$ decreases, $\beta$ being fixed. For $\mu = 1/8$, it is so wide that the presence of M can be established by relatively low-accuracy numerical methods. Because of its easy detectability in a rather large interval of $\mu$, the existence of the self-intersection point M is not a fortuitous incident, but rather a characteristic feature of the full phase portrait of (4-27).

Consider now the exterior main antecedent branch $\mathcal{A}$ of (4-27). In contrast to $\mathcal{C}$, which is simple, in the sense that all consequents of a point $(x_n, y_n)$ on $\mathcal{C}$ are also on $\mathcal{C}$, the antecedent invariant curve $\mathcal{A}$ is multiple (cf. Fig. 4-15). In fact, every point of $\mathcal{A}$ possesses three distinct antecedents near (1,0). The main antecedent curve $\mathcal{A}$ coexists therefore with two excess antecedent curves $\mathcal{A}_1$ and $\mathcal{A}_2$, which start at the rather distant excess antecedent points $A_1 \approx (1, 31.6)$ and $A_2 \approx (1, -31.6)$ of (1,0), respectively. As $\mathcal{A}$ recedes from (1,0), it approaches the excess antecedent curve $\mathcal{A}_2$ and attains the latter at the point N ≈ (11.2, 18.3), which is

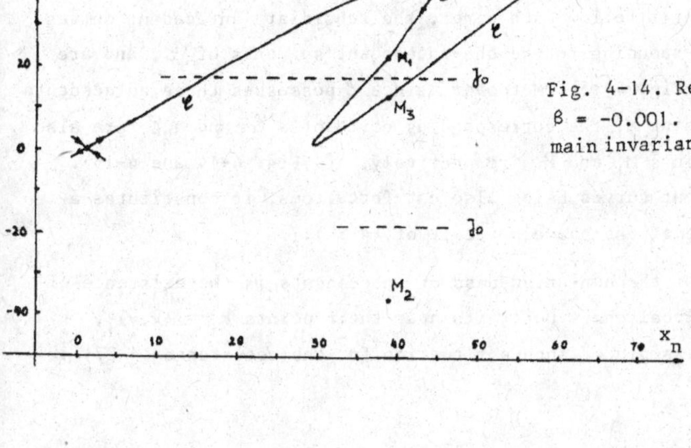

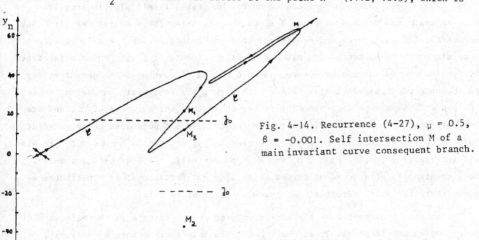

Fig. 4-14. Recurrence (4-27), $\mu = 0.5$, $\beta = -0.001$. Self intersection M of a main invariant curve consequent branch.

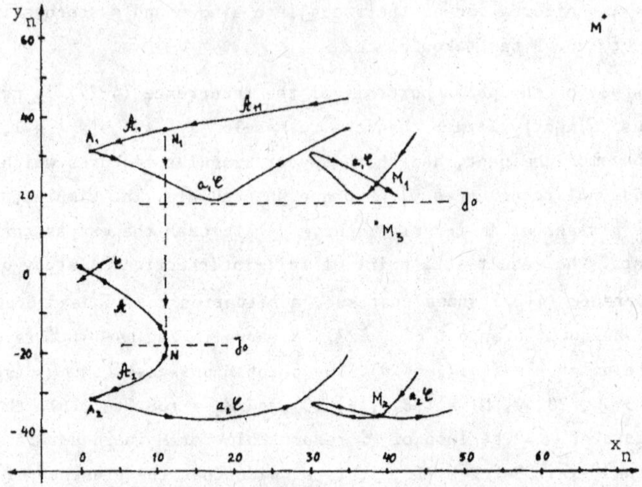


Fig. 4-15. Recurrence (4-27), $\mu = 0.5$, $\beta = -0.001$. Discontinuities and self-intersections of antecedent invariant curves. N : point of alternance. $N_1$ : point of alternance without arrow reversal. $A_1$, $A_2$ : excess antecedents of the main saddle (1,0).

not a fixed point of (4-27). Since N is located on the critical curve $J_o$, it is a point of alternance. Let $N_1$ be a point on the excess antecedent curve $\mathcal{A}_1$ corresponding to the point N. From Fig. 4-15, it is seen that $N_1$ is not located on the critical curve $J_o$, it is therefore not a point of alternance like N. Beyond $N_1$, the three antecedent curves $\mathcal{A}$, $\mathcal{A}_1$, $\mathcal{A}_2$ have a unique continuation : the antecedent curve $\mathcal{A}_{11}$, which joins smoothly with $\mathcal{A}_1$ at $N_1$. The main antecedent curve $\mathcal{A}$ of (4-27) is therefore only piecewise continuous. It consists of a continuous segment linking (1,0) and N, followed by a jump from N to $N_1$, and then by the continuous segment $\mathcal{A}_{11}$, etc.

The discontinuity of antecedent curves manifests itself also indirectly. Consider again the consequent curve $\mathcal{C}$ shown in Fig. 4-14. It is easily verified that some parts of $\mathcal{C}$ have three and others only one antecedent. Since the main antecedents of points of $\mathcal{C}$ are by definition also on $\mathcal{C}$, the segments of $\mathcal{C}$ with three antecedents involve two excess antecedent curves, say $a_1\mathcal{C}$ and $a_2\mathcal{C}$ . Whenever three-antecedent and one-antecedent segments of $\mathcal{C}$ follow each other, the subsidiary antecedent curves $a_1\mathcal{C}$ and $a_2\mathcal{C}$ have gaps corresponding to the one-antecedent segments of $\mathcal{C}$, and are therefore discontinuous (cf. Fig. 4-15). Moreover, since $\mathcal{C}$ possesses three antecedents near the self-intersection point M, the corresponding parts of $a_1\mathcal{C}$ and $a_2\mathcal{C}$ are also self-intersecting (at the points $M_1$ and $M_2$, respectively, cf. Fig. 4-14 and 4-15). The discontinuity of antecedent curves being also not fortuitous, it constitutes a second characteristic feature of the phase portrait of (4-27).

Another consequence of the non-uniqueness of antecedents is the existence of cycles having non-identical local phase portraits near their points $P_i = (x,y)_i$, $i = 1, 2, \ldots, k$, $k > 1$. An example of such a situation for the recurrence (4-27) is the external cycle saddle of order $k = 2$ :

$$P_1 = (53.710\ 278\ 920,\ 44.721\ 359\ 550), \quad P_2 = (35.825\ 446\ 913,\ 0) \quad ,$$

$\mu = 0.5$, $\beta = -0.001$. The point $P_1$ has three antecedents : $P_2$, $P_{10} = (x, y)$ and $P_{11} = (x, -y)$, where $x = 35.825\ 446\ 876$, $y = 31.622\ 776\ 603$, whereas the only antecedent of $P_2$ is $P_1$. The non-conservative term $\beta y^3$ is responsible for many other external cycles, some of which are listed in Table 4-5. The external fixed point, the cycle

| k | x | y | $\lambda_1$ | $\lambda_2$ |
|---|---|---|---|---|
| 1 | 89.465 | 44.721 | 0.7416 | -6.742 |
| 2 | 53.710 | 44.721 | 0.3482 | -14.36 |
| 2 | 17.911 | 44.721 | 45.60 | 0.5482 |
| 3 | 45.542 | 44.950 | 0.07935 | -19.22 |
| 3 | 63.805 | 47. 285 | 48.24 | 0.2633 |
| 4 | 46.236 | 46.244 | 0.2094 | -3.201 |
| 4 | 42.197 | -10.490 | 0.1761 | -284.2 |

Table 4-5. External cycles of the recurrence (4-27), $\mu = 0.5$, $\beta = -0.001$.

of order two already mentioned, and the type 1 saddle of order $k = 3$ have no bifurcations as $\mu$ increases ; in the limit $\mu \to +1$, one of their eigenvalues approaches monotonically the bifurcation value +1. The cycle saddle 3/1 of type 2 and the cycles of order $k > 4$, possess on the contrary numerous bifurcations. The evolution of the saddle 3/1 of type 2 is summarized in Table 4-6. The sequence of bifurcations is as follows : saddle → stable node → stable focus → stable node. At the saddle → node bifurcation there is a coalescence with a stable node of order $k = 6$. The evolution of the bifurcated node of order six is described in Table 4-7. It possesses, as

Table 4-6. Recurrence (4-27), $\beta = -0.001$. Evolution of the external saddle 3/1 of type 2 as a function of $\mu$.

| $\mu$ | x | y | $\lambda_1$ | $\lambda_2$ | Remarks |
|---|---|---|---|---|---|
| 0.5 | 45.5422 | 44.9497 | 0.07935 | -19.22 | |
| 0.7 | 68.3374 | 45.1084 | 0.1143 | -7.102 | |
| 0.8 | 96.2771 | 45.2506 | 0.1317 | -3.036 | |
| 0.81 | 100.6800 | 45.2689 | 0.1324 | -2.697 | |
| 0.82 | 105.5730 | 45.2881 | 0.1326 | -2.371 | (1) |
| 0.83 | 111.0439 | 45.3085 | 0.1321 | -2.0561 | |
| 0.85 | 124.1876 | 45.3527 | 0.1272 | -1.460 | |
| 0.86 | 132.1793 | 45.3769 | 0.1213 | -1.178 | |
| 0.866 | 137.5574 | 45.3921 | 0.1157 | -1.012 | |
| 0.8663 | 137.8338 | 45.3929 | 0.1154 | -1.004 | |
| 0.8667 | 138.2110 | 45.3939 | 0.1149 | -0.9928 | (2) |
| 0.867 | 138.4956 | 45.3947 | 0.1146 | -0.9846 | |
| 0.87 | 141.4128 | 45.4026 | 0.1107 | -0.9028 | |
| 0.89 | 164.9790 | 45.4592 | 0.04170 | -0.3426 | |
| 0.893 | 169.2835 | 45.4684 | 0.00610 | -0.2376 | |
| 0.89330 | 169.7275 | 45.4693 | 0.00074 | -0.2254 | |
| 0.89335 | 169.8017 | 45.4695 | -0.00021 | -0.2233 | (3) |
| 0.894 | 170.7732 | 45.4715 | -0.01456 | -0.1941 | |
| 0.895 | 172.2916 | 45.4746 | -0.05382 | -0.1321 | |
| 0.895240 | 172.6604 | 45.4754 | -0.08780 | -0.09266 | |

Table 4-6 continued

| $\mu$ | x | y | $\|\lambda\|$ | $\varphi$ | (4) |
|---|---|---|---|---|---|
| 0.895245 | 172.6681 | 45.4754 | 0.09012 | 3.085 | |
| 0.896 | 173.8396 | 45.4778 | 0.1067 | 2.442 | |
| 0.9 | 180.3453 | 45.4906 | 0.1689 | 1.792 | |
| 0.92 | 222.7859 | 45.5611 | 0.3384 | 1.040 | |
| 0.94 | 294.0287 | 45.6450 | 0.4508 | 0.6616 | |
| 0.96 | 437.8879 | 45.7488 | 0.5495 | 0.3352 | |
| 0.973 | 647.7279 | 45.8337 | 0.6180 | 0.01839 | |
| 0.973055 | 649.0513 | 45.8340 | 0.6184 | 0.002974 | |
| | | | $\lambda_1$ | $\lambda_2$ | (5) |
| 0.973060 | 649.1718 | 45.8341 | 0.6212 | 0.6156 | |
| 0.974 | 672.6726 | 45.8411 | 0.6717 | 0.5794 | |
| 0.994 | 2951.7884 | 46.0440 | 0.9533 | 0.6630 | |
| 0.997 | 5945.7162 | 46.0988 | 0.9785 | 0.7374 | (6) |

(1) $\lambda_1 \to$ max, $\lambda_2 < -1$ ; (2) $\lambda_2 \to -1$, $0 < \lambda_1 < 1$, bifurcation of a cycle of order $k = 6$ ; (3) $\lambda_1 \to 0$, $-1 < \lambda_2 < 0$ ; (4) $\lambda_1 \to \lambda_2$, bifurcation node $\to$ focus ($\varphi = \pi$) ; (5) $\varphi \to 0$, bifurcation focus $\to$ node ($\lambda_1 = \lambda_2$), (6) $\lambda_1 \to +1$, $0 < \lambda_2 < 1$.

Table 4-7. Recurrence (4-27), $\beta = -0.001$. Evolution of the external cycle of order $k = 6$ as a function of $\mu$.

| $\mu$ | x | y | $\lambda_1$ | $\lambda_2$ | Remarks |
|---|---|---|---|---|---|
| 0.85005 | 127.480 | 46.2632 | -0.1009 | -0.5544 | |
| 0.85405 | 130.104 | 46.1415 | -0.1363 | -0.1517 | |
| 0.854055 | 130.107 | 46.1414 | -0.1386 | -0.1489 | |
| | | | $\|\lambda\|$ | $\varphi$ | (1) |
| 0.854060 | 130.111 | 46.1412 | 0.1436 | 3.124 | |
| 0.8541 | 130.138 | 46.1399 | 0.1426 | 3.027 | |
| 0.856 | 131.416 | 46.0761 | 0.09277 | 2.137 | |
| 0.85767 | 132.554 | 46.0156 | 0.03486 | 0.2904 | |
| 0.857684 | 132.563 | 46.0151 | 0.03414 | 0.04070 | |
| | | | $\lambda_1$ | $\lambda_2$ | (2) |
| 0.857686 | 132.565 | 46.0150 | 0.03769 | 0.03073 | |
| 0.85769 | 132.567 | 46.0148 | 0.04075 | 0.02807 | |
| 0.858 | 132.780 | 46.0031 | 0.09928 | 0.00092 | |
| 0.85801 | 132.787 | 46.0027 | 0.1006 | 0.00059 | (3) |
| 0.85903 | 132.801 | 46.0019 | 0.1033 | -0.00005 | |
| 0.8581 | 132.849 | 45.9219 | 0.3197 | -0.1325 | (4) |
| 0.862 | 135.533 | 45.8282 | 0.5268 | -0.00781 | |
| 0.8635 | 136.547 | 45.7442 | 0.6840 | -0.00140 | |
| 0.8640 | 136.876 | 45.7119 | 0.7370 | 0.00095 | (5) |
| 0.8645 | 137.196 | 45.6764 | 0.7903 | 0.00338 | |
| 0.866 | 138.023 | 45.5258 | 0.9521 | 0.01098 | |
| 0.8663 | 138.093 | 45.4672 | 0.9849 | 0.01255 | |
| 0.8667 | none | none | | | (6) |

(1) $\lambda_1 \to \lambda_2$, bifurcation node $\to$ focus ($\varphi = \pi$) ; (2) $\varphi \to 0$, bifurcation focus $\to$ node ($\lambda_1 = \lambda_2$) ; (3) $\lambda_2 \to 0$, $0 < \lambda_1 < 1$ ; (4) $\lambda_2 \to$ min, $\lambda_1 > 0$ ; (5) $\lambda_2 \to 0$, $0 < \lambda_1 < 1$ ; (6) $\lambda_1 \to +1$, $0 < \lambda_2 < 1$.

expected, a higher bifurcation rate than the cycle which gave rise to it. On first glance the properties of the external cycles of the recurrence (4-27) do not appear to be substantially different from those of the external ones, except for the apparently minor fact that each point of an external cycle may have a different number of antecedents. This impression is, however, completely misleading. Like in the case of recurrences of order one, the existence of non-unique antecedents constitutes a strong source of "structural" complexity, which in many cases amounts to the presence of an additional stochasticity generating mechanism. This additional stochasticity becomes immediately apparent when one tries for example to determine the global shape of the invariant curves traversing the external saddles listed in Table 4-5.

Consider first the external fixed point (x, y), x > 0, y > 0, which admits an odd-symmetry companion (-x, -y). When $\mu = 0.5$ and $\beta = -0.001$, the invariant curves traversing (x, y), which is then a saddle of type 2, are shown in Fig. 4-16. The consequent-branches and the outward directed antecedent branch are relatively simple compared to the inward directed antecedent branch. The latter is extremely complicated, possessing many points of alternance and many points of discontinuity. In order to simplify the interpretation of Fig. 4-16, the single-valued antecedent branches are drawn differently from the triple-valued ones (— — — and -·-·-) . The inward directed antecedent branch is single-valued when it starts at (x, y). It terminates at a point of alternance M, whose companion M' is on the $y_n = -\bar{A} \approx -18.3$ branch of the critical curve $J_o$. One possible continuation of the branch (x, y) → M involves a jump from M to M', and then follows the segment $M' \to M_o$, $M_o$ being a point of alternance whose companion $M'_o$ is on the $(y_n = +\bar{A})$-branch of $J_o$. Another (initially smoother) continuation of (x, y) → M procedes from M to $M'_o$, and then jumps from $M'_o$ to $M_o$. Both

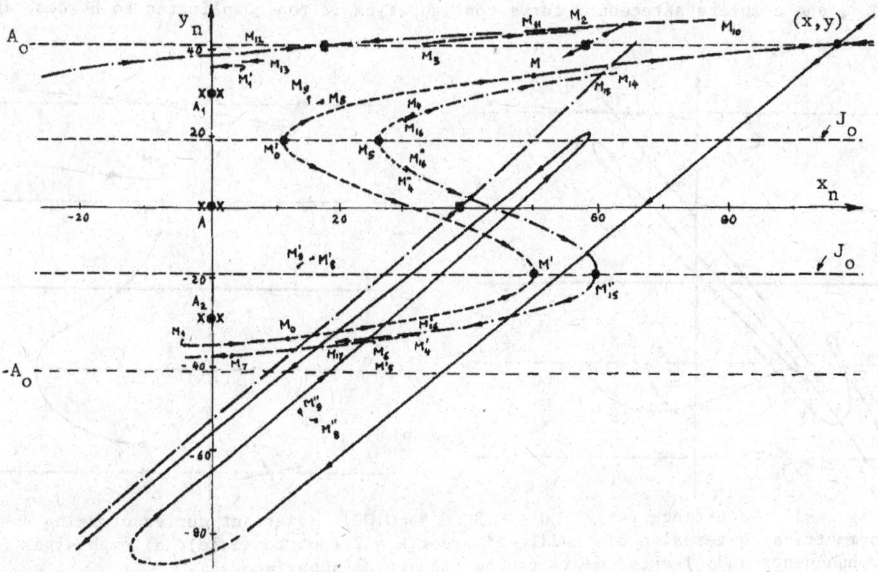


Fig. 4-16. Recurrence (4-27), $\mu = 0.5$, $\beta = -0.001$. Invariant curves crossing an external saddle : — — —unique antecedents, -·-·- non-unique antecedents. xox : Main fixed points (A) and their excess antecedents $(A_1)$ and $(A_2)$. ▪ : points of a cycle of order two. • : points of alternance. $A_o \approx 44.721$.

antecedent paths $(x, y) \to M_o$ bypass the excess antecedent segment $M'_o \to M'$, which constitutes visually a continuous link between the points of alternance $M'_o$ and $M'$. From $M_o$ the antecedent curve continues on a single-valued segment to the point $M_1$, from which it jumps to the point $M_2$, and then while still on a single-valued segment, it continues to the point of alternance $M_3$. The points $M_1$ and $M_2$ are points of discontinuity and not points of alternance, because they do not correspond to a change of the number of antecedents. The evolution of the inward directed antecedent branch beyond the point $M_3$ defies a concise description without the use of auxiliary figures showing explicitly the sequence of various jumps. A rough picture of this evolution can be obtained from the sequence of jump-points $M_i$, $i = 4, \ldots, 17$.

The first cycle of order $k = 2$ is a saddle of type 1 having the oddly symmetrical points $P_1 = (x, y)$ and $P_2 = (-x, -y)$. For $\mu = 0.5$, $\beta = -0.001$ the qualitative shape of consequent invariant curves passing through $P_1 \approx (17.9, 44.7)$ are shown in Fig. 4-17a (not to scale, in order to display more clearly the rather complicated loop structure). The inward consequent branch $\mathcal{C}$ forms several self-intersecting loops before leaving the vicinity of the points $P_1$ and $P_2$. The excess antecedent branches $a_1\mathcal{C}$ and $a_2\mathcal{C}$ of $\mathcal{C}$ are also self-intersecting. Moreover, $a_1\mathcal{C}$ and $a_2\mathcal{C}$ appear to have segments of partial overlap with $\mathcal{C}$. A part of the corresponding antecedent branches is shown in Fig. 4-17b. The single-valued outward branch reaches a point of alternance $M$, from which it jumps to $M'$, $M''$ or $M'''$ on the triple valued continuations. The latter terminate at the points of alternance $M_o$ and $M'_o$, where there occurs again a jump, etc. Combined with the antecedents of the excess antecedent branches $a_1\mathcal{C}$ and $a_2\mathcal{C}$, the complete antecedent curve configuration is too complicated to be described in a few pages.

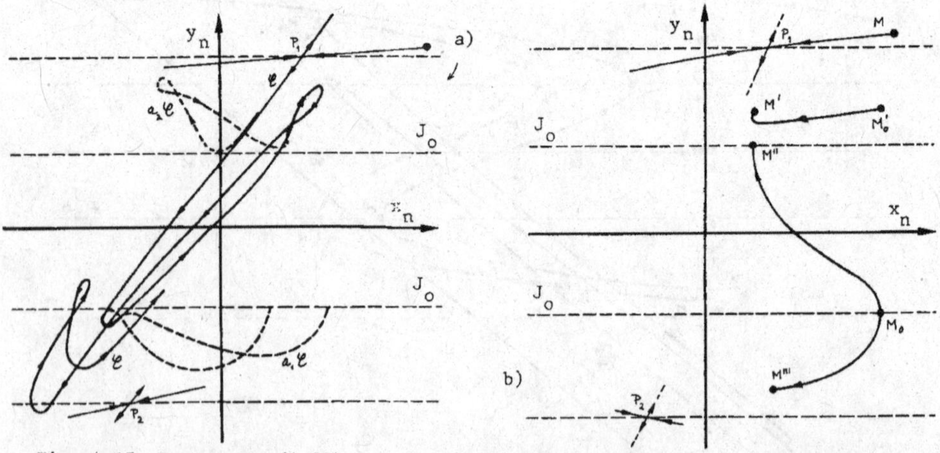


Fig. 4-17. Recurrence (4-27), $\mu = 0.5$, $\beta = -0.001$. Invariant curves crossing a symmetrical external cycle saddle of order $k = 2$ (not to scale). a) emphasizes consequents and b) antecedents (to be thought as superimposed).

The second cycle of order $k = 2$ is double, i.e. there exist two such cycles with complementary symmetry. The qualitative shapes of the invariant curves crossing

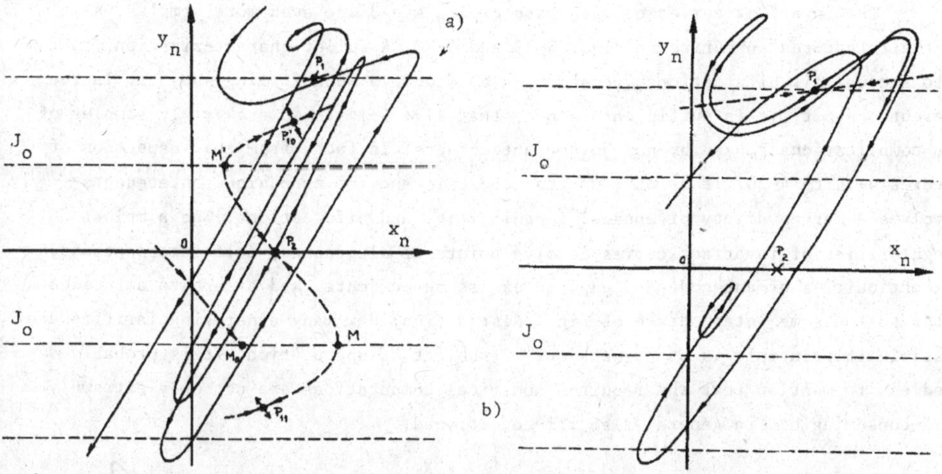


Fig. 4-18. Recurrence (4-27), $\mu = 0.5$, $\beta = -0.001$. Invariant curves crossing an unsymmetrical external cycle saddle of order $k = 2$ (not to scale). The point $P_2$ has two excess antecedents $P_{10}$ and $P_{11}$ whereas $P_1$ has none. (to be thought as superimposed).

the two unsymmetrical points $P_1$ and $P_2$ are shown in Fig. 4-18a and b. The consequent branches possess numerous self-intersection points, and the antecedent branches possess points of alternance and points of discontinuity.

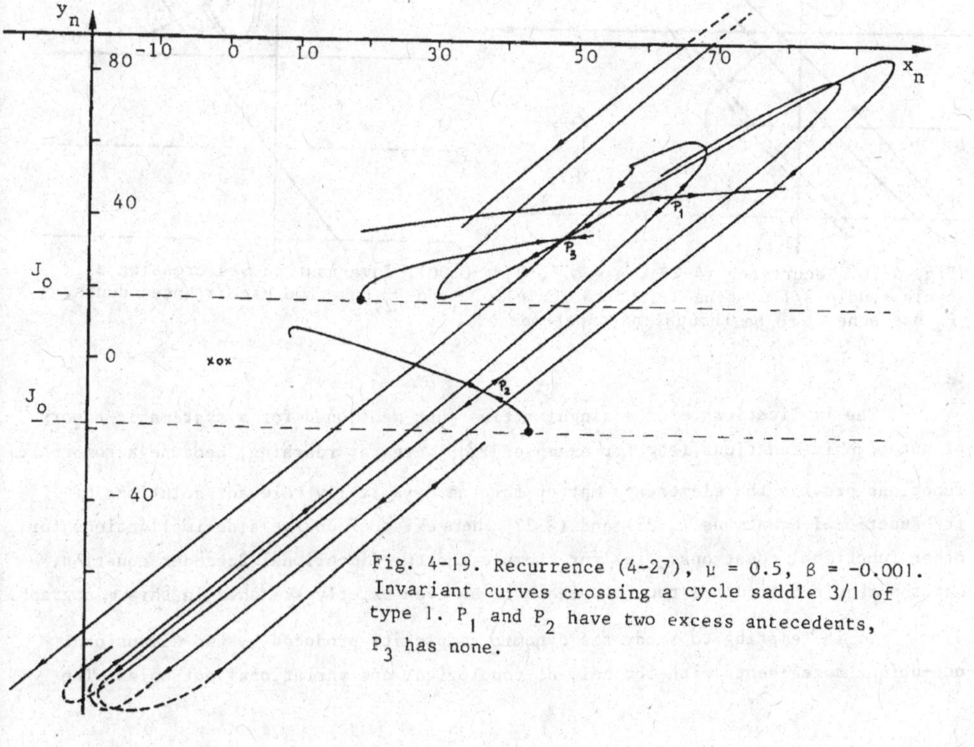


Fig. 4-19. Recurrence (4-27), $\mu = 0.5$, $\beta = -0.001$. Invariant curves crossing a cycle saddle 3/1 of type 1. $P_1$ and $P_2$ have two excess antecedents, $P_3$ has none.

The invariant curves crossing the cycles k = 3 are even more complex, some of their features appearing in Figs. 4-19 and 4-20. A fuller characterization, taking into account all bifurcations, constitutes by itself a book filling subject. In the present context, it is sufficient to note that Fig. 4-14 to 20 are merely samples of the complications caused by non-unique antecedents. In fact, from the inspection of Figures 4-14 to 4-20, it is obvious that the existence of non-unique antecedents involves a large variety of unusual singularities and bifurcations. The simplest singularities of invariant curves involve points of alternance, self-intersections, discontinuities of antecedents, gaps in excess antecedents, and alternate antecedents paths to the same intermediate of final destination. How many other singularities are possible in principle is at present only a subject of speculation. It is probably needless to mention that the required numerical computations are not only extremely time-consuming but in general also ill-conditioned.

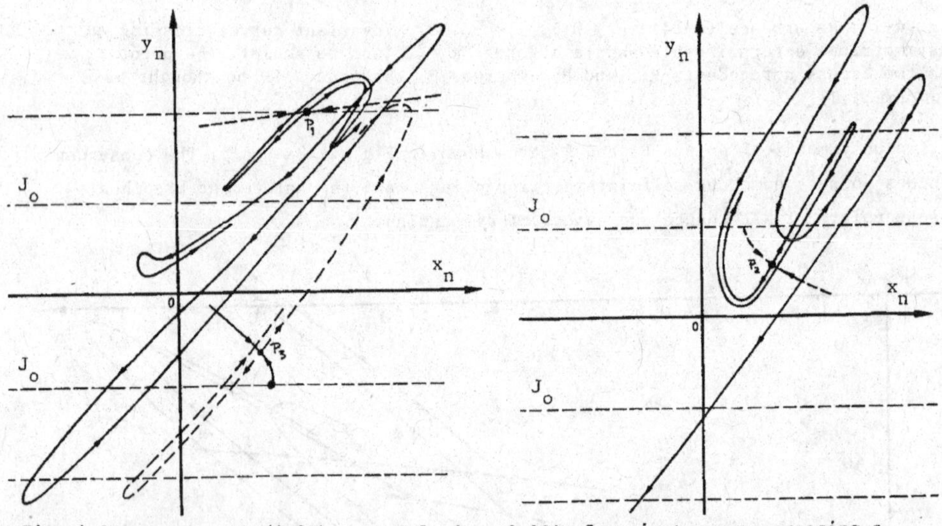


Fig. 4-20. Recurrence (4-27), $\mu$ = 0.5, $\beta$ = -0.001. Invariant curves crossing a cycle saddle 3/1 of type 2 (not to scale). $P_1$ and $P_2$ have two excess antecedents, $P_3$ has none. ( to be thought as superimposed).

The implications of the singularities just mentioned for a systematic theory of automorphic functions (see for example [F3] ) are far reaching, because automorphic functions provide the elementary matter for the dynamically relevant solutions of the functional equations (2-25) and (2-27). There exist of course also implications for other functional equations, like for example the two-dimensional Schröder equation, but the required study is too voluminous to be even briefly sketched in this monograph.

It is tempting to study the dynamic complexity produced by two-dimensional non-unique antecedents with the help of topological and variational entropies. The latter appears especially promising because only discrete consequent half-trajectories are utilized, and those are unique by definition.

Exploratory numerical computations turned out to be disappointing, because the successive iterated of $h_v$, as well as of its mean value, failed to converge inside the stochastic regions of the recurrence (4-27). Close to the main fixed point only the rather trivial result was found that $h_v$ is slightly negative.

The numerical study of the two-dimensional Perron-Frobenius equation must be at present limited to some very promising particular cases, because it is extremely time-consuming (even on rather large computers). The extra insight it would offer in the multiple antecedent situation just described appears to be marginal.

## Chapter V

## STOCHASTICITY IN STRONGLY NON-CONSERVATIVE RECURRENCES

### 5.0 Introduction

It could be surmised that as the amount of damping increases in a recurrence, the complexity of the corresponding dynamics diminishes, and the phase plane portrait approaches an orderly one. This intuitive reasoning finds a partial confirmation in the disappearance of cycles of low ratio k/r, as was repeatadly observed in chapter IV. There exists, however, a physically, biologically, etc. important class of recurrences in which stochasticity persists in spite of the presence of strong damping. The unexpected richness of shapes of invariant curves traversing external cycles, which have an unequal number of antecedents at each point , suggests another complexity generating mechanism, distinct from that induced by first order stochastic "source" recurrences. It is obvious that an imbedding of a first order recurrence into one of second order does not necessarily conserve the number of antecedents, i.e. the second order recurrence may possess unique antecedents whereas the first-order recurrence does not. Uniqueness of antecedents is therefore no intrinsic obstacle to the existence of stochasticity in the presence of strong damping. Research in third and fourth order continuous dynamic systems suggests that the presence of stochasticity is a rather common phenomenon (for specific examples see [R3-9]). This observation implies that non linear second (or higher) order recurrences involve as rule stochastic rather than orderly dynamics. The examples discussed in this chapter are therefore in no way physically, biologically, etc. pathological. Quite to the contrary, they provide samples of "natural" properties. Whether these examples are generic in a more profound sense, like for example that of the Van der Pol oscillator, is too early to tell.

### 5.1. Effect of strong damping on an initially conservative recurrence

When a small damping term is added to a conservative recurrence, its effect on cycles of $k/r > k_c$, where $k_c$ is a certain threshold, is also small. The positions of cycle points and the corresponding $\lambda$ are somewhat changed. In particular, the relation $\lambda_1 \cdot \lambda_2 = 1$ is no longer satisfied. As the damping is increased, the threshold $k_c$ increases and the eigenvalues of the surviving cycles deviate more and more from $\lambda_1 \cdot \lambda_2 = 1$. In order to illustrate this behaviour quantitatively, consider the linearly damped bounded quadratic recurrence with unique antecedents

$$(4\text{-}20) \qquad \begin{aligned} &x_{n+1} = (1+\alpha)y_n + F(x_n), \qquad y_{n+1} = -x_n + F(x_{n+1}) \\ &F(x) = \mu x + 2(1-\mu)x^2/(1+x^2), \quad -1 < \mu < 1, \qquad \alpha < 0 \end{aligned}$$

The evolution of the eigenvalues of three cycles is shown in Table 5-1 for $\mu = 1/8$

Table 5-1. Recurrence (4-20), $\mu$ = 1/8. Evolution of the eigenvalues of three typical cycles. Effect of damping.

| $\alpha$ | 6/1 | | 6/1 | | 9/1 | |
|---|---|---|---|---|---|---|
| | $\lambda_1$ | $\lambda_2$ | $\lambda_1$ | $\lambda_2$ | $\lambda_1$ | $\lambda_2$ |
| 0 | 15.86 | 0.0632 | -17.40 | -0.0575 | 185.9 | 0.0054 |
| -0.2 | 14.80 | 0.0598 | -16.11 | -0.0550 | 168.3 | 0.0050 |
| -0.4 | 13.53 | 0.0579 | -14.35 | -0.0545 | 149.0 | 0.0046 |
| -0.6 | 12.01 | 0.0574 | -12.13 | -0.0569 | 128.1 | 0.0045 |
| -0.8 | 10.15 | 0.0597 | -9.363 | -0.0648 | 105.2 | 0.0045 |
| -1.0 | 7.596 | 0.0700 | -5.726 | -0.0928 | 78.74 | 0.0049 |

as a function of $\alpha$. This evolution is typical for other cycles and other values of $\mu$. The absolutely larger eigenvalue diminishes rapidly with $\alpha$, while the absolutely smaller one changes relatively slowly. Both eigenvalues evolve in the direction of a simpler cycle distribution, i.e. as the damping increases a larger and larger part of the phase plane portrait becomes orderly. A similar conclusion is provided by the observation of invariant curves, because absolutely smaller eigenvalues imply less initial spreading of singular invariant curves, and thus fewer saddles with stochasticity generating invariant curve intersections.

A bifurcation leading to the disappearance of main homoclinic points.

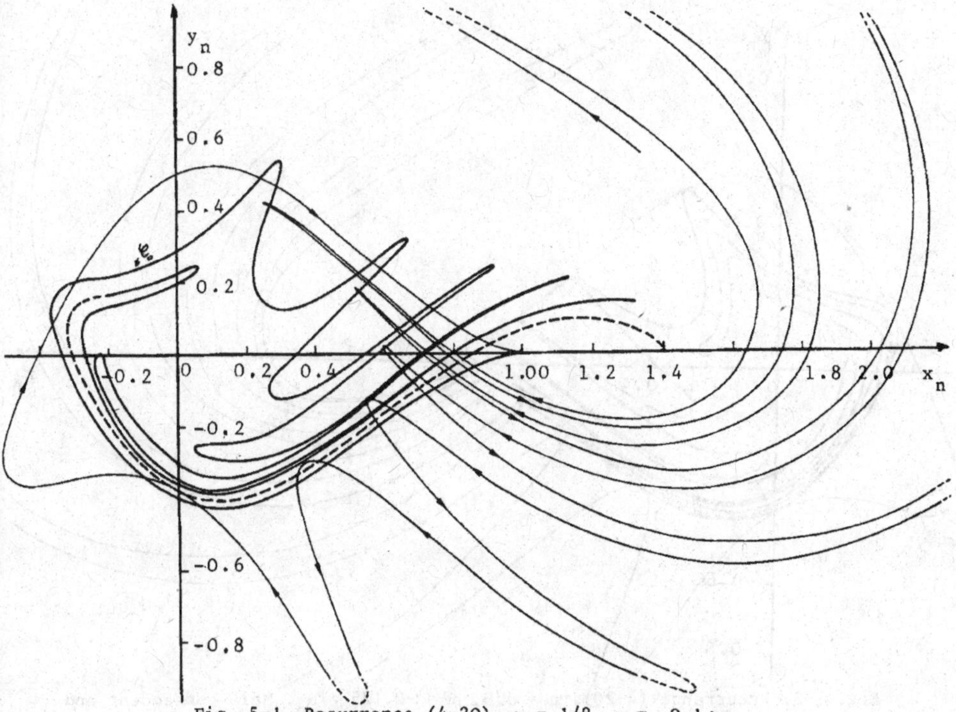


Fig. 5-1. Recurrence (4-20), $\mu$ = 1/8, $\alpha$ = -0.1 > $\alpha_c$

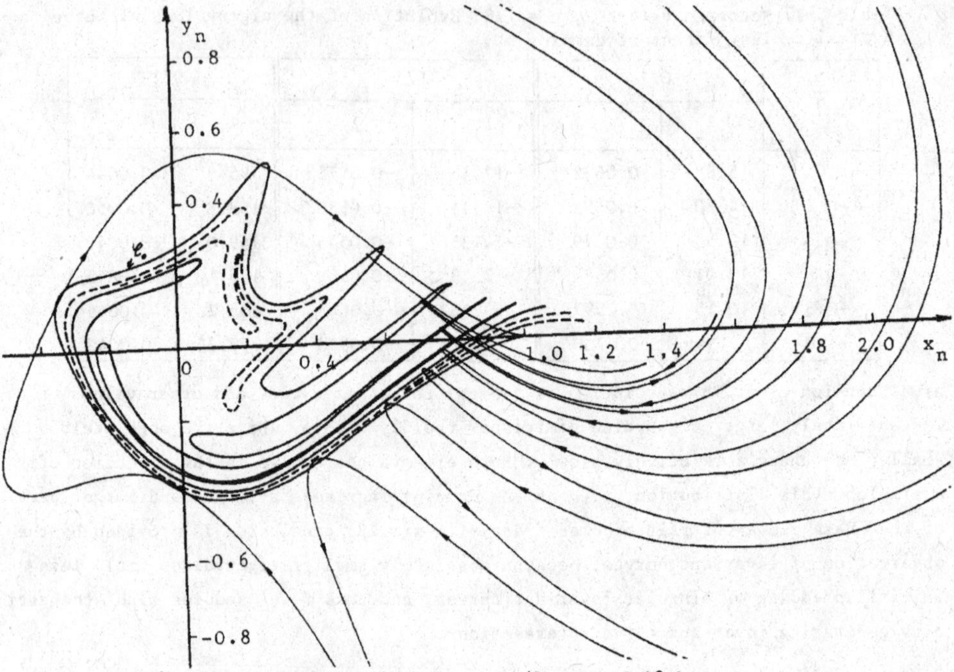


Fig. 5-2. Recurrence (4-20), $\mu = 1/8$, $\alpha = -0.12 \gtrsim \alpha_c$

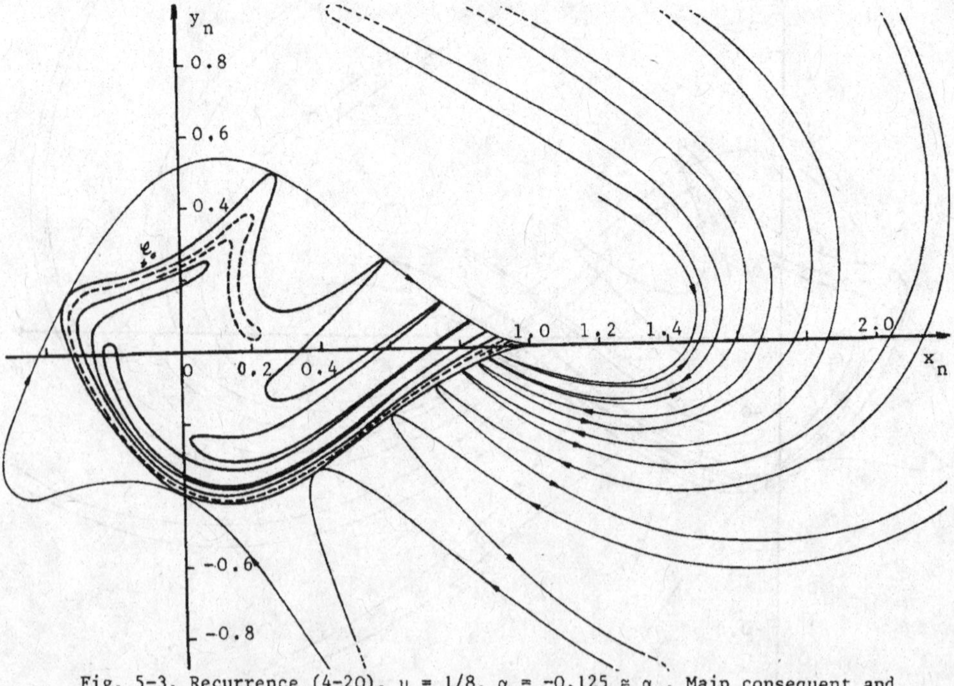


Fig. 5-3. Recurrence (4-20), $\mu = 1/8$, $\alpha = -0.125 \simeq \alpha_c$. Main consequent and antecedent curves are tangent to each other.

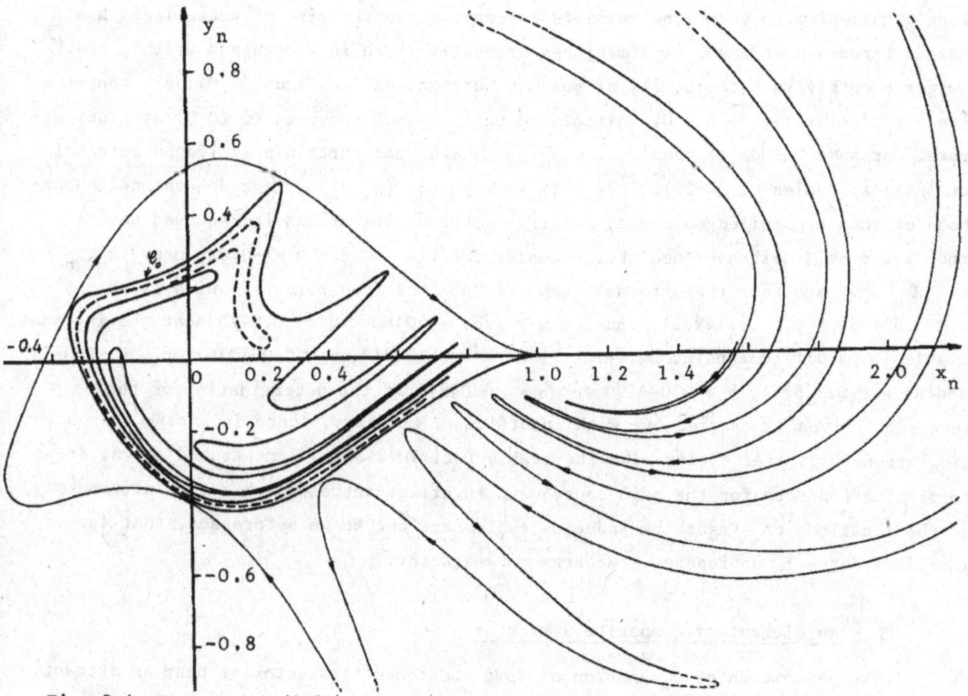


Fig. 5-4. Recurrence (4-20), $\mu = 1/8$, $\alpha = -0.13 < \alpha_c$. Main homoclinic points have ceased to exist.

As an illustration, consider the main invariant curves of (4-20). An increase of damping manifests itself at first as a decrease of the entanglement of consequent and antecedent branches. In such a process a stage is soon reached where these branches no longer intersect. There exists thus a damping threshold $\alpha_c = \alpha_c(\mu)$ beyond which main homoclinic points no longer exist, i.e. the stochasticity is destroyed. The corresponding bifurcation is illustrated in Figs. 5-1 to 5-4. When $\alpha = \alpha_c$, the loops of the consequent and antecedent branches are tangent to each other (Fig. 5-3). This "critical" property is observed in other recurrences, and of course also in differential equations (see for example [S 5]) and the references therein). In some recurrences, the disappearance of main homoclinic points is accompanied by a change of final destination, or attractor, for one branch, say $\mathscr{C}_o$, of the main consequent invariant curve. This is for example so for the linearly damped quadratic recurrence

(3-6a)
$$x_{n+1} = (1+\alpha)y_n + F(x_n), \quad y_{n+1} = -x_n + F(x_{n+1}) \quad ,$$
$$F(x) = \mu x + (1-\mu)x^2, \quad -1 < \mu < 1, \quad \alpha < 0 \quad .$$

Whereas for $\alpha > \alpha_c$ the branch $\mathscr{C}_o$ diverges to infinity, for $\alpha < \alpha_c$ it converges to the main fixed point (0,0). In the case of the bounded quadratic recurrence (4-20), the change of attractor does not coincide with the tangency bifurcation $\alpha = \alpha_c$. It takes place earlier. It is observed in fact (see Fig. 5-1 to 5-3) that $\mathscr{C}_o$ converges to (0,0) as soon as $\alpha < 0$. When the damping is small ($-1 \ll \alpha < 0$), $\mathscr{C}_o$ has an extremely long transient with very long multiple S-loops. After the size of these loops has passed through a maximum, it diminishes gradually (both in length and width), and $\mathcal{C}_o$ merges smoothly into the family of spirals surrounding the focus (0,0). All consequent trajectories $\{x_n, y_n\}$, $n \geq 0$, initialized on $\mathcal{C}_o$ do not converge to (0,0) at a uniform rate. For some values of $\alpha$ many $\{x_n, y_n\}$ stagnate near certain priviledged external or internal cycles of (4-20), i.e. once some points $(x_n,y_n) \in \mathcal{C}_o$ reach the neighbourhood of such a priviledged cycle, a large number of iterations is required before they leave this neighbourhood. For example, for $\mu = 1/8$ and $\alpha = -0.02$, many $\{x_n,y_n\}$ $\in \mathcal{C}_o$ stagnate near the external stable focus 10/2 (one point of which is at $x \simeq 5.534455$ ; $y \simeq 0.114972$), and for $\mu = 1/8$, $-0.0480 < \alpha < -0.0475$ near the internal stable focus 5/1 (one point of which is at $x \simeq 0.329717$, $y \simeq 0.097026$ for $\alpha = -0.0475$ and at $x \simeq 0.329744$, $y \simeq 0.096770$ for $\alpha = -0.048$). If the determination of the invariant curves is carried out with insufficient accuracy, there is a risk of concluding erroneously that cycles like the stable foci 10/2 and 5/1 mentioned above, are terminal attractors for the main consequent invariant curve branch $\mathcal{C}_o$, or alternately, if the locations of stagnation-inducing cycles are not known beforehand, that the invariant curve $\mathcal{C}_o$ approaches some strange attractor.

## 5.2 An elementary composite attractor

The phenomenon of stagnation of some consequent trajectories near an attractive cycle, i.e. the almost "capture" of a passing particle, is typical for non-conservative recurrences with unique antecedents. When the Jacobian determinant satisfies the inequality $0 < J(x_n,y_n) < 1$, an increase of damping leads usually to a decrease of stochasticity, or equivalently, to an increase of orderiliness. Something entirely different happens when $-1 < J(x_n,y_n) < 0$, or when the antecedents are non-unique. In the latter case, one example, showing the shapes of invariant curves, was given in Chapter 4. Another, qualitatively distinct example, is furnished by the quadratic recurrence

$$(5\text{-}1) \qquad x_{n+1} = y_n,\ y_{n+1} = y_n - c\,x_n + x_n^2, \qquad 0 < c \leqslant 8/5 \quad ,$$

whose inverse is double-valued

$$(5\text{-}2) \qquad y_n = x_{n+1},\ x_n = c/2 \pm \sqrt{(c/2)^2 + y_{n+1} - x_{n+1}} \quad .$$

The recurrence (5-1) admits two main fixed points, (0,0) and (c,c). The point (0,0) is a stable node for $0 < c \leqslant 1/4$, a stable focus for $1/4 < c < 1$, and an unstable focus for $c > 1$. The point (c,c) is a saddle of type 2 for all admissible values of c.

As c is made to increase from $c = 0$, a complicated chain of bifurcations takes place [C6 -8]. At the stability → instability transition $c = 1$ of the focus (0,0), there occurs first a bifurcation of a type described by equations (2-63)-(2-72). It leads to the appearance of an asymptotically stable closed regular invariant curve $\mathcal{L}$, surrounding the unstable focus (0,0). When $c - 1 > 0$ is quite saml1, the shape of $\mathcal{L}$ is

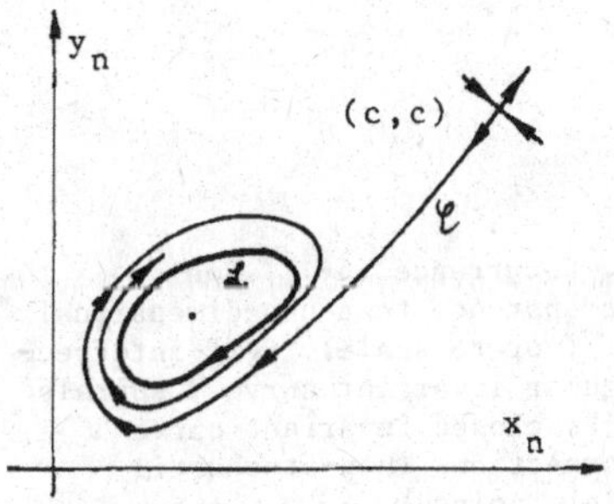


Fig. 5-5. Recurrence (5-1). Monotonic approach to a one-dimensional attractor (not to scale). The singular invariant curve $\mathcal{C}$ spirals towards the closed regular invariant curve without intersections.

an oval resembling an ellipse. One branch of the consequent invariant curve traversing the saddle (c, c) diverges to infinity whereas the other, designated in Fig. 5-5 by $\mathcal{C}$, winds monotonically arround $\mathcal{L}$ and approaches it asymptotically without intersections. The length S of $\mathcal{L}$ and the area $\mathcal{A}$ enclosed by $\mathcal{L}$ increase simultaneously with c-1. The shape of $\mathcal{L}$ deviates at first little from that of an oval, and then, it develops (rather suddenly) concave parts. The qualitative change of the shape of $\mathcal{L}$ takes place when the ovals would otherwise intersect the critical curve $J_1$ : $y_n - x_n - c^2/4 = 0$. $\mathcal{L}$ becomes simultaneously tangent to $J_o : 2x_n - c = 0$ and $J_1$ at two distinct points when $c = c_1 \simeq 1.187735$. Since an intersection of $\mathcal{L}$ with $J_1$ does not occur, and a contact between whole segments of invariant curves and critical curves is excluded in principle (the corresponding definitions are mutually exclusive, except in highly degenerate recurrences), the growth of S requires a local abandon of convexity, and thus the appearance of concave parts. For $1 < c < c_1$, the curve $\mathcal{L}$ is not only invariant with respect to consequents but also with respect to a single antecedent branch. For $c > c_1$, the invariance of $\mathcal{L}$ with respect to antecedents involves both branches of (5-2), i.e. some segments of $\mathcal{L}$ are mapped onto $\mathcal{L}$ via the (+)-part, and others via the (-)-part of (5-2). $\mathcal{L}$ is therefore not completely regular for $c > c_1$, because its antecedents contain points of alternance without arrow reversal.

The tangency bifurcation $c = c_1$ of $\mathcal{L}$ involves also a sudden change of the shape of the singular invariant curve $\mathcal{C}$ originating at the saddle (c,c). For $c > c_1$ the curve $\mathcal{C}$ possesses self-intersections. Moreover, while spiraling towards the attractive curve $\mathcal{L}$, $\mathcal{C}$ intersects the latter (see Fig. 5-6), i.e. if forms heteroclinic points. Although the self-intersections of $\mathcal{C}$, and the intersections of $\mathcal{C}$ and $\mathcal{L}$ constitute a complexity generating process, the approach of $\mathcal{C}$ towards $\mathcal{L}$ is essentially elementary. It can be thought of as a monotonic approach of a spiral in a three-dimensional space to a (planar or non-planar) closed curve. The self-intersections of $\mathcal{C}$, as well as the intersections of $\mathcal{C}$ and $\mathcal{L}$, appear then as virtual intersections due a projection from three to two dimensions.

The closed invariant curve $\mathcal{L}$ is the sole attractor of the recurrence (5-1) when $1 < c < c_2 = 1.37833405...$ At $c = c_2$, there appears spontaneously a double cycle of order k = 7, and rotation number r = 1, which splits into two simple cycles, a

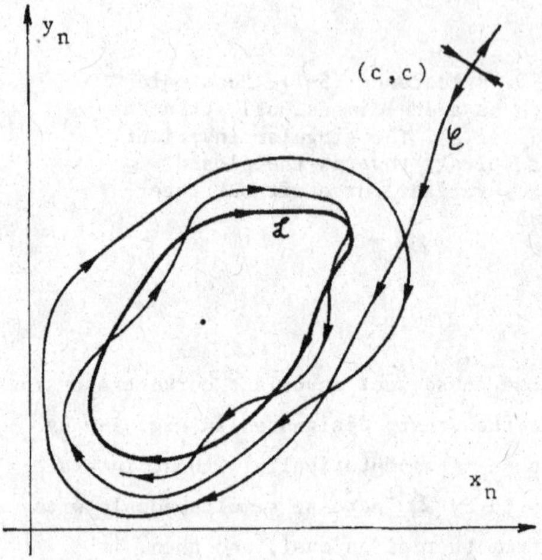


Fig. 5-6. Recurrence (5-1). Non-monotonic approach to a one-dimensional attractor (not to scale). Self-intersecting singular invariant curve $\mathscr{C}$ spirals towards the closed invariant curve $\mathscr{L}$ with intersections (non-stochastic heteroclinic points).

Coexistence of an asymptotically stable closed invariant curve $\mathscr{L}$ with a cycle pair saddle-stable focus (of order k = 7).

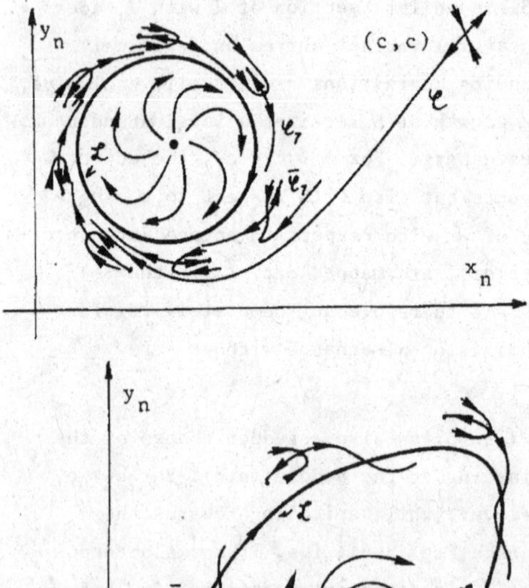


Fig. 5-7a. Extrapolated from properties of weakly non-conservative recurrences (not to scale).

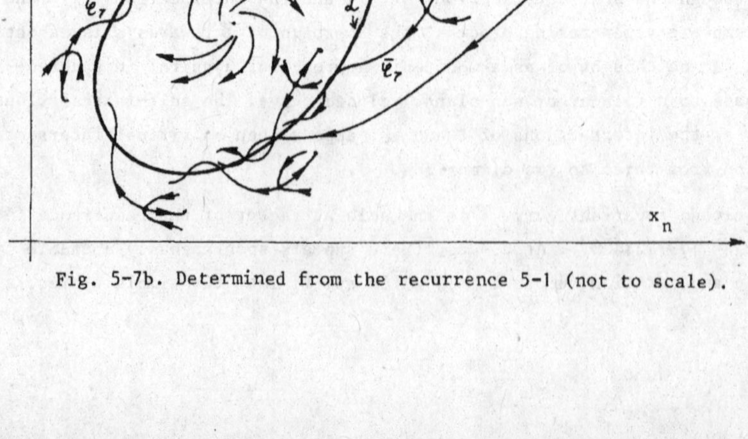


Fig. 5-7b. Determined from the recurrence 5-1 (not to scale).

stable node and a saddle of type 1, when $c > c_2$. The invariant curve $\mathcal{L}$ is still attractive and it coexists with the stable nodes 7/1 in the interval $c_2 < c < \bar{c}_2 \simeq 1.378632$. One might think that during this coexistence the singularity configuration would be as shown in Fig. 5-7a : the nodes and saddles 7/1 are outside of $\mathcal{L}$, one set of consequent invariant curves ($\mathcal{C}_7$) approaches the nodes 7/1, and another ($\bar{\mathcal{C}}_7$) spirals monotonically towards $\mathcal{L}$. The actual situation is, however, quite different, because one pair of points of the cycles 7/1 happens to be inside $\mathcal{L}$, and the other six pairs outside. The invariant curve sets $\mathcal{C}_7$ and $\bar{\mathcal{C}}_7$ approach $\mathcal{L}$ as shown in Fig. 5-7b.

The existence of the closed invariant curve $\mathcal{L}$ ceases at $c = \bar{c}_2$, when $\mathcal{L}$ becomes involved in a non-elementary bifurcations, all details of which have net yet been identified. The end-result is shown in Fig. 5-7c. Apparently $\mathcal{L}$ loses its influence domain as it comes closer to the invariant curves $\bar{\mathcal{C}}_7$ (Fig. 5-7d). When $c-\bar{c}_2 > 0$ is small, there exists only one attractor : the nodes 7/1.

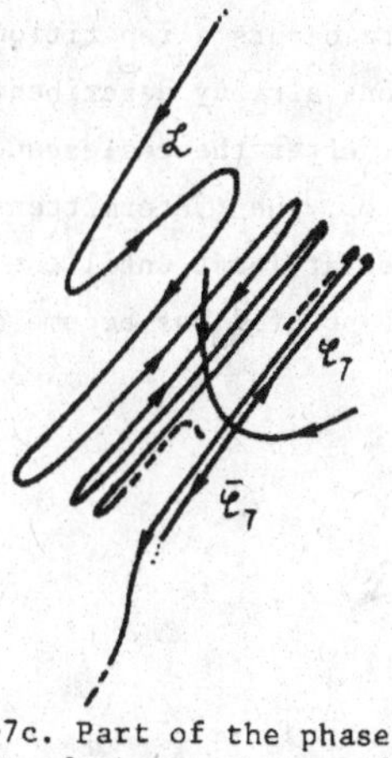

Fig. 5-7c. Part of the phase plane portrait of the recurrence (5-1) after the disappearance of $\mathcal{L}$ (not to scale). The dashed segments lead to other saddles of order k = 7. The foci 7/1 are asymptotically stable.

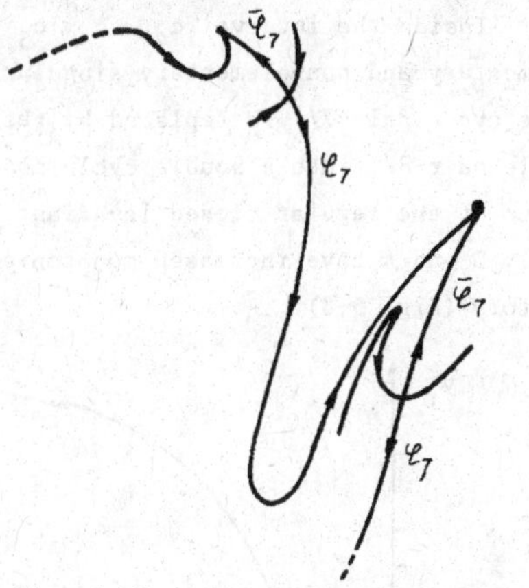

Fig. 5-7d. Part of Fig. 5-7b (not to scale) just before the disappearance of $\mathcal{L}$.

The closed invariant curve $\mathcal{L}$ reappears (increased in size, as expected), at $c = c_3 \simeq 1.4009916$, apparently via the inverse of the bifurcation which led to its disappearance at $c = \bar{c}_2$. The cycle pair 7/1 exists inside the interval $c_2 < c < c_3 \simeq 1.4018035$, except that with an increase of c the nodes change briefly into foci and then back again into nodes. At $c = c_3$, the nodes merge with the saddles, produce a double cycle node-saddle 7/1, and disappear. In contrast to $c_2 < c < \bar{c}_2$, the whole interval $\bar{c}_3 < c < c_3$ is not an interval of coexistence of $\mathcal{L}$ and the stable cycle 7/1, because inside several sub-intervals of $\bar{c}_3 < c < c_3$, $\mathcal{L}$ disintegrates into a cycle pair of a rather high order k. For example k = 92 when $1.401499748 < c < 1.4015024$, k = 394 when 1.400998435 < c < 1.4010..., and k = 211 when c ≈ 1.4009950. These cycles were found with the help of sensitivity coefficients. They appear and disappear as node-saddles. There exists also considerable numerical evidence that for some c the closed invariant curve $\mathcal{L}$ is replaced by a cyclic set of attractive disjoint curve segments $\mathcal{L}_s$, analoguous to the invariant segments of a first order recurrence. Such stochastic curve segments are observed, for example, for c slightly larger than 1.400995. The bifurcation responsible for the appearance of the $\mathcal{L}_s$ is still unknown. When $\mathcal{L}$ coexists with the stable cycle 7/1, six pairs of the cycle points 7/1 are inside $\mathcal{L}$, and a single pair outside. The resulting configuration is therefore a natural continuation of that encountered earlier, the temporary disappearance of $\mathcal{L}$ making room for a continuous displacement of the cycle points 7/1. After the cycles 7/1 cease to exist (by merging into a node-saddle), $\mathcal{L}$ reappears more definitely (with an increased S and $\mathcal{A}$). Apparently $\mathcal{L}$ is the sole attractor of (5-1) till c reaches the bifurcation value $c_4 \approx 1.481125$, where a double cycle (node-saddle) 8/1 comes into existence.

Inside the interval $c_4 < c < c_5 \approx 1.503775$, there occurs a repetition of the elementary and non-elementary singularity configurations already described, except that the cycle pair 7/1 is replaced by the cycle pair 8/1. After the coalescence of the cycle pair 8/1 into a double cycle node-saddle at $c = c_5$, the (intermittent) existence of the regular closed invariant curve $\mathcal{L}$ continues at least until $c = c_6 \approx 1.504821$. S and $\mathcal{A}$ have increased monotonically, but the shape of $\mathcal{L}$ has become quite oscillatory (Fig. 5-8).

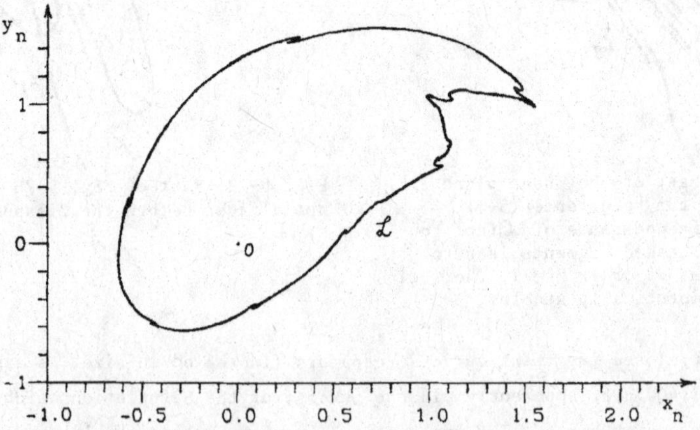


Fig. 5-8. Recurrence (5-1), c = 1.504. Shape of the closed invariant curve $\mathcal{L}$.

Beyond $c_5$ the singularity structure of the recurrence (5-1) is described schematically in Fig. 5-9. The details of the "macro"singularity link $c_6 \to c_9$ of Fig. 5-9 is given in Fig. 5-10a, where for brevity the following notations are used : S = saddle, N = node, F = focus, s = asymptotically stable, u = unstable. Whenever necessary, the order k of a cycle is indicated by a subscript. The evolution of the cycle $sN_9$ is different from that of $sN_7$ and $sN_8$. Although the cycle pair $sN_9$, $S_9$ appears at $c = c_6$, exists inside the interval $c_6 < c < c_9$ and disappears at $c_9$ by becoming a node-saddle, $sN_9$ turns at first into a stable focus and then into an unstable one. The stability transition $sF_9 \rightarrow uF_9$ is accompanied by two bifurcations, one of which gives rise to nine closed regular invariant curves $\mathcal{L}_9$, each analoguous to that depicted in Fig. 5-5, and the other to a double cycle (node-saddle 36/4), and then to the cycle pair $sN_{36}$, $S_{36}$. The cycle pair $sN_{36}$, $S_{36}$ undergoes several bifurcations (Fig. 10b), and disappears at $c = c_8$ by merging into a node-saddle. Bifurcation of $sN_{36}$ or $S_{36}$ into cycles of order $k = 72$ or higher are likely, but the study of their existence was not pursued due to essentially numerical reasons. An important property of the recurrence (5-1) is the coexistence inside the interval $c_7 < c < c_8$ of at least two stable cycles. This situation is somewhat similar to that of an almost conservative recurrence.

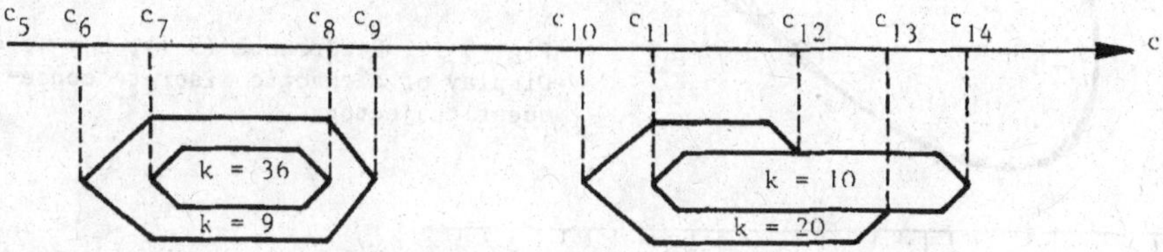


Fig. (5-9). Recurrence 5-1. Bifurcation schemes represented by macro-links. $c_6 = 1.504821$, $c_7 = 1.512312$, $c_8 = 1.524278$, $c_9 = 1.525788$, $c_{10} = 1.534227$, $c_{11} = 1.544742$, $c_{12} = 1.544742$, $c_{13} = 1.554000$, $c_{14} = 1.560957$.

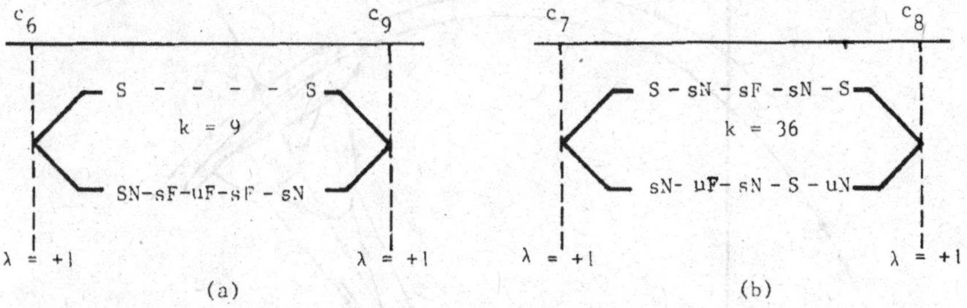


Fig. 5-10. More detailed macro-links of Fig. 5-9.

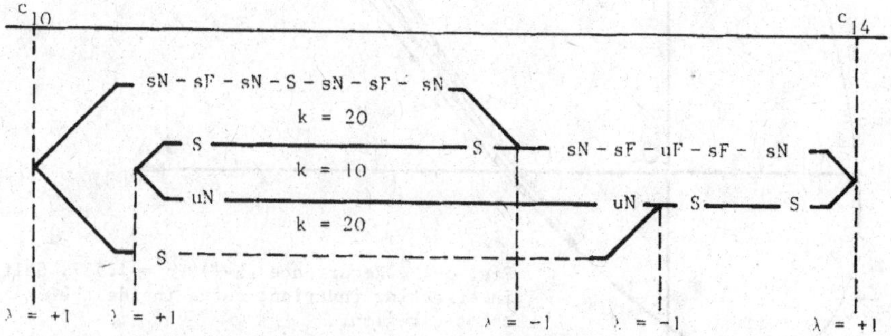


Fig. 5-11. A more detailed macro-link of Fig. 5-9.

Between $c_9$ and $c_{10}$ there exist stable cycles of a high order, but again for numerical reasons, their properties could not be studied with a reasonable effort. The final macro-singularity link of Fig. 5-9 is considerably more complicated. Some details are given in Fig. 5-11. At $c = c_{10}$, there appears a node-saddle of order $k = 20$ and rotation number $r = 2$, which splits into a cycle pair $sN_{20}$, $S_{20}$ as c exceeds $c_{10}$. The disappearance of this cycle pair is, however, completely different.

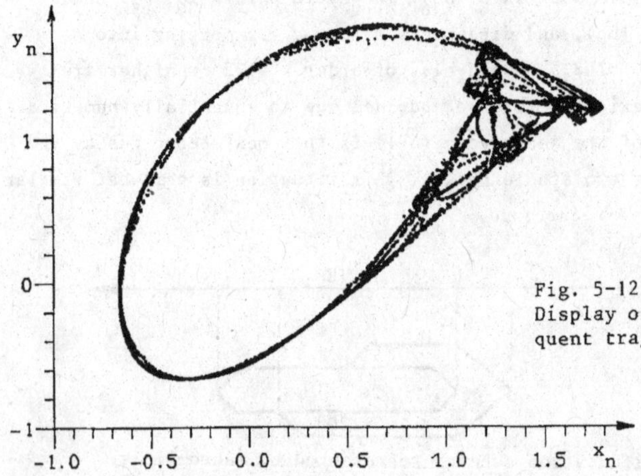


Fig. 5-12. Recurrence (5-1), c = 1.54. Display of a chaotic discrete consequent trajectory.

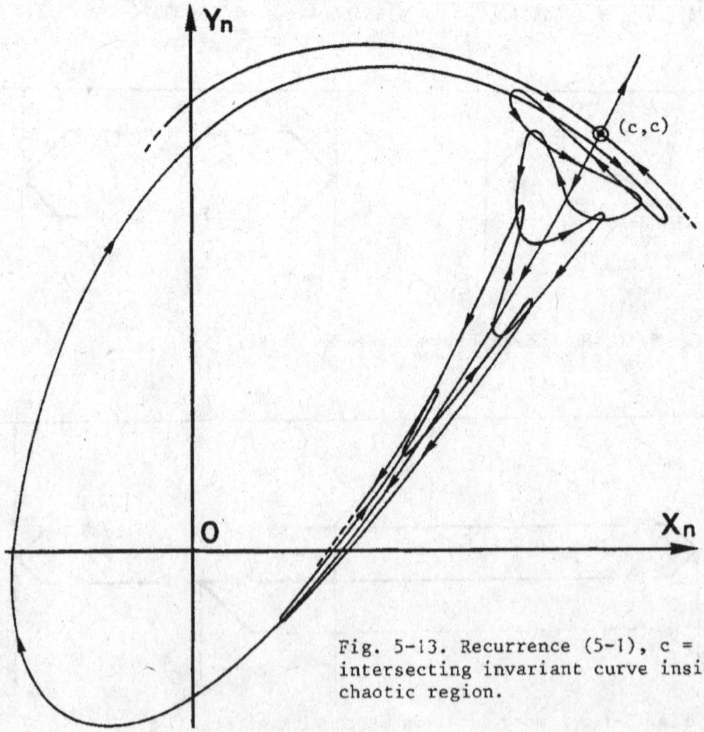


Fig. 5-13. Recurrence (5-1), c = 1.557. Self-intersecting invariant curve inside the chaotic region.

The $sN_{20}$ traverses several bifurcations releasing each time a higher-order cycle (the first of which is apparently 40/4) and then merges with a saddle 10/1 at $c = c_{12}$. The saddle 20/2 exists in a longer interval $c_{10} < c < c_{13}$ ($> c_{12}$), and merges at $c = c_{13}$ with a node 10/1 (see Fig. 5-11). The two cycles of order k = 10 appear at $c = c_{11} > c_{10}$ as a node-saddle and disappear in the same way at $c = c_{14}$. No stable point-singularities were found when $c > c_{14}$.

From an observation of discrete trajectories $\{x_n, y_n\}$, $n \geq 0$ it was concluded that a chaotic no-escape region D (of finite area) exists when $1.53 \lesssim c \lesssim 1.55$ (see Fig. 5-12). For c = 1.557 a sample self-intersecting invariant curve entering D, and then remaining indefinitly inside D is shown in Fig. 5-13.

The study of the chaotic regime by means of the variational entropy turned out to be inconclusive. For c = 1.535 one finds for example $h_v = 0.331$ with the initial conditions ($x_n = 0.8$, $y_n = 0.2$) and $h_v = 0.317$ with ($x_n = 1.1$, $y_n = 0.8$). For c = 1.549 the same initial conditions yield $h_v = 0.337$ and $h_v = 0.345$, respectively. The fact that $h_v > 0$ implies only that for the initial conditions chosen the transient to the elementary or composite attractor inside D is rather long, but it gives no information whatever on the nature of this attractor.

### 5.3 Chaos in a predator-prey recurrence

Another recurrence whose complexity is attributable to non-unique antecedents is the predator-prey recurrence

$$x_{n+1} = x_n \exp\left[c(1-x_n/K)\right] . \exp(-a\, y_n) \ ,$$
(2-6)
$$y_{n+1} = b\, x_n \left[1 - \exp(-a\, y_n)\right] \ ,$$

where c is the intrinsic growth rate, K the carrying capacity and a, b, coupling constants. Two fixed points of (2-6) can be found by inspection : (0,0) and (K,0). For the biologically relevant parameter ranges they are both unstable, so no further mention will be made of them. A third fixed point $A = (\bar{x}, \bar{y})$, $\bar{x}, \bar{y} > 0$, is given by a solution of the algebraic equations

(5-3) $$c(1-x/K) - ay = 0, \qquad y - bx\left[1 - \exp(-ay)\right]$$

which, unfortunately, can only be solved numerically. For this reason, in what follows the parameters will be restricted to the particular values K = 10, b = 1, and $0 < c < 5$. The value of a is fixed indirectly by means of an auxiliary parameter $\gamma = \bar{x}/K$, related to a by

(5-4) $$a = \frac{c(1-c)}{\gamma K} . \left[1 - \exp(c\gamma - c)\right]^{-1} ,$$

which is a measure of the extent to which the predator can depress the prey below its carrying capacity (cf. [B 2 ]). When $\gamma = 0.4$, and $0 < c < c_o \approx 0.706952$, the fixed point A is a stable focus. $c_o$ is a root of the algebraic equation

Table 5-2a. Recurrence (2-6). Bifurcation chain generated by a saddle 5/1. (1), (2), (3) : $\lambda_2 = -1$.

| c | k | x | y | $\lambda_1$ | $\lambda_2$ | $p_1$ | $p_2$ | |
|---|---|---|---|---|---|---|---|---|
| 2.18018343 | 5 | 10.732538 | 1.259975 | 1.0003 | -0.81258 | 0.21131 | -0.27947 | |
| 2.20 | 5 | 10.287144 | 1.158971 | 1.5625 | -0.79076 | 0.24693 | -0.24399 | |
| 2.24 | 5 | 9.985692 | 1.068495 | 1.9174 | -0.84209 | 0.25646 | -0.22243 | |
| 2.28 | 5 | 9.807633 | 0.997228 | 2.1113 | -0.91620 | 0.25380 | -0.20746 | |
| 2.31955 | 5 | 9.682230 | 0.934500 | 2.2233 | -0.99991 | 0.24640 | -0.19435 | (1) |
| 2.32 | 5 | 9.681110 | 0.933971 | 2.2245 | -1.0010 | 0.24629 | -0.19420 | |
| 2.5044 | 5 | 9.269490 | 0.682669 | 1.9297 | -1.4523 | 0.18960 | -0.13144 | |
| 2.3196 | 10 | 9.718980 | 0.927480 | 4.49448 | 0.99957 | 0.24364 | -0.19159 | |
| 2.36 | 10 | 10.597680 | 0.730962 | 5.3361 | 0.62181 | 0.19212 | -0.11580 | |
| 2.40 | 10 | 10.937161 | 0.655392 | 5.6827 | 0.21550 | 0.18256 | -0.09115 | |
| 2.44 | 10 | 11.203921 | 0.603801 | 5.9966 | -0.21508 | 0.17998 | -0.05676 | |
| 2.48 | 10 | 11.437201 | 0.565388 | 6.2690 | -0.66528 | 0.18179 | -0.06764 | |
| 2.50 | 10 | 11.545597 | 0.549551 | 6.3830 | -0.89681 | 0.18419 | -0.06437 | |
| 2.5085 | 10 | 11.590304 | 0.543379 | 6.4256 | -0.99641 | 0.18552 | -0.06318 | (2) |
| 2.5090 | 10 | 11.592911 | 0.543015 | 6.3112 | -1.0023 | 0.18561 | -0.06311 | |
| 2.5090 | 20 | 11.635391 | 0.540319 | 41.293 | 0.99082 | 0.18170 | -0.06380 | |
| 2.513 | 20 | 11.791709 | 0.528800 | 40.990 | 0.80080 | 0.16942 | -0.06566 | |
| 2.528 | 20 | 12.013580 | 0.509491 | 39.668 | 0.05925 | 0.15633 | -0.06759 | |
| 2.538 | 20 | 12.109117 | 0.500358 | 38.618 | -0.45923 | 0.15244 | -0.06860 | |
| 2.548 | 20 | 12.188625 | 0.492462 | 37.430 | -0.99508 | 0.15020 | -0.06977 | |
| 2.54809 | 20 | 12.189289 | 0.492395 | 37.418 | -0.99997 | 0.15018 | -0.06978 | (3) |
| 2.5482 | 20 | 12.190099 | 0.492313 | 37.404 | -1.0060 | 0.15017 | -0.06980 | |

Table 5-2b. Recurrence (2-6). Bifurcation chain generated by a node 5/1. (1), (2), (3), (4) : $\lambda_2 = -1$.

| c | k | x | y | $\lambda_1$ | $\lambda_2$ | $p_1$ | $p_2$ | |
|---|---|---|---|---|---|---|---|---|
| 2.18018343 | 5 | 10.732994 | 1.260072 | 0.99969 | -0.81264 | 0.21126 | -0.27951 | |
| 2.20 | 5 | 11.104020 | 1.348086 | 0.40846 | -0.89921 | 0.15856 | -0.33333 | |
| 2.24 | 5 | 11.231903 | 1.416665 | -0.01646 | -0.95478 | 0.11754 | -0.40158 | |
| 2.32 | 5 | 11.098448 | 1.526232 | -0.53173 | -0.90879 | 0.07421 | -0.94189 | |
| 2.36 | 5 | 10.932706 | 1.582591 | -0.73400 | -0.82000 | 0.06189 | 0.28019 | |
| 2.40 | 5 | 10.718746 | 1.643820 | -0.68764 | -0.91829 | -0.13706 | 0.05192 | |
| 2.418520 | 5 | 10.604265 | 1.674610 | -0.60946 | -0.99998 | -0.19109 | 0.04902 | (1) |
| 2.42 | 5 | 10.594690 | 1.677154 | -0.60271 | -1.0064 | -0.19443 | 0.04882 | |
| 2.5044 | 5 | 9.917704 | 1.857251 | -0.05269 | -1.3577 | -0.30404 | 0.46551 | |
| 2.418526 | 10 | 10.607628 | 1.674785 | 0.99999 | 0.37141 | 0.04828 | -0.19120 | |
| 2.42 | 10 | 10.746561 | 1.681601 | 0.97388 | 0.37488 | 0.07106 | -0.19456 | |
| 2.46 | 10 | 11.528822 | 1.651036 | 0.30663 | -0.02898 | -0.19055 | -0.23926 | |
| 2.48 | 10 | 11.799213 | 1.624379 | 0.21995 | -0.83705 | -0.17863 | -0.29372 | (2) |
| 2.48387 | 10 | 11.843415 | 1.620041 | 0.19289 | -1.0003 | -0.17728 | -0.30108 | |
| 2.5044 | 10 | 12.031549 | 1.602742 | 0.01628 | -1.8624 | -0.17304 | -0.32963 | |
| 2.48387 | 20 | 11.822445 | 1.626356 | 0.99871 | 0.03700 | 0.30127 | 0.18373 | |
| 2.488 | 20 | 11.233094 | 1.817841 | 0.36889 | 0.02069 | -0.30797 | -0.28843 | |
| 2.498 | 20 | 10.074583 | 2.231221 | 0.16852 | -0.77398 | -0.36612 | -0.33490 | |
| 2.498982 | 20 | 9.968058 | 2.272971 | 0.08125 | -1.0001 | -0.37746 | -0.34258 | (3) |
| 2.5044 | 20 | 9.631132 | 2.417525 | -0.2507 | -2.8165 | -0.44517 | -0.36966 | |
| 2.498982 | 40 | 9.973295 | 2.271177 | 0.99977 | 0.00658 | -0.34268 | -0.37683 | |
| 2.5 | 40 | 10.370219 | 2.139227 | 0.19930 | -0.23465 | -0.33867 | -0.34391 | |
| 2.501 | 40 | 10.565799 | 2.076044 | 0.03432 | -0.99611 | -0.32990 | -0.33895 | (4) |
| 2.501005 | 40 | 10.566667 | 2.075765 | 0.03400 | -1.0004 | -0.32986 | -0.33893 | |
| 2.5044 | 40 | 11.008936 | 1.936289 | -0.0260 | -4.8189 | -0.30792 | -0.33170 | |

Table 5-3. Recurrence (2-6). Some cycles independent of the family $k = 5.2^m$, $m = 0, 1, 2, \dots$ .

| c | k | x | y | $\lambda_1$ or $\rho$ | $\lambda_2$ or $\varphi$ | $p_1$ | $p_2$ |
|---|---|---|---|---|---|---|---|
| 2.8 | 1 | 4 | 3.254504 | $\rho$=1.27821 | $\varphi$=1.46702 | — | — |
| 2.8 | 4 | 3.312430 | 5.577332 | 2.1280 | -2.2505 | -0.55700 | 99.000 |
| 2.8 | 4 | 17.03128 | 0.844316 | -0.2800 | -2.3289 | -0.09633 | 0.18422 |
| 2.8 | 8 | 3.180551 | 4.247864 | 3.3152 | -6.1749 | -0.05278 | -4.09873 |
| 2.8 | 8 | 7.917645 | 3.515393 | -0.7794 | -8.8692 | -1.77531 | -0.47665 |
| 2.8 | 9 | 3.829778 | 1.808717 | 4.9127 | -4.6008 | 0.26724 | -2.01417 |
| 2.8 | 9 | 10.32207 | 1.597601 | $\rho$=4.0586 | $\varphi$=1.8852 | — | — |
| 2.8 | 11 | 0.626571 | 2.901126 | 6.8988 | -6.1536 | 1.80342 | 18.5582 |
| 2.8 | 11 | 0.482222 | 2.622240 | -2.3767 | -5.3358 | 2.68556 | 1.84582 |
| 2.8 | 11 | 0.567141 | 2.673334 | $\rho$=6.50627 | $\varphi$=0.76271 | — | — |
| 2.8 | 11 | 0.381109 | 2.136189 | 7.2258 | 0.0352 | 72.9319 | 4.87704 |
| 2.8 | 12 | 3.485239 | 3.567009 | 7.8712 | -13.7547 | 0.19017 | -1.89137 |
| 2.8 | 12 | 4.831342 | 4.948113 | -2.4775 | -20.4779 | -12.7961 | -0.42644 |
| 2.8 | 12 | 3.201826 | 6.305706 | 13.3090 | 7.9791 | 24.6873 | -0.53797 |
| 2.8 | 12 | 15.189918 | 1.135507 | 28.4793 | -1.1860 | -0.43033 | -0.10295 |
| 2.8 | 13 | 4.318845 | 3.460041 | 13.6358 | -12.2642 | -0.39927 | 1.41824 |
| 2.8 | 13 | 3.791522 | 4.430569 | $\rho$=7.29776 | $\varphi$=1.89920 | — | — |
| 2.8 | 13 | 4.621747 | 3.473271 | $\rho$=12.1489 | $\varphi$=1.49976 | — | — |
| 2.8 | 13 | 3.766182 | 4.842982 | 7.6521 | -8.1386 | -2.72469 | 0.82067 |
| 2.8 | 14 | 4.363598 | 3.478267 | 10.9485 | -14.9376 | 0.09972 | -1.37150 |
| 2.8 | 14 | 4.526641 | 3.492576 | -8.2747 | -14.9965 | 0.08747 | -0.79053 |
| 2.8 | 15 | 2.104981 | 2.892633 | 19.7328 | -16.1313 | 0.55963 | -1.75215 |
| 2.8 | 15 | 3.007712 | 2.902543 | $\rho$=17.0089 | $\varphi$=1.48191 | — | — |
| 2.8 | 15 | 1.047865 | 1.578465 | -0.2540 | -39.2925 | 65.6279 | 0.01395 |
| 2.8 | 15 | 2.099552 | 0.382328 | 8.7335 | -20.4340 | 1.08212 | 0.00785 |

(5-5)
$$c(1-c)\left[1 - c\gamma \exp(c\gamma - c)\right]/\left[1 - \exp(c\gamma - c)\right] = 1 \quad .$$

At $c = c_o$ the recurrence (2-6) undergoes a bifurcation : when $c > c_o$, the fixed point A becomes an unstable focus and there appears around it a regular closed invariant curve $\mathcal{L}$. The length S of $\mathcal{L}$ increases simultaneously with the difference $c-c_o$, and the shape of $\mathcal{L}$ deviates more and more from an oval (cf. Fig. 2-15), similarly to the $\mathcal{L}$ of the recurrence (5-1). The regular curve $\mathcal{L}$ ceases to exist when its expansion brings it too close to a cycle. The cycle in question is either a node or a saddle of order $k = 5$ and rotation number $r = 1$, which came into existence as a node-saddle at $c = c_1 = 2.180183$. When $c > c_1$ both cycles 5/1 undergo a large number of bifurcations of the type $\lambda = +1$ and $\lambda = -1$, producing a large number of "satellite" cycles. The "macro"-bifurcation sheme resembles somewhat that of Fig. 5-11, except that it involves more elements. The first three bifurcations are summarized in Tables 5-2a and 5-2b (cf. [G13], [C 5]). When $c > 2.5044$ there exist cycles whose orders are relatively prime with respect to $k = 5$. Some of these cycles are listed in Table 5-3. Like in the case of the recurrence (5-1), two or more stable cycles coexist in several (sub-) intervals of c.

By computing a large number of discrete trajectories $\{x_n, y_n\}$, $n \geqslant 0$ of (2-6), it was observed that there exists a finite number N, such that all points $(x_n, y_n)$, $n > N$ remain inside a bounded region D of the phase plane, regardless of the choice of the initial point $(x_o, y_o)$. D constitutes therefore a no-escape region. As c increases beyond $c_1$, the points $(x_n, y_n)$, $n > N$ fill D in a more and more chaotic manner (see for example Fig. 5-14a and b).

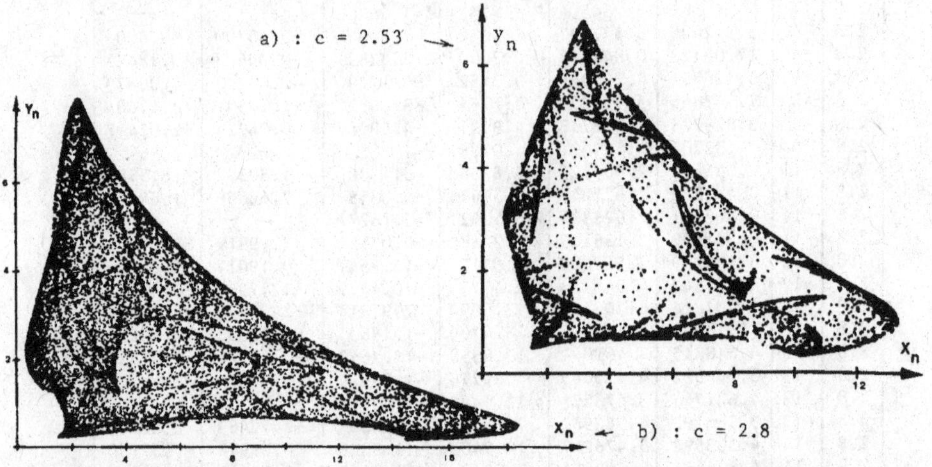


Fig. 5-14. Recurrence (2-6). Display of a chaotic discrete consequent trajectory.

By comparing the boundary $\mathcal{C}$ of D to the shapes of critical curves $J_n$, $n > 0$ it was found that $\mathcal{C}$ is composed of a rather small number of segments of the $J_n$ (cf. Fig. 5-15). The critical curves $J_o$ and its consequent $J_1$ have relatively simple explicit expressions :

(5-6) $$J_o \; : \; cx/K - \exp(ay) = 0 \quad ,$$

(5-7) $$J_1 \; : \; K\, c^{-1}.\ \exp\left[c-1 - cy/(bK)\right] = 0 \quad .$$

The curve $J_1$ separates the phase-plane into two regions where the consequent $J_1(x_n, y_n)$ of the Jacobian determinant $J(x_n, y_n)$ is positive and negative, respectively ; a point $(x_n, y_n)$ of the first region possesses two antecedents and a point of the second

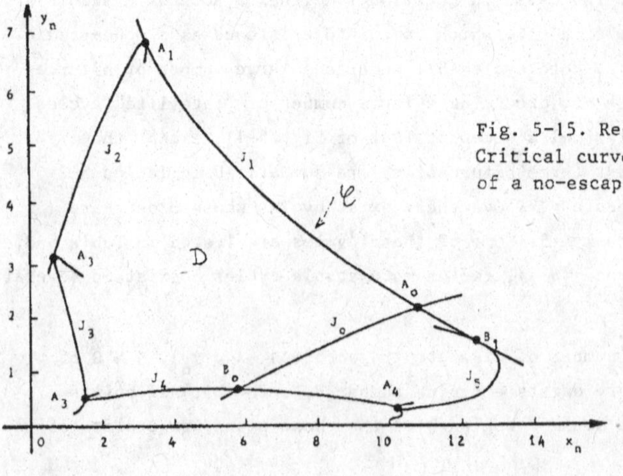


Fig. 5-15. Recurrence (2-6), c = 2.53. Critical curves constituting the boundary of a no-escape region.

none. For the parameters chosen the curves $J_o$ and $J_1$ of the recurrence (2-6) admit a point of intersection $A_o = (x_o, y_o)$ :

(5-8) $$x_o = (K/c)\ \exp\ (ay_o), \quad y_o = Kb(c-1)/(baK - c) \ .$$

The consequent $A_1 = (x_1, y_1)$ of A is a point of first-order contact between $J_1$ and $J_2$. In fact $A_1$ belongs to both $J_1$ and $J_2$, because $J_2$ is a consequent of $J_1$. $J_2$ and $J_1$ are tangent at $A_1$, because otherwise the curve $J_2$ would extend into the region $J_1(x_n, y_n)<0$, which is excluded by definition. The curve $J_2$ separates the region $J_1(x_n, y_n)>0$ into three subregions where the number of antecedents of rank two of $(x_n, y_n)$ is none, two and four. A similar separation is produced by the curves $J_n$, $n > 2$. Analoguously to $A_1$ and $A_o$, the consequents $A_m = (x_m, y_m)$, $m > 1$, of $A_o$ are fisrt-order contact points of the curves $J_m$ and $J_{m+1}$. Moreover, the consequent of the intersection point of the curves $J_o$ and $J_{m+1}$ is a first-order contact point of the curves $J_1$ and $J_{m+2}$. For the set of numerical parameters given before, there exists an additional intersection point $B_o = (\bar{x}_o, \bar{y}_o)$ of the curves $J_o$ and $J_4$, whose consequent $B_1 = (\bar{x}_1, \bar{y}_1)$ is a point of first-order contact of the curves $J_1$ and $J_5$. The domain D is defined by the curve segments $J_m$, $m = 1, 2, \ldots, 5$, joining the points $A_1$, $A_2$, $A_3$, $A_4$ and $B_1$. For $c = 2.53$ the coordinates of these points are given in Table 5-4. Examining the possible loca-

Table 5-4. Recurrence (2-6), c = 2.53. Coordinates of the junction points of the $J_n$ in Fig. 5-15.

| | $A_1$ | $A_2$ | $A_3$ | $A_4$ | $B_1$ |
|---|---|---|---|---|---|
| x | 3.21931 | 0.63852 | 1.50841 | 8.34541 | 12.5 |
| y | 6.85849 | 3.1044 | 0.49729 | 1.47858 | 1.45 |

tions of antecedents of a point $(x_n, y_n)$ inside D, the antecedent separation property of the curves $J_m$, $m = 1, 2, \ldots, 5$, implies that all consequents of $(x_n, y_n)$ remain inside D. D is, however, smaller than the composite influence domain of the recurrence (2-6). In fact, if an initial point $(x_o, y_o)$ is chosen outside of D, its consequents enter D after a finite number of iterations.

During the computation of the points displayed in Fig. 5-14, it was noted that the images obtained look substantially the same for the same number of points displayed, regardless of whether the first 1000, 2000, or more initial points were omitted. The consequent trajectories $(x_n, y_n)$, $n_o < n < n_1$, $n_1 - n_o \geqslant 5000$ of the recurrence (2-6) are therefore repetitive in a suitably defined sense when $2.5 \leqslant c \leqslant 2.8$, without being periodic. The repetitiveness suggests the existence of collective invariants, one example of which is the mean mass-density (cf. Chapter I). In order to determine the mean mass density inside the region of no escape D shown in Fig. 5-14b, D was partitioned into elementary cells by the use of the curvilinear coordinates $u = xy$ and $v = \log(y/x)$. The curves $u = \alpha_i$ = const., $v = \beta_j$ = const., $i, j = 1, 2, \ldots, 500$, divide D into 2500 approximately equal cells $D_{ij}$. A unit mass was assumed to exist at the initial point $(x_o^i, y_o^j)$ chosen by different methods inside each cell, and the corresponding discrete consequent trajectories were computed. The number of mass points $r_n(i, j)$ visiting each cell $D_{ij}$ was determined as a function of n. As expected, the numbers $r_n(i, j)$ were found to oscillate in an irregular manner. The amplitude of these oscillations turned out, however, to be so moderate that the mean sums

$$r_{N,m} = \frac{1}{N} \sum_{n=m}^{m+N} r_n(i, j) \tag{5-9}$$

showed little variation after a few hundred iterations. It was noted that the asymptotic values of $r_{N,m}$ are independent of the choice of the initial point inside each cell $D_{ij}$, and of the number m of "transient" points omitted. Hence, it may be assumed that there exists a limit density

$$r(i, j) = \lim_{N\to\infty} r_{N,m}(i, j) \tag{5-10}$$

independent of m, and of the subdivision of D into elementary cells $D_{ij}$. Expression (5-10) represents therefore a macroscopic invariant of the recurrence (2-6). For $N \simeq 700$, the larger values of r(i, j) are indicated by the darker areas in Fig. 5-16.

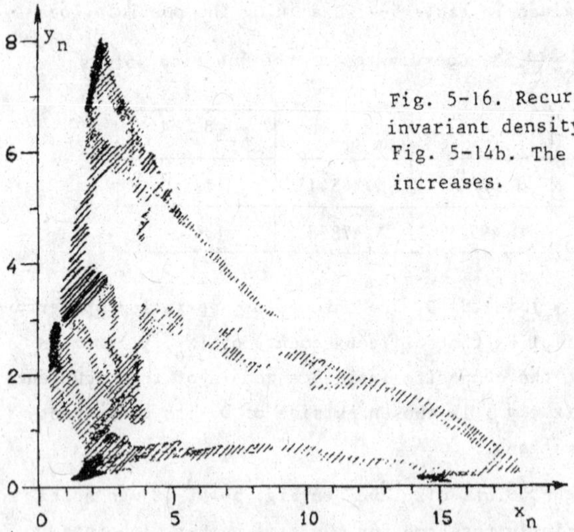


Fig. 5-16. Recurrence (2-6), c = 2.8. Approximate invariant density inside the chaotic region of Fig. 5-14b. The dark area shrinks slowly as N increases.

Contrary to the appearance of the discrete point displays (Fig. 5-14), the no-escape region D is therefore not a two-dimensional attractor. Whether the attractor inside D is simple or composite, one- or zero-dimensional, is still unknown. Analytical methods of solving the two-dimensional Perron-Frobenius equation

$$\iint_D w(x, y)\, dx\, dy = \iint_{D_i} w(x, y)\, dx\, dy \quad ,$$

where $D_i$ is the mapping of D into itself by means of (non-unique) antecedents, are insufficiently worked out to be of any help in this problem.

### 5.4 Chaos in a non-conservative recurrence with unique antecedents

A quadratic recurrence similar to (5-1), which possesses a chaotic behaviour inside a no-escape region D, is :

(5-11) $$x_{n+1} = y_n - a\,x_n^2 + 1, \qquad y_{n+1} = b\,x_n, \qquad a = 1.4,\ b = 0.3 \quad .$$

D exists not only bor the isolated parameter values given above, but in a certain range around them. The unique inverse of (5-11) is obtained by rearrangement :

(5-12) $$x_n = y_{n+1} / b, \qquad y_n = x_{n+1} + a\,x_n^2 - 1, \qquad b > 0 \quad .$$

A continuity argument suggests that the complexity generating mechanism of the recurrence (5-11) is induced by that of the single-variable degenerate recurrence

(5-13) $$x_{n+1} = -a\,x_n^2 + 1$$

with non-unique antecedents, into which (5-11) degenerates as $b \to 0$. The conjectured continuous dependence on b is confirmed by the analysis of the cycles of (5-11), a partial list of which is given in Table 5-5. These cycles have no identifiable rotation number. The rotation sequence is, however, typical of first-order recurrences, and it evolves smoothly into that of (5-13). The coordinates of two fixed points, and of one cycle of order two are known analytically.

(5-14) $$\begin{aligned} k = 1 &: ax^2 + cx - 1 = 0, \qquad c = 1-b, \qquad y = bx \quad , \\ k = 2 &: a^2x^2 - a\,c\,x + c^2 - a = 0, \qquad y = b(1 - ax^2)/c \quad . \end{aligned}$$

The expressions (5-14) facilitates a highly accurate determination of invariant curve germs.

The existence of a no-escape region D has been inferred from the observation of many discrete consequent trajectories $\{x_n, y_n\}$, $n \geqslant 0$, and in particular from the boundedness of $|x_n|$ and $|y_n|$ for large n [H 7]. A display of the points $(x_n, y_n)$ of $\{x_n, y_n\}$ produces no orderly pattern, and if it is assumed that the transient part of $\{x_n, y_n\}$ is rather short ($n < \bar{n} < 100$, [H 7]), then $\{x_n, y_n\}$, $n > \bar{n}$ appear to constitute a strange attractor of the Cantor type. The determination of $\{x_n, y_n\}$ was carried out with different accuracies, and with the same accuracy on different computers (see for example [C14]). The resulting points of $\{x_n, y_n\}$ are somewhat shifted, but the qualitative picture remains the same : the points $(x_n, y_n)$ show no tendency to accumulate when $\bar{n}$ is reasonably large ($\bar{n} \simeq 10^4$).

Unfortunately, long transients constitute a rather common property of non-conservative recurrences, even if it is known beforehand that the points $(x_n, y_n)$ approach asymptotically a point-singularity (for example when $\{x_n, y_n\}$ is initialized inside D, and D is known to contain a single attractive cycle). It is therefore impossible to infer the existence of a strange attractor from the sole observation of discrete trajectories of a finite length. Moreover, when a discrete trajectory

Table 5-5. Cycles of the recurrence (5-11).

| k | x | y | $\lambda_2$ | $\lambda_1$ | $p_2$ | $p_1$ |
|---|---|---|---|---|---|---|
| 1 | 0.631354 | 0.189406 | 0.15595 | -1.9237 | 0.15595 | 1.9237 |
| 1 | -1.131354 | -0.339406 | -0.09203 | 3.2598 | -3.2598 | 0.09203 |
| 2 | -0.475800 | 0.292740 | -0.02 | -3.01 | -1.2115 | -0.12074 |
| 2 | 0.975800 | -0.142740 | -0.029899 | -3.0101 | 2.4846 | 0.24763 |
| 4 | 0.217762 | 0.191458 | -0.000937 | -8.6394 | 0.71015 | -0.18421 |
| 6 | 0.485336 | 0.132573 | -0.000026 | -27.515 | 1.4729 | -0.27297 |
| 6 | -0.415159 | 0.311418 | $2.5920 \times 10^{-5}$ | 28.125 | -1.0567 | -0.094185 |
| 7 | 0.818035 | 0.154744 | $-6.1187 \times 10^{-6}$ | 35.743 | 2.7160 | -0.19022 |
| 7 | 0.694135 | 0.042436 | $1.2156 \times 10^{-5}$ | -17.991 | 2.2045 | -1.0890 |
| 7 | 0.851159 | 0.142703 | $8.4142 \times 10^{-6}$ | -25.992 | 3.0517 | -0.19048 |
| 7 | 1.000005 | 0.111205 | $-1.2401 \times 10^{-5}$ | 17.635 | 2.3810 | -0.23650 |
| 8 | 0.557519 | 0.110936 | $-7.2965 \times 10^{-7}$ | -89.920 | 1.7052 | -0.33233 |
| 8 | -0.389522 | 0.281492 | $-8.1815 \times 10^{-7}$ | -80.193 | -0.98473 | -0.12454 |
| 8 | -0.388170 | 0.310400 | $7.4522 \times 10^{-7}$ | 88.041 | -0.98454 | -0.095094 |
| 8 | -0.233652 | 0.295100 | $-1.4290 \times 10^{-6}$ | -45.914 | -0.56404 | -0.10022 |
| 8 | -0.140442 | 0.284806 | $1.9613 \times 10^{-6}$ | 33.453 | -0.30596 | -0.10415 |
| 8 | -0.002188 | 0.270009 | $-1.9510 \times 10^{-6}$ | -33.628 | 0.080240 | -0.11005 |
| 8 | 0.100420 | 0.259370 | $1.2654 \times 10^{-6}$ | 51.847 | 0.36939 | -0.11450 |
| 9 | 0.682265 | 0.178767 | $-1.2723 \times 10^{-7}$ | 154.70 | 2.0994 | -0.16539 |
| 9 | 0.617191 | 0.089500 | $1.4896 \times 10^{-7}$ | -132.13 | 1.9062 | -0.42548 |
| 9 | 0.692494 | 0.170378 | $1.8668 \times 10^{-7}$ | -105.44 | 2.1399 | -0.16052 |
| 9 | 0.679890 | 0.102065 | $-5.5643 \times 10^{-7}$ | 35.374 | 2.1409 | -0.39271 |
| 9 | 0.687523 | 0.099016 | $6.0282 \times 10^{-7}$ | -32.651 | 2.1263 | -0.40678 |
| 10 | 0.638229 | 0.173086 | $-1.8624 \times 10^{-8}$ | -317.06 | 1.9509 | -0.15215 |
| 10 | -0.488159 | 0.318206 | $2.4157 \times 10^{-8}$ | 244.44 | -1.2479 | -0.091833 |
| 10 | 0.619725 | 0.128797 | $2.9287 \times 10^{-8}$ | 201.62 | 1.9024 | -0.31257 |
| 10 | -0.427269 | 0.312541 | $-2.4529 \times 10^{-8}$ | -240.73 | -1.0890 | -0.093782 |
| 10 | 0.631026 | 0.125143 | -3.0848 | -191.42 | 1.9413 | -0.32208 |
| 10 | 0.658631 | 0.170069 | $1.9924 \times 10^{-8}$ | 296.37 | 2.0204 | -0.15553 |
| 10 | 0.707114 | 0.167122 | $-4.9848 \times 10^{-8}$ | -118.46 | 2.2028 | -0.16366 |
| 10 | 0.720878 | 0.164854 | $4.9885 \times 10^{-8}$ | 118.37 | 2.2301 | -0.16595 |
| 11 | 0.644350 | 0.184934 | $-2.9966 \times 10^{-9}$ | 591.17 | 1.9678 | -0.16003 |
| 11 | 0.624400 | 0.121691 | $9.4445 \times 10^{-9}$ | -187.57 | 1.9201 | -0.32189 |
| 12 | 0.588053 | 0.132827 | $-1.2955 \times 10^{-9}$ | -410.21 | 1.7977 | -0.29250 |
| 12 | -0.465705 | 0.290831 | $-7.2563 \times 10^{-10}$ | -732.44 | -1.1854 | -0.12072 |
| 12 | 0.211043 | 0.193227 | $7.6648 \times 10^{-10}$ | 693.36 | 0.69067 | -0.18298 |
| 12 | 0.282118 | 0.237830 | $7.0523 \times 10^{-10}$ | 753.60 | 0.89032 | -0.12497 |
| 12 | 0.528063 | 0.204903 | $-4.5961 \times 10^{-10}$ | -1156.4 | 1.5902 | -0.14457 |
| 12 | 0.540419 | 0.193182 | $-3.1905 \times 10^{-9}$ | -166.57 | 1.6325 | -0.14036 |

Table 5-5. continued.

| k | x | y | $\lambda_2$ | $\lambda_1$ | $p_2$ | $p_1$ |
|---|---|---|---|---|---|---|
| 12 | -0.390652 | 0.309083 | $-6.8300\times10^{-10}$ | -778.19 | -0.99105 | -0.094988 |
| 12 | -0.432188 | 0.314554 | $-7.2740\times10^{-10}$ | -730.69 | -1.1022 | -0.093707 |
| 12 | 0.586206 | 0.133368 | $1.3069\times10^{-9}$ | 406.65 | 1.7916 | -0.29123 |
| 12 | -0.486054 | 0.319558 | $7.1313\times10^{-10}$ | 745.24 | -1.2428 | -0.092060 |
| 12 | 0.309917 | 0.234155 | $3.1057\times10^{-9}$ | 171.12 | 0.97225 | -0.12679 |
| 12 | 0.564106 | 0.199572 | $5.0042\times10^{-10}$ | 1062.1 | 1.7061 | -0.14829 |
| 12 | 0.328101 | 0.232013 | $-6.6299\times10^{-10}$ | -801.60 | 1.0255 | -0.12806 |
| 12 | 0.572544 | 0.198144 | $-7.4209\times10^{-10}$ | -716.13 | 1.7332 | -0.14920 |
| 13 | 0.614596 | 0.181610 | $-2.1231\times10^{-10}$ | 751.04 | 1.8712 | -0.14997 |
| 13 | 0.645450 | 0.187136 | $-7.2191\times10^{-11}$ | 2205.6 | 1.9709 | -0.15775 |
| 13 | 0.580474 | 0.140457 | $3.2475\times10^{-10}$ | -490.93 | 1.7714 | -0.28561 |
| 13 | 0.631694 | 0.179866 | $1.1022\times10^{-10}$ | -144.60 | 1.9274 | -0.15179 |
| 13 | 0.656152 | 0.183839 | $2.1561\times10^{-10}$ | -739.43 | 2.0076 | -0.16081 |
| 13 | 0.599652 | 0.178848 | $1.2201\times10^{-10}$ | -1305.8 | 1.8236 | -0.14599 |
| 13 | 0.670445 | 0.173877 | $-3.3143\times10^{-10}$ | 481.05 | 2.0597 | -0.15722 |
| 13 | 0.614903 | 0.190489 | $-1.1534\times10^{-10}$ | 1381.6 | 1.8699 | -0.15543 |
| 13 | 0.665337 | 0.183974 | $7.9524\times10^{-11}$ | -2003.3 | 2.0387 | -0.16035 |

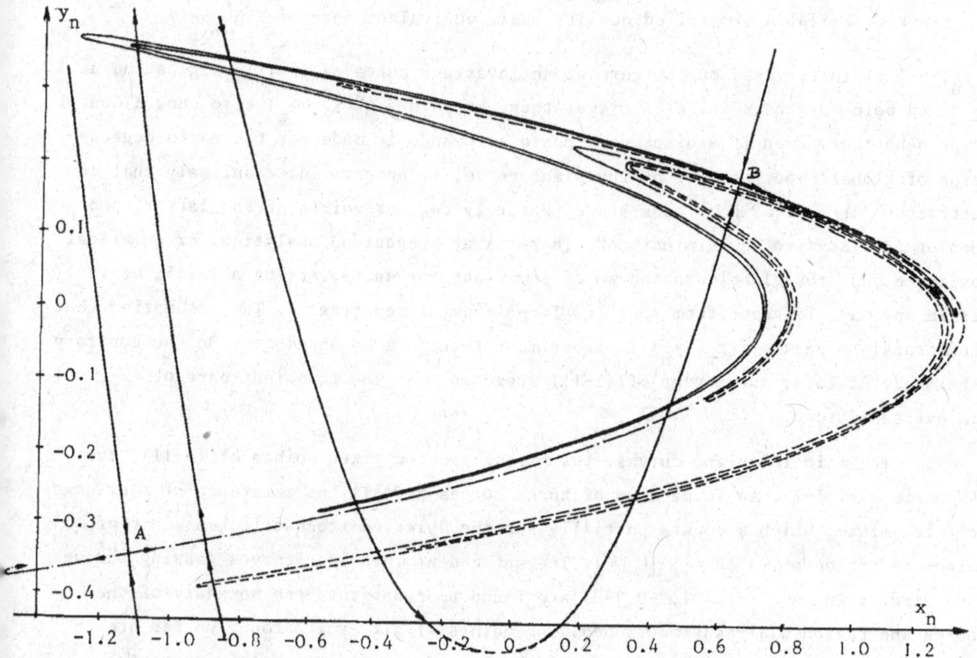

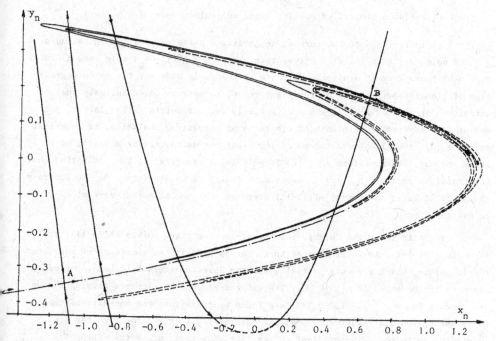

Fig. 5-17. Invariant curves traversing two fixed points of the recurrence (5-11).

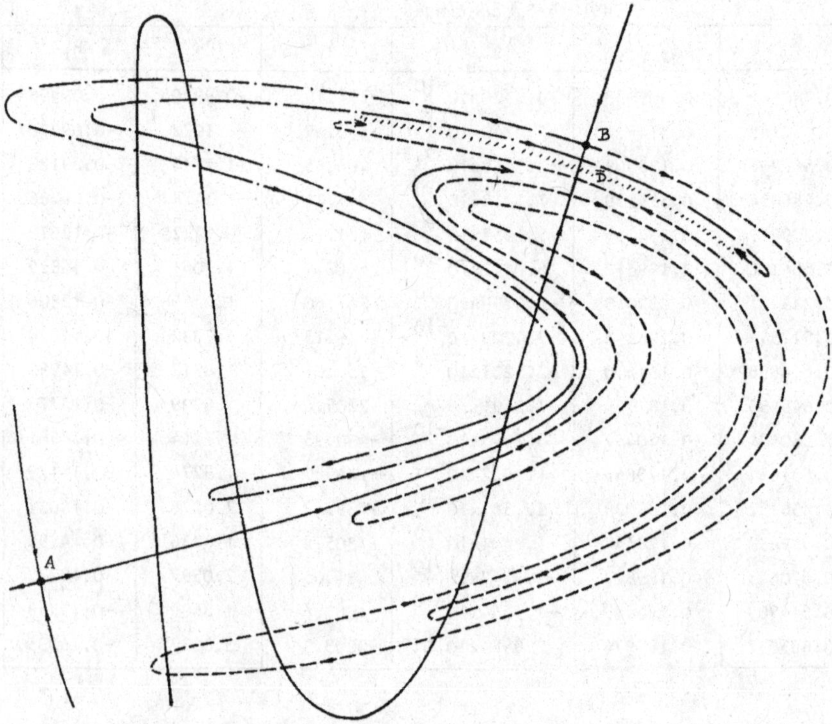


Fig. 5-17a. A simplified qualitatively equivalent form of Fig. 5-17.

$\{x_n, y_n\}$ is initialized on the germ of an invariant curve of (5-11) originating at a fixed point or the point of a cycle, then the points of $\{x_n, y_n\}$ also show a Cantor type behaviour. Even if a disproportionate allowance is made for the rapid degradation of global smoothness of an invariant curve, it appears quite unlikely that an attractor (strange of not) consists of entirely regular points of the latter, and not of its "asymptotic termination". There is at present no analytical or numerical evidence that the global smoothness of invariant curves traversing a saddle of (5-11) is in any way different from that of otherpolynomial recurrences. The assumption that the transient part of $\{x_n, y_n\}$ is short must therefore be abandoned. On the contrary, the study of invariant curves of (5-11) suggests that the transient part of $\{x_n, y_n\}$ is extremely long.

The main invariant curves, traversing the two fixed points of (5-11), are shown in Fig. 5-17. An inspection of these curves reveals the existence of many homoclinic points, which accounts partially for the observed stochastic behaviour of discrete trajectories [H 7], [C 14]. The antecedent invariant curves passing through the fixed point A ≃ (-1.131, -0.339) are found to constitute the boundary of the no-escape region D already mentioned. The points of all cycles found so far are located close to the main invariant curves. The invariant curves traversing the latter follow the outline of the main invariant curves. This property is not abnormal in view of the fact that all cycles found so far are saddles, i.e. at the parameter values a = 1.4, b = 0.3 these cycles are far from their bifurcational origin, where some of them were stable nodes.

The global properties of the main invariant curves can be used to estimate the part of D inside of which an (eventually composite) attractor of (5-11) is located. Fig. 5-17 can be redrawn without any qualitative loss by straightening out unessential S-loops, so that the existence of a smaller no-escape region $\bar{D}$ (Fig. 5-18) becomes apparent. The estimate of the size of $\bar{D}$ can be refined by adding to Fig. 5-18 some consequent invariant curves emanating from cycles. This was done for several low order saddles listed in Table 5-5. The area of $\bar{D}$ appears to shrink more and more as longer invariant curve segments are computed. Because of essentially numerical limitations, it is not possible to establish that the area of $\bar{D}$ shrinks indefinitely. It appears, however, that the consequent invariant curves approach asymptotically an attractor located at most on a one-dimensional manifold.

In order to obtain some additional information about the possible singularity structure inside $\bar{D}$, the variational entropy $h_v$ was determined for several initial conditions $(x_o, y_o)$ and max $n \approx 10^4$. It was found that $h_v = 0.623$. The only information furnished by this result is that the elementary or composite attractor inside $\bar{D}$ has a very complicated influence domain structure, or in other words, that the transient leading to it is rather long.

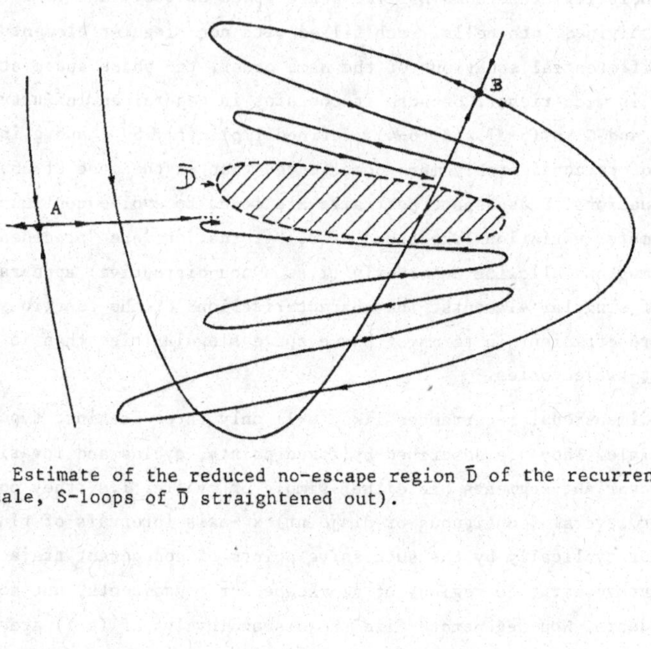


Fig. 5-18. Estimate of the reduced no-escape region $\bar{D}$ of the recurrence (5-11). (not to scale, S-loops of $\bar{D}$ straightened out).

## Chapter VI

## CONCLUSION AND SOME OPEN PROBLEMS

The study of particular first and second order recurrences of form

(6-1) $$x_{n+1} = f(x_n, y_n, c),\ y_{n+1} = f(x_n, y_n, c), \quad n = 0, \pm 1, \pm 2, \ldots$$

and

(6-2) $$x_{n+1} = g(x_n, y_n, c), \quad y_{n+1} = f(x_n, y_n, c),\ n = 0, \pm 1, \pm 2, \ \ldots$$

with sufficiently smooth right-hand sides, shows that it is possible to describe their general solutions

(6-3) $$x_n = F(x_c, n, c)$$

and

(6-4) $$x_n = G(x_o, y_o, n, c), \quad y_n = F(x_o, y_o, n, c)$$

in terms of phase space portraits. For any fixed value of the parameter c, the phase portrait of (6-1) or (6-2) is characterized by singular elements, or for short singularities, corresponding to sets of initial points $(x_o)$, or $(x_o, y_o)$ for which some salient evolution property of $(x_n)$ or $(x_n, y_n)$ becomes apparent as either $n \to +\infty$ or $n \to -\infty$. The simplest singularities describe static and dynamic steady states, and the boundaries of their influence domains. The phase space of (6-1) and (6-2) can in principle be subdivided into cells, each filled with non-singular elements. In contrast to analoguous differential equations of the same order, the phase space structure of (6-1) and (6-2) is much richer, because it contains in general an infinity of cells. The functions F and G in (6-3) and (6-4), defined implicitly by f and g in (6-1) and (6-2), are therefore considerably more complicated than in the case of analoguous differential equations. Phase space portraits are found to evolve continuously with respect to parameter variation like those of c. Bifurcations are "preceded" by smooth structure deformations allowing an orderly (i.e. a non-disruptive) appearance or disappearance of singular elements. The characterizations of the functions F and G is obviously more efficient in terms of phase space singularities than in terms of (discrete) point-trajectories.

In one-dimensional recurrences like (6-1) only three distinct types of steady states are possible. They are described by fixed points, cycles and invariant segments, respectively. Invariant segments are either simple or cyclic, i.e. they consist either of one or several (contiguous or disjoint) $x_n$-axis intervals of finite size, visited simply or cyclically by the successive points of consequent trajectories. Invariant segments constitute regions of no-escape for consequents, but not necessarily for antecedents. Non-degenerate fixed points and cycles of (6-1) are described by simple roots of the algebraic equations

(6-5) $$x_{n+k} = x_n, \qquad k = 1, 2, \ldots$$

Their relationship to neighbouring points is conveniently described in terms of the notion of stability or instability in the sense of Liapunov. A non-degenerate fixed point or cycle of (6-1) is thus either asymptotically stable (attractive) or unstable (repulsive). Since the function f in (6-1) is assumed to be smooth, local stability or instability of a point-attractor (i.e. of a fixed point or of a cycle) can be deduced from its eigenvalue $\lambda = f'(x, c)$, $' = d/dx$. $|\lambda| < 1$ implies local stability and $|\lambda| > 1$ instability. Multiple roots or one-dimensional loci of roots of (6-5) require a special examination. Another characteristic property of a cycle of order $k > 2$ is its rotation sequence, defined by the order of successively appearing consequents compared to the positive direction of the $x_n$-axis.

Every asymptotically stable fixed point or cycle of (6-1) possesses a total influence domain, which in general consists of an immediate and a subsidiary part. Since the immediate influence domain is always one-dimensional (i.e. it consists of an interval of non-zero length), point-attractors are always internal points of their immediate influence domains. The subsidiary part of the influence domain, if it exists, is given by excess antecedents of the immediate influence domain. The boundary of the total influence domain consists of points belonging to at least one of the following sets : a) unstable fixed points (simple or multiple) ; b) points of unstable cycles (simple or multiple) ; c) excess antecedents of a) and b) ; d) accumulations points of b) and c). Discrete trajectories initialized on a point of d) may appear to be highly unusual in an abstract context, but dynamically they are not different from those initialized on a), b) or c), and carry thus no additional qualitative or quantitative information.

As a rule invariant segments of (6-1) have also an influence domain, but exceptionally they are internally closed, i.e. there exist no phase space points whose consequents become internal points of the invariant segments after a finite number of steps. Invariant segments are therefore either asymptotically or "neutrally" stable. They may occur as isolated entities or as parts of composite influence domains.

Multiple roots of the algebraic equations (6-5) play an important role in the bifurcations of the recurrence (6-1). When f is smooth such roots represent cycles with the eigenvalue $\lambda = 1$ or $\lambda = -1$. The simplest bifurcation of (6-1) occurs at $\lambda = +1$ and it involves the emergence of a new cycle pair from the domain of complex numbers by the coalescence of two complex-conjugate roots into a real double root of (6-5). When this double root splits into two simple ones, it gives rise to two cycles of the same order, one asymptotically stable and the other unstable. The value $\lambda = -1$ characterizes the destabilization of an asymptotically stable cycle (or vice-versa), and is accompanied by the emergence (or disappearance) of a cycle pair of double order. New cycles of (6-1) appear therefore either spontaneously (via the splitting of roots of (6-5) of even multiplicity), or they are spun off by other cycles (via the splitting of roots of (6-5) of odd multiplicity). Invariant segments are produced by a qualitatively different bifurcation : the coalescence of an unstable fixed point (or a point of an unstable cycle) with a consequent of an extremal point of the function f in (6-1), i.e. with a critical point in the sense of Julia-Fatou. But every coalescence of a critical point and an unstable cycle point does not necessarily yield an invariant segment. The result is usually only a minor modification of the subsidiary part of the influence domain of some asymptotically stable cycle. Since there exist no unstable invariant segments, the coalescence of a critical point with an asymptotically stable fixed point or cycle point changes merely the sign of the eigenvalue of the latter, i.e. inside the immediate influence domain a monotonic transient becomes an oscillatory one, or vice versa.

In a recurrence of form (6-1) with a single extremum (and convexity), there exists for a fixed c only one attractor in the finite part of the phase space. It is a fixed point, a cycle or a cyclic set of invariant segments, and exceptionally a single internally closed invariant segment. If f possesses several extrema, the coexistence of several attractors is not only possible, but it is more a rule than an exception. Any motion described by a consequent trajectory initialized inside the total influence domain of a point-attractor is essentially orderly from a dynamic point of view, because consequents of a sufficiently high rank of the initial point will enter the immediate influence domain and then approach the point attractor exponentially (with oscillations when the point attractor has a negative eigenvalue). The traversal of the subsidiary influence domain may be, however, quite long, i.e. it may require a large, although a finite number of iterations. The impression of chaos produced by this irregularly wandering transient is especially strong when the influence domain of an attractor is entirely contained inside the composite influence domain (of a finite size) of several attractors. At a given (finite) level of accuracy there exists thus considerable uncertainty about the future evolution of the motion, and in particular it may be difficult or even impossible to predict on which specific attractor the motion will terminate (i.e. into which specific immediate influence domain it will eventually enter). Such a type of unpredictability is a prominent feature of many devices used for entertainment or gambling (for example, coins, dice, lottery wheels, bingo boards, roulettes, etc.). In each case the motion is completely orderly and deterministic, but the subsidiary influence domains of the various attractors are highly interlaced, and the size-ratios between immediate and subsidiary influence domains are rather low. The device is unbiased when these ratios and the sizes of the total influence domain are the same for all attractors with the same gaming return.

Another impression of chaos arises during the observation of a motion described by a discrete consequent trajectory evolving inside an invariant segment, or inside a set of cyclic invariant segments. For simplicity consider that the recurrence (6-1) admits for $c = c_o$ a single invariant segment X. By definition X is a no-escape interval of finite length for all consequent trajectories $\{x_n\}$, $n > 0$ if $x_o$ is inside X. When $f(x_n, c_o)$ is sufficiently smooth with respect to $x_n$, then in general $x_n = F(x_o, n, c_o)$ turns out to be continuous and at least continuously differentiable with respect to $x_o$. The derivative $S_n = \frac{\partial x_n}{\partial x_o} = \frac{\partial F}{\partial x_o}$ is therefore a meaningful sensitivity coefficient of the dynamic steady state motion inside X. Differentiating (6-1) with respect to $x_o$ it is easily seen that $S_n$ satisfies the recurrence

(6-6) $$S_{n+1} = \left[\frac{\partial}{\partial x_n} f(x_n, c_o)\right] . S_n \quad ; \qquad S_o = 1 \quad .$$

The discrete "trajectory" $\{S_n\}$, $n > 0$, can thus be determined simultaneously with the discrete trajectory $\{x_n\}$, $n > 0$, for any fixed $x_o$, without the knowledge of the solution (6-3) of (6-1). If $x_o$ is an internal point of the influence domain of a point-attractor, then obviously $S_n \to 0$ as $n \to +\infty$. If $x_o$ is an internal point of an invariant segment X, then just as obviously $\max |S_n| \to \infty$ as $n \to +\infty$. These asymptotic properties can be used to discriminate numerically between an invariant segment and the total influence domain of a point-attractor of an unknown order. Since the growth of $\max |S_n|$ is easily detectable, the discrimination is quite sharp, but the values of n required may be large. As an illustration, consider the Myrberg recurrence with $c = -2$. Letting $x = 2y$, one obtains instead of (1-21) and (1-29)

(6-7) $$y_{n+1} = 2y_n^2 - 1, \qquad Y : -1 < y_n < +1$$

and

(6-8) $$y_n = F(y_o, n) = \cos\,(2^n . \text{arc cos } y_o) \quad .$$

The sensitivity coefficient inside the invariant segment Y is therefore

(6-9) $$S_n = \frac{\partial y_n}{\partial y_o} = \frac{\partial}{\partial y_o} F(y_o, n) = -\frac{2^n}{\sqrt{1-y_o^2}} \sin\,(2^n . \text{arc cos } y_o) \quad ,$$

and its modulus increases rapidly with n. Hence two initially close discrete trajectories $\{y_n\}$, $n > 0$, cease to be close as n increases. Although the geometrical distance between $y_o$ and $y_n$ is always finite ($y_o$ and $y_n$ remain inside Y), the "effective" distance, not subjected to the modulo-$2\pi$ reduction, grows without limit as $n \to \infty$. The dynamic situation is somewhat analoguous to the large-amplitude motion of an ideal pendulum (described by $\ddot{y} + \sin y = 0$), whose oscillation period is a function of the amplitude. The effective separation between two initially close oscillations is represented by the phase difference, which increases of course indefinitly with time, although the geometrical distance between the extremal points [max y(t)] and [min y(t)] cannot exceed twice the length of the pendulum. Contrary to the case of the pendulum an impression of chaos arises in the case of (6-7) because at a given (in practice necessarily finite) level of accuracy it is impossible to predict the location of $y_n$ from $y_o$ when n is large. The mathematical source of this chaos is the loss of continuity of F and $\frac{\partial F}{\partial y_o}$ with respect to $y_o$ as $n \to \infty$. For $|y_n| > 1$, the solution of the recurrence (6-7) is orderly

(6-8a) $$y_n = F(y_o,n) = \frac{1}{2}\left[(y_o+\sqrt{y_o^2-1})^\alpha + (y_o-\sqrt{y_o^2-1})^\alpha\right], \qquad \alpha = 2^n \quad ,$$

but both $y_n$ and $S_n$ increase indefinitly as $n \to \infty$.

When a recurrence like (6-1) is considered together with a functional equation, then a reasonably smooth solution of the latter represents a collective invariant of the former, because it relates structurally neighbouring values of $x_n$ and removes the dependence on n. Mathematically the choice of functional equations is unlimited, but only a few of them happen to possess a dynamically meaningful content. This is in particular so for the Perron-Frobenius equation

(6-10) $$w(x) = \frac{d}{dx}\int_{s_o}^{s} w(\xi)d\xi \quad , \qquad s = f_{-1}(x, c) \ , \ x \in X \ ,$$

s being the generally multivalued inverse of f, the Schröder equation

(6-11) $$v(f(x, c)) = \lambda\, v(x, c) \quad ,$$

and its formal inverse, the Picard equation

(6-12) $$u(\lambda x) = f(u(x))$$

In fact, let $z = u(x)$, which implies $x = u^{-1}(z)$, then (6-12) becomes

(6-11a) $$\lambda u^{-1}(z) = u^{-1}(f(z)) \quad ,$$

i.e. it is of the same form as (6-11), provided $v(x) = u^{-1}(x)$. In order to illustrate the relationship between the general solution F of the recurrence (6-1) and the solution w, v and u of (6-10), (6-11) and (6-12), respectively, assume that x represents internal points of an invariant segment X. Assume furthermore that (6-11) and (6-12) admit smooth solutions on X for at least one suitably determined value of $\lambda$, i.e. assume that (6-11) and (6-12) represent eigenvalue-type problems defined on X. In the particular case of the recurrence (6-7), it is easily verified that $\lambda = 2$, and that the Schröder and Picard functional equations

(6-13) $$v(2y^2 - 1) = 2\, v(y) \quad ,$$

(6-14) $$u(2y) = 2u^2(y) - 1 \quad ,$$

admit the solutions

(6-15) $$v(y) = C \ \arccos y \quad ,$$

(6-16) $$u(y) = \cos y \quad ,$$

respectively, where C is an arbitrary constant. In order to verify (6-15) directly, note that

$$\arccos \alpha + \arccos \beta = \arccos\,(\alpha\beta - \sqrt{(1-\alpha^2)(1-\beta^2)})$$

and let $\alpha = \beta$. The corresponding solution of the Perron-Frobenius equation is (cf.

(1-44)) :

(6-17) $$w(y) = A/\sqrt{1 - y^2} \quad ,$$

where A is also an arbitrary constant. Combining (6-8), (6-15), (6-16) and (6-17), there results (subject to $F(y_o, 0) = y_o$)

(6-18) $$y_n = F(y_o, n) = u(\lambda^n v(y_o))$$

and (cf. (1-45)),

(6-19) $$w(y) = B \frac{d}{dy} v(y), \qquad B = CA \quad .$$

Hence, the existence of any one of u, v, w and F implies the existence of all others. As can be seen by inspection of (6-13) - (6-17), the formulation (6-18) is more general than that expressed in terms of the solution of the Schröder equation alone (cf. 2-80a))

(6-20) $$y_n = v^{-1}(\lambda^n v(y_o)) \quad ,$$

because the solutions of (6-13) and (6-14) exist and are single valued in essentially different intervals of y. These intervals are $Y_v : -1 < x < +1$ for (6-13) and $Y_u = -\infty < y < +\infty$ for (6-14). Moreover the inverse (6-15) of (6-16) is not unique, it represents a sort of principal inverse having a maximally extended independent variable range. The corresponding minimal interval of (6-16) is $Y_{um} : 0 < y < \pi$. The minimality of $Y_{um}$ is conditioned by the property that $Y_{um}$ is sufficient to cover the full range of $Y_v$. $Y_{um}$ is much smaller than $Y_u$ or the natural (single period) interval $Y_{un} : -\pi < y < +\pi$ of $u = \cos y$. In general the existence interval (whether maximal minimal or "natural") of a solution of (6-12) is not known. It has to be determined simultaneously with the (generalized) eigenvalue $\lambda$. The existence interval of a solution of (6-11) coincides usually with the invariant segment X, but in some cases it may comprise the whole influence domain of X.

The solutions (6-15) and (6-16) of (6-13) and (6-14) are not only smooth, they are analytic. If it is however attempted to determine their Mc Laurin series

(6-15a) $$v(y) = \frac{\pi}{2} - \sum_{n=0}^{\infty} a_n \frac{y^{2n+1}}{2n+1}, \quad a_o = 1, \; a_n = \frac{1}{2}\cdot\frac{3}{4}\cdot\frac{5}{6}\cdot\ldots\cdot\frac{2n-1}{2n}, \; n > 0 \; ,$$

and

(6-16a) $$u(y) = \sum_{n=0}^{\infty} b_n y^{2n}, \quad b_o = 1, \quad b_n = (-1)^n / (2n)!, \quad n > 0 \; ,$$

directly from the corresponding functional equations, great difficulties are encountered. For example, successive differentiations of (6-13) and (6-14) lead to indeterminacies, and the method of undetermined coefficients triggers a combinatorial explosion. The usual method of power series inversion (see for example p. 259 of [P 3]) does not apply, because $a_o \neq 0$ in v(y) implies then that u(y) contains a non-vanishing linear term in y (which of course it does not). The knowledge that u(y) is a periodic function of a known period is also of limited help, because the insertion of a Fourier series with undetermined coefficients into (6-14) yields a system of simultaneous algebraic equations which cannot be solved by the Kantorovich method of reduction (see for example [K 1] and Chapter 14 of [K 2]). There is therefore substantial need of gaining more insight into the fundamental properties of the functional equations (6-13) and (6-14), and a fortiori into those of (6-11) and (6-12). The usual study of (6-11) near an asymptotically stable fixed point of the recurrence (6-1), say $x = 0$ with the eigenvalue $0 < \lambda < 1$, is motivated by mathematical convenience ant not by any intrinsic subject-matter. In fact, the invariant segment Y of the recurrence (6-7) possesses an infinity of point-singularities, none of which is stable. As the example (6-7) shows, the (generalized) eigenvalue problems associated with the Schröder and Picard equations are not only meaningful, but they may admit even analytic solutions.

When the existence of an invariant segment X (or of a set of k cyclic segments $X_j$, $j = 1, 2, \ldots, k$) of the recurrence (6-1) has been established for some $c = c_o$ (for example, by means of the bifurcation involving a favourable coalescence of unstable point-singularities with critical points), then it is usually possible to determine the approximate invariant density $w(x)$ in a routine manner, for example by means of the stationary mean-mass distribution algorithm, based on equation (1-41). Once $w(x)$ is known with sufficient accuracy, an approximate solution $v(x)$ of the Schröder equation (6-11) requires only one quadrature (see for example pp. 271-325 of [K 4] for numerical algorithms). Knowing $v(x)$, it is possible to estimate the value of $\lambda$, and simultaneously determine the accuracy of all preceding computations, by rewritting (6-11) in the form

(6-21) $$\lambda = v(f(x, c_o)) \,/\, v(x, c_o) \simeq \text{constant} \quad , \quad x \in X \quad .$$

The next step towards the general solution of (6-1) is, however, not straightforward at all, because even a single valued $u(x)$ is usually not a simple inverse of $v(x)$, i.e. the curve $x = v(y)$ is not the y-image of one piece of the curve $y = u(x)$, but the superposition of y-images of many pieces of not necessarily equal sizes. The simplest kind of superposition takes place only if $u(x)$ is a periodic function of x, but this simplifying structural property cannot be postulated a priori. Its existence for a given recurrence and invariant segment must be ascertained independently. The Picard equation (6-12) is therefore not trivially equivalent to the Schröder equation (6-11), in spite of the formal equivalence expressed by (6-11a).

As an illustration of such a situation (fundamentally more complicated than that of the recurrence (6-7)), consider the two cyclic invariant segments of the Myrberg recurrence (1-21) for $c = c_o \simeq -1.543689013$ (cf. Fig. 1-11). One of these segments can be described by the iterated recurrence

(6-22) $$x_{n+2} = x_n^4 + 2c_o\, x_n^2 + c_o^2 + c_o, \qquad X : \bar{x} < x_n < -\bar{x}$$

where $\bar{x} = \frac{1}{2}(1 - \sqrt{1 - 4c_o}) \simeq -0.839286755$. The change of scale $x = \bar{x}y$ reduces (6-22) to

(6-23) $$y_{n+2} = -1 + (2+a)\, y_n^2 - a\, y_n^4, \qquad Y : -1 < y_n < +1 \quad ,$$

where $a = a_o \simeq 0.591195485$. If (6-23) is interpreted as a description of a simple invariant segment instead of a cyclic one, then $y_{n+2}$ may be replaced by $y_{n+1}$, and (6-23) becomes a smooth family of first order recurrences

(6-24) $$y_{n+1} = -1 + (2+a) y_n^2 - a\, y_n^4, \qquad Y : -1 < y_n < +1 \quad ,$$

a being a free parameter, into which (6-7) is imbedded for $a = 0$. The invariant density $w(y)$ of (6-24) is given by the (unsymmetrical) curve segment cutting the w-axis in Fig. 1-18. The integral of $w(y)$ is a single valued monotonic function of y, which is quantitatively but not qualitatively different from (6-15). At least when $|a| < 1$, the eigenvalue of the Schröder equation (6-11) is $\lambda = 2$, independently of a. The corresponding Picard equation to be solved is therefore

(6-25) $$u(2y) = -1+(2+a)\, u^2(y) - a\, u^4(y), \qquad Y_u : -\infty < y < +\infty, \quad |a| < 1 \, .$$

All efforts to show that (6-25) possesses a non-constant periodic solution have failed so far. The unavailability of a smaller natural or minimal interval replacing $Y_u$ renders a direct numerical exploration of the solutions (6-25) highly impractical. If one tries the successive iterations

(6-26) $$u_{n+1}(y) = \left[\alpha + \sqrt{\alpha^2 - (1+u_n(2y))/a}\right]^{\frac{1}{2}} \quad ,$$
$$u_o(y) = \cos y, \; \alpha = (2+a)/2a \qquad , \quad 0 < y < 4\pi \quad ,$$

it is found that they converge to a constant, and are therefore uninformative. The iterations

(6-27) $$u_{n+1}(2y) = -1 + (2+a) u_n^2(y) - a\, u_n^4(y), \qquad u_o(y) = \cos y, \quad y \in Y_u \, ,$$

can be studied analytically. $u_n(y)$ is of the form

(6-28) $$u_n(y) = \cos y + P_{\bar{m}}(y), \qquad P_{\bar{m}}(y) = \sum_{m=0}^{\bar{m}} b_m \cos \frac{m}{2^{n-1}} y, \qquad \bar{m} = \max m,$$

where the integer $\bar{m}$ grows very rapidly with n. The constants $b_m$ are polynomials in "a" with rather small coefficients (less than unity in modulus). Inside the index-interval $\frac{1}{2} < m/2^{n-1} < \frac{3}{2}$, the $b_m$ start with linear terms in a, and outside of it with quadratic or higher power terms. For $|a| < 1$ and $n \to \infty$ the trigonometric polynomials $u_n(y) = \cos y + P_{\bar{m}}(y)$ converge therefore to a bounded function $u_\infty(y)$, which is smooth, oscillatory but non-periodic. If $u_\infty(y)$ is the only natural extension to $a \neq 0$ of the periodic solution $u(y) = \cos y$, then the dynamically relevant inverse of $v(y) = \frac{1}{B} \int w(y)\, dy$ is an almost-periodic function (approaching cos y uniformly as $a \to 0$). The lack of periodicity of $u_\infty(y)$ is a possible explanation of the failure of regular conjugate transformations to account for the dissymetry of the invariant density $w(y)$. The chaos generated by an invariant segment appears thus as a consequence of an orderly sampling (at exponentially or at least monotonically increasing abscissae) of an almost-periodic function (and only exceptionally of a periodic one).

This conclusion corroborates the results of pseudo-randomness tests applied to discrete trajectories (of finite length).

Once a solution of the Picard functional equation (6-25) has been found, the general solution of the second order recurrence (6-23) takes the form

$$y_n = u\left[(C_1 + (-1)^n C_2)(\sqrt{2})^n\right] , \qquad n \geqslant 2 \quad ,$$

(6-29)

$$C_1 = \frac{1}{2} v(y_o) + \frac{\sqrt{2}}{4} v(y_1) , \quad C_2 = \frac{1}{2} v(y_o) - \frac{\sqrt{2}}{4} v(y_1),$$

where $y_o$ is an initial point inside Y, and $y_1$ its consequent evaluated by means of the first order recurrence (1-21) with $c = c_o$. Contrary to appearances, the general solution (6-29) of (6-23) depends thus only on one free parameter (the initial point $y_o$). This is not necessarily so when $a \neq a_o$, because the appropriate parametric family of first order recurrences, possessing two cyclic invariant segments for all a, $|a| < 1$, is not yet known.

When the dependence of the solution of the Picard equation on an arbitrary function is properly taken into account, then the relationship existing between the general solution (6-3) of (6-1) and the Schröder and Picard functional equations permits to construct many recurrences with analytically known particular solutions, using the Myrberg example (6-7), (6-8) and (6-13) - (6-16) as a guide (cf. [S8]). Some simple examples are :

(1) $$\begin{cases} x_{n+1} = 4x_n^3 - 3x_n \quad , \\ x_n = \cos (3^n \text{ arc cos } x_o) \quad \text{or ch } (3^n \text{ arg ch } x_o) \quad , \end{cases}$$

(2) $$\begin{cases} x_{n+1} = 16\, x_n^5 - 20\, x_n^3 + 5x_n \quad , \\ x_n = \cos (5^n \text{ arc cos } x_o) \quad \text{or } x_n = \sin (5^n \text{ arc sin } x_o) \quad , \end{cases}$$

(3) $$\begin{cases} x_{n+1} = 64\, x_n^7 - 112\, x_n^5 + 56\, x_n^3 - 7\, x_n \quad , \\ x_n = \cos (7^n \text{ arc cos } x_o) \quad , \end{cases}$$

(4) $$\begin{cases} x_{n+1} = 2x_n \,/\, (1 - x_n^2) \quad , \\ x_n = \text{tg } (2^n \text{ arc tg } x_o) \quad , \end{cases}$$

(5) $$\begin{cases} x_{n+1} = (3x_n - x_n^3) \,/\, (1 - 3x_n^2) \quad , \\ x_n = \text{tg } (3^n \text{ arc tg } x_o) \quad , \end{cases}$$

(6) $$\begin{cases} x_{n+1} = (5x_n - 10\, x_n^3 + x_n^5) \,/\, (1 - 10\, x_n^2 + 5\, x_n^4) \quad , \\ x_n = \text{tg } (5^n \text{ arc tg } x_o) \quad , \end{cases}$$

(7) $$\begin{cases} x_{n+1} = 2x_n \sqrt{1 - x_n^2} \quad , \\ x_n = \sin (2^n \text{ arc sin } x_o) \quad , \end{cases}$$

$$(8)\quad \begin{cases} x_{n+1} = -4x_n^3 + 3\,x_n , \\ x_n = \sin\,(3^n \text{ arc } \sin x_o) , \end{cases}$$

$$(9)\quad \begin{cases} x_{n+1} = 4x_n^3 + 3\,x_n , \\ x_n = \text{sh}\,(3^n \text{ arg sh } x_o) , \end{cases}$$

$$(10)\quad \begin{cases} x_{n+1} = (x_n^3 + 3x_n)\;/\;(3x_n^2 + 1) , \\ x_n = \text{th}\,(3^n \text{ arg th } x_o) , \end{cases}$$

$$(11)\quad \begin{cases} x_{n+1} = \left[-1 + 2x_n^2 + k^2(1-2x_n^2+x_n^4)\right] / \left[1-k^2(1-2x_n^2+x_n^4)\right] , \\ x_n = \text{cn}\,(2^n.z_o,\,k)\;,\; z_o = \text{am}\,(x_o,\,k),\quad 0 < k^2 < 1 , \end{cases}$$

$$(12)\quad \begin{cases} x_{n+1} = \left[x_n^4 + (1-k^2)(1-2x_n^2)\right] / (k\,x_n)^2 , \\ x_n = \text{dn}\,(2^n.z_o,\,k),\quad z_o = \text{am}\,(x_o,\,k)\;,\; 0 < k^2 < 1 . \end{cases}$$

The cases 1) - 3) are "fractional" iterates of the Myrberg recurrence (6-7), while 7), 8) are two of its symmetrical counterparts. 11) and 12) are straightforward extensions of (6-7) and 4) to elliptic functions (of modulus k).

A two dimensional example of the same structural type ($s_{n+1} = (2s_n-1)^2 > 0$, $r_{n+1} = r_n + \varphi$), is :

$$(13)\quad \begin{cases} x_{n+1} = (4s_n - 4 + 1/s_n)(x_n - y_n.\text{tg}\,\varphi)\cos\varphi\;,\; s_n = +\sqrt{x_n^2 + y_n^2} , \\ y_{n+1} = (4s_n - 4 + 1/s_n)(y_n + x_n.\text{tg}\,\varphi)\cos\varphi ,\qquad 0 < \varphi < \pi , \end{cases}$$

where $\varphi$ is a free parameter. The solution of (13) inside the no-escape region bounded by the circle $\mathcal{C}: x_n^2 + y_n^2 = 1$ is

$$(13a)\quad \begin{cases} x_n = \frac{1}{2}\left[1 + \cos\,(2^n \text{ arc } \cos s_o)\right] / r_n\;,\quad r_n = \text{tg}\left[n\varphi + \text{arc tg}\,(y_o/x_o)\right] , \\ y_n = r_n\,x_n . \end{cases}$$

The circle $\mathcal{C}$ is a complex singularity of (13), in fact it is simultaneously an invariant curve and a critical curve ($J_2$). The chaotic behaviour inside $\mathcal{C}$ is especially pronounced when $\varphi$ is incommensurable with $\pi$.

A somewhat different two-dimensional example is

$$(14)\qquad x_{n+1} = y_n,\; y_{n+1} = \frac{5y_n - x_n + 6x_n\,y_n}{6 + 16x_n - 8y_n} ,$$

which admits a general solution in spite of a point of indetermination of $y_{n+1}$ :

$$(14a)\qquad x_n = \frac{(1/3)^n(3x_o-6y_o) + (1/2)^n(6y_o - 2x_o + 4x_o\,y_o)}{1 + 3x_o - 2y_o - (1/3)^n.(3x_o - 6y_o)} ,\quad y_n = x_{n+1} .$$

When the general solution (6-3) of (6-1) is such that there exists a natural extension of n to fractional and continuous values, then F(x, n, c) constitutes also an explicit expression of all iterates of $f(x, c)$, except possibly for $n \to \pm\infty$. Moreover, when F is not only continuous but also continuously differentiable with respect to n, $|n| < \infty$, then the two expressions

$$\text{(6-30)} \qquad \frac{dx}{dn} = \frac{\partial}{\partial n} F(x_o, n, c), \qquad x - F(x_o, n, c) = 0$$

are equivalent to the non-autonomous differential equation

$$\text{(6-31)} \qquad \frac{dx}{dn} = X(x, n, c) \qquad ,$$

provided the "arbitrary constant" $x_o$ can be eliminated from (6-30). In such a case, the recurrence (6-1) and the differential equation (6-31) have the same solution (and of course, the same singularity structure in phase space). If $x_o$ and n occur in F in such a way that an insertion of a root $x_o = x_o(x, n, c)$ of $x - F(x_o, n, c) = 0$ into $\frac{\partial}{\partial n} F(x_o, n, c)$ leads to a simultaneous elimination of $x_o$ and n, then (6-31) simplifies into the autonomous differential equation

$$\text{(6-32)} \qquad \frac{dx}{dn} = \bar{X}(x, c) \qquad .$$

Since the root $x_o = x_o(x, n, c)$ is in general neither unique nor a smooth function of x and n, the equivalence between (6-1) and (6-32) is highly exceptional. As an illustration of the difficulties obstructing the passage from (6-1) to (6-32) in the case of an invariant segment, consider the particular recurrence (6-7) and its general solution (6-8). Since F depends on n via $2^n$, $y_n$ is unambiguously defined for all real n (in the case of (6-8a) and $y_o < -1$ it is necessary to pass to a polar form), i.e. it represents not only integer but also fractional and continuous iterates of $f(y) = 2y^2-1$. This is so in spite of the fact that the recurrence (6-7) admits inside the invariant segment $Y : -1 < y < +1$ both a fixed point and a cycle of order two (as well as another fixed point and many other cycles). The example (6-7) illustrates again the unnecessarily restrictive nature of the framework of contemporary iteration theory (cf. section 2.5 , chapter II).

Let $y = 2^n$ arc cos $y_o$. Although the two expressions

$$\text{(6-30a)} \qquad \frac{dy}{dn} = -(\log 2)\, z.\, \sin z, \qquad y = \cos z \quad ,$$

derived from (6-8), depend formally on $y_o$ and n only via the combination z, the elimination of z involves the inverse of a modulo-$2\pi$ operation, and leads to the discontinuous non-autonomous differential equation

$$\text{(6-31a)} \begin{cases} \dfrac{dy}{dn} = -(\log 2)(m\pi + \text{arc cos } y).\, \sin(m\pi + \text{arc cos } y), \qquad y(0) = y_o \quad , \\ m = \text{integer verifying the equation :} \qquad \text{arc cos } y = -m\pi + 2^n \text{ arc cos } y_o, \\ y_o \in Y, \quad y \in Y, \qquad Y : -1 < y < +1, \qquad -\infty < n < +\infty \quad . \end{cases}$$

No further reduction to the form (6-32) is possible.

Due to the higher dimensionality the phase space portrait of the recurrence (6-2) is much more complex than that of (6-1), because zero- , one- and two-dimensional singularities are possible in principle. As a consequence, the functions F and G, eq. (6-4), describing the general solution of (6-2) are inherently more complex than the function F of (6-3). Like in the case of (6-1), the simplest zero-dimensional singularities of (6-2) are fixed points and cycles, defined by real roots of the algebraic equations

(6-33) $$x_n = g_k(x_n, y_n, c), \quad y_n = f_k(x_n, y_n, c) \qquad k = 1, 2, \dots \quad .$$

In addition to their order k, cycles are characterized by two geometric features associated with the relative locations of their points : a) rotation number or rotation sequence and b) number of excess antecedents at each cycle point. One-dimensional singularities of (6-2) correspond to certain solutions of the functional equation of automorphic functions

(6-34) $$w(g(x, y, c), \; f(x, y, c)) = w(x, y, c) \quad .$$

A curve $\mathcal{C}$ defined by $w(x, y, c) = C(x_o, y_o, c)$ where C is a known constant, is said to be an invariant curve of the recurrence (6-2) (analytic in the sense of Poincaré, cf. Chapter II), if $\mathcal{C}$ is also represented parametrically in terms of n by the general solution (6-4). Because of possible non-uniqueness of w, it is often convenient to replace locally the linear functional equation (6-34) and $\mathcal{C}$: $w(x, y, c) = C$ by the non linear functional equation

(6-35) $$f(x , \theta(x,c), c) = \theta(g(x, \theta(x,c), c)), \qquad y_o = \theta(x_o, c)$$

and $\mathcal{C}$: $y = \theta(x, c)$. Other types of non-uniqueness, and occasionally an apparent lack of existence of $\mathcal{C}$, are removed by replacing f, g in (6-34) and (6-35) by their consequents $f_k$, $g_k$ or their antecedents $f_{-k}$, $g_{-k}$. $\mathcal{C}$ is called a consequent curve when $k > 1$, and an antecedent curve when $-k < -1$. If desired, the solutions of (6-33) can be considered as a degenerate, "strongly" discontinuous automorphic functions, because like $w \equiv 0$, (6-33) is a trivial solution of (6-34). The simplest one-dimensional singularities are (continuous) loci of cycle points and invariant curves crossing a point singularity (or terminating at the latter). A slightly more complicated one constitutes an analogue of invariant segments occuring in first order recurrences. Two-dimensional singularities of (6-2) are still a subject of debate, no uncontrived (non-trivial) examples having been found so far.

Let $(\bar{x}, \bar{y})$ be a non-stochastic point-singularity of the recurrence (6-2), i.e. let $(\bar{x}, \bar{y})$ be either a fixed point or a cycle point not having another point-singularity in its infinitesimal neighbourhood. This is usually (but not always) so when $(\bar{x}, \bar{y})$ is a simple root of (6-33) for a finite value of k. $(\bar{x}, \bar{y})$ is then either an accumulation point of some discrete trajectories $\{x_n, y_n\}$ or it is an isolated point. In the first case $(\bar{x}, \bar{y}) = \lim (x_n, y_n)$ as either $n \to +\infty$ or $n \to -\infty$, and occasionally $n \to +\infty$ for some initial points $(x_o, y_o)$ and $n \to -\infty$ for others, whereas in the second case there is no such relationship. If one has $(\bar{x}, \bar{y}) = \lim (x_n, y_n)$ only for $n \to +\infty$, then an accumulation point $(\bar{x}, \bar{y})$ is said to be asymptotically stable, otherwise it is said to be unstable. An isolated $(\bar{x}, \bar{y})$ is either simply stable or unstable (in the sense of Liapunov). Because of the definition of an invariant curve, it is more efficient to characterize the stability or instability of a point-singularity in terms of invariant curves rather than in terms of discrete trajectories. By analogy with second order differential equations the simplest point-singularities of the recurrence (6-2) are therefore centres, foci, (simple, critical or bi-critical) nodes and saddles, according to the shape of the invariant curves in their immediate neighbourhood. A centre is an isolated point, whereas a focus, node and saddle are accumulation points of discrete trajectories. A focus and a node is asymptotically stable when it is approached by consequent invariant curves, and unstable when it is approached by antecedent invariant curves. A saddle is always unstable. In addition to zero-dimensional attractors (asymptotically stable nodes, foci or multiple fixed points and cycles) there exist one-dimensional ones. Three distinct types have been observed so far : continuous loci of fixed points or cycles, stochastic curve segments and closed regular invariant curves (not traversing any point-singularity). Continuous loci of point-singularities are easily identified, because their points satisfy simulteneously the albebraic equations (6-33) and the functional equations (6-34), (6-35). A distinction between stochastic curve segments and closed invariant curves becomes very difficult when the number of the former is so large that their configuration resembles a jagged closed curve. The difference consists then in the smoothness properties and in the existence of possible gaps. Whether attractive closed stochastic curves are possible in principle is not yet known. One bifurcation leading to the appearance of regular closed invariant curves involves analytic solutions of the functional equations (6-34), (6-35), at least when the latter are expressed in appropriate coordinates. Short arc segments of a closed invariant curve are then mapped onto each other by means of smooth functions, and close to the bifurcation (in the sense of a parametric distance) these functions represent almost-translations. As the (parametric) distance from the bifurcation increases, the shape of the closed invariant curves is found to distort rapidly, and it is likely that the corresponding solutions of (6-34), (6-35) become less smooth, i.e. instead of being analytic they possess merely a finite number p of continuous derivatives. This conjectured decrease of smoothness cannot be studied analytically, because the required constructive algorithms of solving the functional equations (6-34), (6-35) are not yet available. The abstract theory of automorphic functions (see for example [F 3] and [M 11]) is also silent on this subject. If the numerically deduced decrease of smoothness does indeed occur, then $p = 1$ is a critical threshold, because below it a curve ceases usually to be rectifiable, and the meaning of the notion "closed curve" becomes debatable. It is therefore possible that a regular closed invariant curve degenerates "smoothly" into an (almost) closed stochastic curve ($p \approx 0$) possessing at first infinitesimal gaps, and that these gaps grow gradually ($p < 0$) till a certain number of disjoint stochastic curve-segments becomes detectable. Closed invariant curves and stochastic segments would thus be described by essentially the same solutions of (6-34), (6-35), whose smoothness depends "continuously" on a parameter, and decreases as the parametric distance from the bifurcation increases.

The degree of smoothness near $p = 1$ can be ascertained numerically by means of sensitivity coefficients. Consider in fact a discrete consequent trajectory $\{x_n, y_n\}$, $n > 0$ of the recurrence (6-2) for a fixed c and a given initial point $(x_o, y_o)$, and let

$$S_{x,n} = \frac{\partial x_n}{\partial x_o}, \quad S_{y,n} = \frac{\partial x_n}{\partial y_o}, \quad T_{x,n} = \frac{\partial y_n}{\partial x_o}, \quad T_{y,n} = \frac{\partial y_n}{\partial y_o} \quad \text{(6-36)}$$

If the general solution (6-4) of (6-2) is continuously differentiable with respect to $x_o$ and $y_o$ (almost everywhere), and there is no evidence so far that in general it is not , then the sensitivity coefficients S and T can be determined simultaneously with the discrete trajectory $\{x_n, y_n\}$, by means of the coupled linear recurrences

$$\begin{aligned} S_{x,n+1} &= \left[g_x(x_n,y_n,c)\right].S_{x,n} + \left[g_y(x_n,y_n,c)\right].T_{x,n}, & S_{x,0} &= 1, \\ S_{y,n+1} &= \left[g_x(x_n,y_n,c)\right].S_{y,n} + \left[g_y(x_n,y_n,c)\right].T_{y,n}, & T_{x,0} &= 0, \\ T_{x,n+1} &= \left[f_x(x_n,y_n,c)\right].S_{x,n} + \left[f_y(x_n,y_n,c)\right].T_{x,n}, & S_{y,0} &= 0, \\ T_{y,n+1} &= \left[f_x(x_n,y_n,c)\right].S_{y,n} + \left[f_y(x_n,y_n,c)\right].T_{y,n}, & T_{y,0} &= 1, \end{aligned} \quad \text{(6-37)}$$

where $f_x$, $f_y$, $g_x$, $g_y$ designate partial derivatives, or by some equivalent alternate means. If $\{x_n, y_n\}$ terminates on a point-attractor, then, like in the case of first-order recurrences, $S \to 0$ and $T \to 0$ as $n \to \infty$. If on the contrary at least one sensitivity coefficients becomes unbounded as $n \to \infty$, then $\{x_n, y_n\}$ terminates on a stochastic one-dimensional attractor or is already located on the latter. Eventually $\{x_n, y_n\}$ may belong to a two-dimensional attractor, if the latter are proven to exist. A necessary condition for the existence of an invariant-curve attractor is thus the doundedness of S and T as $n \to \infty$, with at least one of these coefficients different from zero. Since the computation of S and T is readily implemented, the differentiation between a regular and a stochastic curve attractor becomes a matter of routine.

Sensitivity coefficients can also be used to determine the smoothness of an attractive closed invariant curve, say $\mathcal{L}$. Assume that the coordinates in (6-2), (6-35) and (6-37) have been so chosen that a solution of (6-35) is a single valued representation of $\mathcal{L}$. Such a choice constitutes theoretically a trivial problem, but practically a difficult one (polar coordinates are satisfactory only exceptionally). If these coordinates are designated by x and y, then $\mathcal{L}$ is described by $y = \theta(x)$ and $\theta$ is a periodic function of x, say of the normalized period $2\pi$. It is not efficient to choose x to be the arc-length of the curve, because such a choice implies the rectifiability of the latter, and introduces thus an unnecessary restriction. Choosing the initial point $(x_o, y_o)$ of $\{x_n, y_n\}$, $n \geqslant 0$, inside the influence domain of $\mathcal{L}$, after a finite (but possibly large) number of steps, the points $(x_n, y_n)$, $n \geq \bar{n}$, $\bar{n}$ = known integer, will lie on $\mathcal{L}$ or close to $\mathcal{L}$ with an error, say $\varepsilon(n)$. The following $2N+1$ points $(x_n, y_n)$ furnish then $2N+1$ samples of the periodic function $\theta(x)$. Evaluating the Fourier integrals

$$a_o = \frac{1}{2\pi}\int_0^{2\pi}\theta(x)dx,\quad a_m = \frac{1}{\pi}\int_0^{2\pi}\theta(x)\cos mx dx,\quad b_m = \frac{1}{\pi}\int_0^{2\pi}\theta(x)\sin mx dx$$

at these samples (using Lagrange interpolation for example), one obtains the following approximation of $\theta(x)$ :

$$y = \theta(x) \approx \theta_N(x) = a_o + \sum_{m=1}^{N}(a_m \cos mx + b_m \sin mx) \quad . \tag{6-38}$$

Consider now the sequences $\{a_m\}$ and $\{b_m\}$. If for m close to N the coefficients $a_m$ and $b_m$ are already typical, and they decrease not slower than $1/m^2$, then $\theta_N(x)$ converges to an at least piecewise continuously differentiable function as $N \to \infty$, and

$$\theta'_N(x) = \sum_{m=1}^{N} m(-a_m \sin mx + b_m \cos mx) \tag{6-39}$$

converges to at least a piecewise continuous one (with at most discontinuities of the first kind and an accompanying bounded Gibb's phenomenon). But an approximation $\theta'_s(x)$ of $\theta'(x)$ can also be determined independently by means of the sensitivity coefficients S and T. Comparing $\theta'_N(x)$ to $\theta'_s(x)$ at a certain number of abscissae (and avoiding, if possible, the proximity of discontinuities), one obtains both an estimate of the difference $\theta(x) - \theta_N(x)$ and of the error $\varepsilon(n)$. The procedure can, of course, be repeated with other values of n and N, and the estimates improved. If $a_m$ and $b_m$ decrease at least as $1/m^3$, then $\theta'(x)$ is continuous, and so forth, for a still faster decrease. A sufficient condition for the rectifiability of $\mathcal{L}$ is a $1/m^2$ decrease of $a_m$ and $b_m$. If $a_m$ and $b_m$ decrease at a slower rate, then $\theta(x)$ may either cease to be continuous, or remain continuous but cease to be of bounded variation.

After a certain number of locally stable singularities have been determined, the next step in the study of the phase portrait of (6-2) is the determination of the corresponding immediate and total influence domains. Like in the case of first order recurrences, a total influence domain may consist of several disjoint parts. The influence domain of a set of attractors may form a composite influence domain with highly intertwined individual boundaries. The qualitative properties of all possible boundary points are not yet known. Since a boundary is necessarily one-dimensional, compared to first order recurrences there appears another point set, composed of regular and singular invariant curve segments. The boundary of an influence domain may thus consist of a wide variety of qualitatively distinct elements, ranging from a single regular invariant curve to an infinite set of singular and stochastic curve segments, separated by simple and composite singular points. The inverse problem of determining the number and the nature of attractors located inside a known no-escape region, is even more difficult, because the list of known attractors is by no means complete. There exists thus always the possibility that one of the interior attractors is of a new (unstudied) type, or if one likes emotional connotations (expressing astonishment), that it is "strange". A boundary of a no-escape region may be composed entirely of critical curve segments, it is thus a different entity than a total or composite (stability) influence domain. No-escape regions exist even in stochastic phase plane regions of conservative recurrences (see for example Fig. 3-33 to 36).

The attractor content of a stochastic no-escape region may be studied by means of the two-dimensional Perron-Frobenius functional equation, but at present there are no efficient algorithms of finding its solution. In the case of conservative recurrences one solution is a constant density corresponding to the Lebesgue measure, but this solution is trivial because it merely restates the area conservation expressed by Liouville's theorem. A non-trivial solution should quantify the filamentation of a closed contour enclosing only a part of the no-escape region, or in other words it should describe the spreading and mean concentration (or dilution) of an initially known mass distribution, similarly to the spreading of a drop of non-diffusing dye placed into a bounded two-dimensional fluid flow.

A phase portrait of a conservative or non conservative recurrence (6-2) is said to be orderly, or to describe orderly dynamics, if it can be subdivided into a finite number of cells, and each cell is characterized by one elementary point-singularity or one isolated closed invariant curve. Elementary point singularities are fixed points and cycles of finite rank and finite multiplicity. Cell boundaries may contain an infinity of regular or singular elements. A necessary condition for the recurrence (6-2) to be (at least locally) stochastic or chaotic, or to describe complex or chaotic dynamics, is the existence of an infinity of cycles not located on cell-boundaries. In order to determine this property, cycles can be ordered into sets characterized by certain asymptotic properties, the simplest of which are growth or boundedness of the ratio $k/r$ (order divided by rotation number), growth or decrease of the eigenvalues (and eventually convergence of eigendirections) and growth or boundedness of the loci of coordinates. In this manner, it was possible to show that homoclinic points, i.e. intersections of consequent and antecedent curves originating at the same point-singularity (fixed point or point of the same cycle), are accumulation points of cycles as $k/r \to \infty$. They are thus cycle points of an infinite order. Intersection of consequent and antecedent invariant curves originating at basically distinct singularities are called heteroclinic points. There exist thus heteroclinic points of different types involving : a) two fixed points, b) points of two cycles of different order (one of them eventually a fixed point), c) two separate cycles of the same order, d) one fixed point, or a point of a cycle, and an isolated regular closed invariant curve, e) one fixed point, or a point a cycle, and a singular invariant curve consisting of a locus of fixed points or cycle points. Only the types a) and c) appear to be accumulation points of cycles, similarly to homoclinic points. In non-conservative recurrences the types b) and d) are generally not generators of stochasticity, they introduce merely some complications ("deterministic" indeterminacy) into the boundaries of influence domains. Intersections of consequent (or antecedent) invariant curves with other consequent (or antecedent) invariant curves are also possible, but only in non-conservative recurrences (example : a consequent invariant curve originating at a saddle and approaching non-monotonically an attractive closed invariant curve). Self-intersections of consequent (or antecedent) invariant curves can occur only in non-conservative recurrences with multivalued inverses. In such recurrences there exists a variety of other intersection points attributable to the existence of excess antecedents : a) points of alternance (junction of two distinct antecedent branches on a critical curve with arrow reversal), "continuation" points (junction of two distinct antecedent branches without arrow reversal), c) intersection of a principal and an excess antecedent branch, d) intersection of a primary or excess antecedent curve with an excess antecedent branch of a consequent curve, etc. For this reason the complete configuration of antecedent curves is much more difficult to determine than the complete configuration of consequent curves ; or in other words, in evolution processes described by recurrences with non-unique functions $f_{-1}$, $g_{-1}$, more than one historical development leads to a given present state, but there exists a unique prediction of the future, subject of course to a perfect knowledge of the functions f, g in (6-2). Conservative recurrences have also their specific complications. Since in contrast to non-conservative recurrences they possess (as a rule) cycles of any order k and rotation number r, the limit configuration of cycle points as $k \to \infty$, $r \to \infty$ but $k/r = k_e$, $0 < k_e < \infty$, may produce non-elementary singularities. When $k_e$ is irrational and relatively small, i.e. the accumulation of cycle points does not take place inside a stochastic region, then such a singularity is either an already known point set (centres with $\varphi = 2\pi/k_e$) or a singular closed invariant curve $\Gamma(k_e)$ (representing an infinitely narrow island structure with an infinity of islands, cf. the curves $\Gamma_3$, $\Gamma_4$ in Fig. 3-1). The shape of the curve $\Gamma(k_e)$ may be quite regular, in the sense that it is describable by a smooth or even an analytic function $w(x_n, y_n) = C$, C = fixed constant. What happens when $k_e$ is irrational but not small, i.e. the corresponding (non-trivial) accumulations take place inside a stochastic region, is not yet fully understood. Rational values of $k_e$ correspond obviously to exceptional cases at centres ; whether they can give rise to one-dimensional accumulations is uncertain, but it seems rather unlikely.

The complete invariant curve configuration near a fixed point (or a cycle point of finite order) can only be determined by global solutions of the functional equation of invariant curves (6-34). If it turns out, however, that this configuration is entirely defined by local solutions, then the fixed point is said to be ordinary (or simple). In the contrary case, it is said to be weakly or strongly stochastic. The adverbs weakly and strongly quantify roughly the number of non-local invariant curves in its neighbourhood. For example, the saddle (1,0) of the conservative recurrence (3-6) is strongly stochastic, and so is the cusp (1,0) of (3-80).

Other conservative examples are most saddles of type 3 (exceptions are saddles of orderly recurrences, like the saddle (0,0) of (2-93) when $a < -1$), and mixed centres. Weakly stochastic point-singularities arise only in non-conservative recurrences, like for example, the cycles 3/1 shown in Fig. 4-10, because their local invariant curve structure is crossed by some invariant curves originating at the 40/13 saddles. For the same parameter values, the saddle (1,0) of the recurrence (4-23) is strongly stochastic. The transition from strong to weak stochasticity, and finally to orderiliness, is smooth in a similar way as the transition from stochastic curve segments to a rectifiable closed invariant curve, and finally to a regular one.

The second order recurrence (6-2) possesses of course a larger variety of bifurcations than the first order recurrence (6-1). The simplest bifurcations of (6-2) involve only cycles. New cycles appear either spontaneously (critical case $\lambda = +1$) or they split off from already existing ones (critical cases $\lambda = 1$, $\lambda = -1$ and exceptional cases $\varphi = 2\pi r/k$). The critical cases $\lambda = +1$, $\lambda = -1$ (including $\varphi \to 0$ and $\varphi \to \pi$) can be subdivided into numerous sub-cases, depending on the exponents of the lowest order terms in the corresponding canonical forms of (6-2), i.e. depending on how many low order terms can be entirely removed from the Mc Laurin developments of the functions f and g (or more commonly of their k-th iterates) by successive applications of almost-identity transformations. The detailed enumeration of the various possibilities would consume too much space to be attempted here. When (6-2) is conservative, the exceptional cases $\varphi = 2\pi r/k$ are of two different types : a) the centre releases a cycle pair of order k and rotation number r (stable resonance), and b) the centre merges with k already existing (simple) saddles and becomes a 2k-branch saddle (unstable resonance). If the exceptional case $\varphi = 2\pi r/k$ occurs on an almost conservative focus ($|\lambda| \approx 1$), then the k/r cycle pair appears (spontaneously) near the focus, or the k saddles move towards the focus but fail to reach it (weak resonance). An elementary bifurcation involving a point-singularity and an invariant curve occurs at the stability change of a focus, the bifurcated singularity being an asymptotically stable or unstable closed invariant curve. Numerous non-elementary bifurcations involving invariant curves have been observed numerically, but their analytical formulation awaits a better understanding of global solutions of the functional equations (6-34) and (6-35).

Since invariant curves are key elements of the phase plane portrait of the recurrence (6-2), it is instructive to examine some particular (non linear) cases possessing at least one analytically known invariant curve. One method of constructing such particular cases is based on the properties of involutions. Two functions, say u(x) and v(x), are said to form a (simple) involution when their iterative product is unity, i.e. when $u(v(x)) = v(u(x)) = x$, and an involution with respect to a function g(x) when either $u(v(x)) = g(x)$ or $v(u(x)) = g(x)$, but usually not both. Like in the case of involutions defined by the Schröder and Picard funntional equations, the g-involutive functions u(x) and v(x) need not have the same x-existence interval. The simplest involution is of course formed by a single-valued function $u(x)$ and its unique and single-valued inverse $v(x) = u^{-1}(x)$.

Consider now a smooth phase plane curve $y = \theta(x)$, where $\theta(x)$ is a known single-valued function of x admitting a unique single-valued inverse $x = \theta^{-1}(y)$. Although the x- and y-existence intervals need not be the same, assume that they are (or take only their common part), and let

$$\theta(y) + \theta^{-1}(y) = f(y) \quad . \tag{6-40}$$

It is easily verified ([L 2]) that the curves $y = \theta(x)$ and $y = \theta^{-1}(x)$ are invariant curves of the conservative recurrence

$$x_{n+1} = y_n, \qquad y_{n+1} = -x_n + f(y_n) \quad . \tag{6-41}$$

In fact, inserting $\theta(x)$ or $\theta^{-1}(x)$ into the functional equation (6-35), and taking into account (6-41) and (6-40), yields a formal identity in x. In order to avoid complications of a purely mathematical nature, assume that $f(y)$ is continuously differentiable, except perhaps at a finite number of points $\bar{y}$. The Jacobian determinant of (6-41) is therefore unity, except at $\bar{y}$, where it is undefined. As an illustration of the use of (6-40), let

$$y = \theta(x) = a - (x+a)^{-1} \qquad \text{which implies } y = \theta^{-1}(x) = -a - (x-a)^{-1}, \tag{6-42}$$

where a is a real constant. Then $f(y) = 2y / (a^2 - y^2)$, and the recurrence (6-41) coincides with one already discussed in Chapter II (eq. 2-92) :

$$x_{n+1} = y_n, \quad y_{n+1} = -x_n + 2y_n / (a^2 - y_n^2) \quad . \tag{6-43}$$

The Jacobian determinant of (6-43) fails to exist at $y = \pm a$. The fixed points of (6-43) are (0,0) if $|a| < 1$, and (0,0), ($\pm b$, $\pm b$), $b = \sqrt{a^2-1}$ if $|a| > 1$. (0,0) is a saddle of type 1 in the first case and a centre ($\varphi$ = arc tg b) in the second. The points ($\pm b$, $\pm b$) are saddles of type 1. If $|a| > 1$, the curves (6-42) pass through the saddles ($\pm b$, $\pm b$), and they are therefore separatrices. The complete family of invariant curves of (6-43) can be determined in a straightforward way. Consider the product

$$(y - \theta(x))(y - \theta^{-1}(x)) = \left[w(x,y) - (a^2-1)^2\right]/(x^2-a^2) \quad , \tag{6-44}$$

where

$$w(x, y) = x^2 y^2 - a^2(x^2 + y^2) + 2\,xy \quad . \tag{6-45}$$

Inserting this $w(x, y)$ into the functional equation (6-34), it is easily seen that $w(x, y)$ is a solution subject to (6-43). Hence, $w(x, y) = C$ describes all invariant curves of (6-43). The singularity $y = \pm a$ of $f(y)$, leading to a singularity of the Jacobian determinant, has no serious consequences. A simple pole of f introduces merely some horizontal and vertical asymptotes into the phase portrait.

Consider now the inverse problem : knowing a non linear $f(y)$ what is the corresponding $\theta(y)$ verifying (6-40) ? From the preceding example it is clear that (6-40) can be interpreted as a special case of the functional equations (6-34) and (6-35). Under normal or regular circumstances, (6-40) admits therefore a general solution depending on one arbitrary constant. But what is a relevant meaning of the adjectives normal or regular of the preceding sentences ? These adjectives have obviously nothing to do with properties like the monotonicity of $\theta(x)$, insuring the uniqueness and single-valuedness of $\theta^{-1}(x)$, because such restrictions lead invariably to some singularity of the function f. When f is highly regular, then $\theta$ and $\theta^{-1}$ are necessarily strongly singular. The quadratic and cubic recurrences (3-6) and (3-59) are striking illustrations of this property. The main invariant curves $y = \theta(x)$ and $y = \theta^{-1}(x)$ traversing the saddle (1,0) are both described by multi-valued functions, having each an infinity of single-valued branches, but in the coordinates of (6-41), their sum is simply $f(y) = 2\left[\mu x + (1-\mu)x^{\alpha}\right]$, $\alpha = 2$ or 3, which is as regular a function as one can possibly wish. When $\mu = 1$, the functions $\theta(x)$, $\theta^{-1}(x)$ and $f(x)$ are linear, whereas when $-1 < \mu < 1$, the functions $\theta(x)$, $\theta^{-1}(x)$ are not only non linear but also (infinitely) multivalued, even if $0 < 1-\mu < \varepsilon$, $\varepsilon$ arbitrarily small. The same singularity-regularity relationship exists for the recurrences (3-80), (3-90), (3-92), (3-93) and (4-27) when $\alpha = \beta = 0$. The problem of solving the functional equation (6-40) with a given smooth $f(y)$ is therefore regular when $y = \theta(x)$ and $y = \theta^{-1}(x)$ describe smooth non-intersecting curves, and not when the functions $\theta(x)$ and $\theta^{-1}(x)$ are single-valued, or otherwise non-singular. An abstract function-theoretical approach to the equation (6-40) is thus precisely the opposite of what is needed to capture the salient features of its solutions. Since no elementary or (classical) transcendental function possesses singularities which are complementary to those of its inverse, there is little probability of constructing a simple recurrence of form (6-41) with analytically known stochastic solutions. Choosing $\theta(x)$'s with singularities other than poles leads as a rule to pathological situations, without any dynamically meaningful content.

Consider for example a $\theta(x)$ with a branch point and two inverses :

(6-46) $$\theta(x) = +\sqrt{a^2 - x^2}, \quad \theta_1^{-1}(x) = +\sqrt{a^2 - x^2}, \quad \theta_2^{-1}(x) = -\sqrt{a^2 - x^2}, \quad a > 0.$$

The choice $\theta(x) + \theta_2^{-1}(x) = 0$ shows that there exists solutions of the homogeneous equation (6-40). The sum

(6-47) $$\theta(x) + \theta_1^{-1}(x) = f(x) = 2\sqrt{a^2 - x^2}$$

leads to a "secularly" perturbed recurrence

(6-48) $$x_{n+1} = y_n, \quad y_{n+1} = -x_n + 2\sqrt{a^2 - y_n^2} .$$

The Jacobian determinant J of (6-48) is formally unity for all x and y, but in reality it is indeterminate for $y = \pm a$, because $f'(\pm a)$ is not finite, and it does not exist for $y^2 > a^2$, because $f(y)$ is not real-valued. The recurrence (6-48) admits two fixed points (b, b) and (−b, −b), $b = a/\sqrt{2}$, with the eigenvalues $\lambda = -1$ and $\lambda = +1$, respectively. The arc $y = \sqrt{a^2 - x^2}$, $0 < x < a$, is a locus of cycle points of order $k = 2$, with the exception of the point (b, b). The points (a, a), (a, −a), (−a, a) and (−a, −a) form a cycle of order $k = 4$. It is easily verified by substitution into (6-35) that $\mathcal{C}$: $y = +\sqrt{a^2 - x^2} > 0$, $-a < x < +a$ is a consequent invariant curve segment of (6-48). This segment traverses the fixed point (b, b). Using the inverse of (6-48) it if found that $\mathcal{A}$: $x = +\sqrt{a^2 - y^2} > 0$, $-a < y < +a$ is an antecedent curve segment, which also traverses (b, b). The overlapping part of $\mathcal{C}$ and $\mathcal{A}$ coincides with the locus of cycle points of order $k = 2$. The missing arc $x < 0$, $y < 0$ of the circle $x^2 + y^2 = a^2$ is not a part of any invariant curve of (6-48). The general solution of (6-47), or of (6-34), (6-48) containing $\mathcal{C}$ and $\mathcal{A}$ is not known. A numerical exploration of (6-48) suggests that, except for fixed points and cycles (and perhaps also the invariant curves $\mathcal{C}$ and $\mathcal{A}$), all discrete consequent and antecedent trajectories have only a finite number of points (with real-valued coordinates). What happens to Liouville's theorem, to invariant curves other than $\mathcal{C}$ and $\mathcal{A}$, and to the solutions of the Perron-Frobenius equation under such circumstances ?

Another type of pathology arises in the case of the example (14), (14a) of p. 243. The recurrence (14) possesses two fixed points (0,0) and (−1,−1), which are stable and unstable nodes, respectively. In spite of the existence of critical curves ($J_0$ : $4x_n(6y_n-1) - (1+8y_n^2) = 0$), the inverse of (14) is unique

(14b) $$x_n = \left[x_{n+1}(5+8y_{n+1}) - 6y_{n+1}\right]/(1+16y_{n+1}-6x_{n+1}), \quad y_n = x_{n+1} .$$

$J_0$ is therefore analoguous to a horizontal inflection point of a curve. Three invariant curves of (14) are known explicitly :

(15) $$\mathcal{C}_1 : y_n = x_n/2, \quad \mathcal{C}_2 : y_n = x_n/(3+2x_n), \quad \mathcal{C}_3 : y_n = (1+3x_n)/2 .$$

$\mathcal{C}_1$, $\mathcal{C}_2$ pass through (0,0) and $\mathcal{C}_2$, $\mathcal{C}_3$ through (−1,−1). What is unusual is that $\mathcal{C}_1$, $\mathcal{C}_2$ and $\mathcal{C}_3$ (and an infinity of other invariant curves) pass through the point $Q = (-\frac{1}{2},-\frac{1}{4})$, which is neither a fixed point nor a point of a cycle, but a point where $y_{n+1}$ in (14) becomes indeterminate (i.e. $y_{n+1}$ takes the form $\frac{0}{0}$). Moreover, the consequent of $(-1, -\frac{1}{2})$ (evaluated by means of (14)), and the antecedent of $(-\frac{1}{4}, -\frac{1}{8})$ (evaluated by means of (14b)), coincide with Q, as do many other points.

Instead of starting from a given involution and the functional equation (6-40), it is also possible to start from a generalized form of (6-44), and then deduce a solution of the functional equation (6-34). To be more specific, let $y = \theta(x)$ and $y = \Psi(x)$ be two solutions of the albegraic equation $v(x, y) - C = 0$, where C is an arbitrary constant. In general, $\Psi(x)$ is of course not an inverse of $\theta(x)$. If $\theta(x)$ and $\Psi(x)$ are used to construct a recurrence of the form (6-41), where (6-40) is replaced by

(6-49) $$\theta(x) + \Psi(x) = f(x) ,$$

it can happen nevertheless that v(x, y) verifies (6-34). As an illustration, consider

(6-50) $$v(x, y) = x^2y^2 + x^2 + y^2 - 2axy, \qquad a \neq 0$$

Two roots of v(x, y) = C are

(6-51) $$y = \theta(x) = \left[ax + \sqrt{a^2x^2 - (1+x^2)(x^2-C)}\right]/(1+x^2) \quad ,$$
$$y = \Psi(x) = \left[ax - \sqrt{a^2x^2 - (1+x^2)(x^2-C)}\right]/(1+x^2) \quad .$$

Hence, $f(x) = 2ax/(1+x^2)$, and

(6-52) $$x_{n+1} = y_n, \quad y_{n+1} = -x_n + 2ay_n/(1+y_n^2) \quad .$$

It is easily verified that (6-50) is a solution of (6-34) subject to (6-52). The phase portrait of (6-52), described by v(x,y) = C is shown in Fig. 2-18. The recurrence (6-52) is quite exceptional because it involves both a non-singular f(y) and a non stochastic phase portrait. When (6-50) is replaced by the slightly more general bi-quadratic polynomial

(6-53) $$v(x,y) = \alpha x^2y^2 + Bxy(x+y) + \beta(x^2+y^2) + Dxy + E(x+y) \quad ,$$

where α, β, B, D, E are arbitrary constants, then

(6-54) $$f(y) = -(B^2y^2 + Dy + E) \ / \ (\alpha y^2 + By + \beta)$$

and v(x, y) = C furnishes again an explicit description of the phase portrait of (6-41). It should be stressed, however, that an arbitrary factorizable v(x, y) yielding a unique f(y) does not satisfy (6-34), (6-41).

Non unitary involutions provide a way of constructing non-conservative recurrences with one analytically known invariant curve. These recurrences have a rather special form, which provides a conceptual relation between invariant curves and iterative square roots. As an illustration, let $\mathcal{C}$: y = θ(x) be a known smooth curve and g(x) a known smooth function related to θ(x) and one of its inverses $\theta^{-1}(x)$ by the involution

(6-55) $$\theta(\theta(x)) = g(x), \qquad \text{(6-55a)} \quad \theta^{-1}(\theta^{-1}(g(x))) = x \quad ,$$

in some interval X of x. Consider a first order recurrence

(6-56) $$x_{n+1} = \theta(x_n)$$

and its second iterate

(6-57) $$x_{n+2} = \theta(\theta(x_n)) = g(x_n) \quad ,$$

which can be also written in the form (6-2) :

(6-58) $$x_{n+1} = y_n, \quad y_{n+1} = g(x_n) \quad .$$

From an inspection of (6-35), it is obvious that (6-55) defines all consequent invariant curves of (6-58), provided such curves exist ; or in other words, the functional equation (6-55) has a one-parameter family of solutions, in which the curve $\mathcal{C}$: $y = \theta(x)$ is continuously imbedded. Moreover, (6-55) describes by definition all iterative square roots (half-iterates) of $g(x)$, only one of which is $\theta(x)$. The others are given by other (consequent) invaraint curves of (6-58). Since smooth invariant curves are in general described by very complicated and strongly singular functions, the problem of the "best" or "simplest" half-iterate has a different answer, depending on whether it is framed in terms of curves or in terms of functions. An inspection of the phase portraits of some known recurrences makes it obvious that this answer coincides only in special cases.

When the recurrence (6-58) has no point singularity in the finite part of the phase plane, then all invariant curves are dynamically non-singular, and the best one can be characterized either by some geometric feature, or by the way its description $\theta(x)$ is expressed in terms of some more elementary (or perhaps better, some less transcendental) functions. For the special case $g(x) = e^x$, the invariant curves of (6-58) lieing above the $x_n$-axis are somewhat simpler than the others. The "last" one not to cross the $x_n$-axis is perhaps the best.

When the recurrence (6-58) does have at least one point-singularity, then the choice of the best half-iterate is in general much easier. If the point-singularity is crossed by a certain number of invariant curves, then at least one of these curves is a good candidate, because it might be the locus of all integer and half-integer discrete (consequent) trajectories initialized on it. A favourable condition for this property exists when the singularity is of type 1. When the singularity is of type 2 or 3, then the successive points of integer discrete trajectories are in general even more scattered. The locus of successive half-integer points is therefore highly discontinuous as a rule, although the corresponding individual invariant curve segments may be quite smooth. Exceptionally this locus consists of a smooth union of distinct curve segments, described by one single-valued function. The notions of invariant curve and type of singular point turn out therefore to be key elements in the study of half-iterates. The case of a fixed point is more complicated than the case of a cycle, because the latter may involve a finite number of invariant curves, whereas the former is either a star-node of type 2 or a focus. To see this, let $\bar{x}$ be a root of $g(x) = x$. $g(x)$ being by assumption a smooth function, $g'(x)$ is unambiguously defined. If $g'(\bar{x}) = a > 0$, then the eigenvalues of the fixed point $x_n = y_n = \bar{x}$ are $\lambda_{1,2} = \pm\sqrt{a}$, whereas if $a = -\alpha^2 < 0$, then $\lambda_{1,2} = \pm i\alpha$. The critical cases $a = 0$, $a = 1$ and $\alpha = 1$ need a special examination.

As an illustration, consider the particularly transparent linear case $g(x) = ax$, having the fixed point $x_n = y_n = 0$. When $a > 0$, then the only smooth "half-iterates" (6-56) of (6-57) are $x_{n+1} = \pm\sqrt{a}\, x_n$. For $+\sqrt{a}$, the points $(x_n, y_n)$, $n = 0, 1, \ldots$ are located on one invariant curve branch $y = \theta(x)$ originating at (0,0), and for $-\sqrt{a}$ alternately on two branches. When $a < 0$, then the half-iterates (6-56) of (6-57) are located on smooth invariant curves, whose description $y = \theta(x)$ involves two distinct single-valued functions : one to pass from n to n+1, and another from n+1 to n+2, i.e. $x_{n+1} = \theta_-(x_n) = bx_n$ and $x_{n+2} = \theta_+(x_{n+1}) = c\,x_{n+1}$, respectively. Combining the two yields $x_{n+2} = \theta_+(\theta_-(x_n)) = c\,b\,x_n$, which coincides with a $x_n$, $a < 0$, provided $cb = a$. There exists thus an infinity of discontinuous half-iterates (depending on one free parameter), none of which is best in any geometrically or analytically meaningful sense. The same parametric family exists also for $a > 0$, but the choice $c = b = \pm\sqrt{a}$ is obviously priviledged.

As another illustration, consider the Myrberg recurrence

(6-59) $$x_{n+1} = \theta(x) = x_n^2 + c, \qquad -2 \leq c < \frac{1}{4}$$

and its second iterate

(6-60) $$x_{n+2} = g(x) = \theta(\theta(x_n)) = (x_n^2 + c)^2 + c, \qquad x_{n+1} = y_n \quad .$$

Since the fixed points of (6-59) are also fixed points of (6-60), the invariant curve $y = x^2 + c$ of (6-60) is singular, because it traverses two fixed points, which are both unstable when $c < -3/4$. The parametric family of invariant curves into which $y = x^2 + c$ is imbedded becomes progressively more complicated as c approaches -2. A description of their detailed properties is not yet available.

As a final illustration, consider the recurrence

(6-61) $$x_{n+2} = a\,|x_n|^m, \quad x_{n+1} = y_n, \qquad a > 0, \qquad m \geq 2 \quad ,$$

whose fixed point (0,0) has the critical eigenvalues $\lambda_{1,2} = 0$. Two priviledged half-iterates of (6-61) are :

(6-62) $$\begin{aligned} &x_{n+1} = b_1\,|x_n|^c, \qquad x_{n+1} = b_2\,|x_n|^{-c}, \qquad c = \sqrt{m} \quad , \\ &b_1 = a^\alpha, \quad b_2 = a^\beta, \qquad \alpha = 1/(1+c), \qquad \beta = 1/(1-c) \quad . \end{aligned}$$

Let $k = \log a$ and $y = \log|x|$, then (6-61) reduces to

(6-63) $$y_{n+2} = m\,y_n + k \quad ,$$

whose general solution (extensible to non-integer values of n) is

(6-64) $$y_n = C_1\,c^n + C_2(-c)^n - k(m-1) \quad ,$$

or in original variables

(6-65) $$x_n = a^{-\gamma} \cdot e^{C_1 c^n} \cdot e^{C_2(-c)^n}, \qquad \gamma = 1/(1-m) \quad ,$$

where $C_1$, $C_2$ are arbitrary constants, defined by the initial values $x_o$ and $x_1$. Since half-iterates like (6-62) depend only on one arbitrary constant, is there a simple rule governing the choice of $x_o$ and $x_1$ or of $C_1$ and $C_2$ in solutions like (6-65), in order to separate the best half-iterates ? In the particular case of (6-61) and (6-62), this choice can be made by inspection : $C_1 \neq 0$, $C_2 = 0$ and $C_1 = 0$, $C_2 \neq 0$, but it would be worthwhile knowing whether it is "generally" valid, and if so, under what conditions ?

All recurrences with stochastic dynamics studied in this monograph involve phenomenologically a confrontation between a linear growth and a strong non linear growth limitation. The resulting chaos is a generic consequence of this confrontation, in the same sense as a limit cycle is a generic consequence of self-excited oscillations, but at present few naturally meaningful examples have been studied in detail.

An evaluation of all stochastic recurrences studied so far confirms the conjecture that stochasticity arises from three distinct causes : a) disintegration of island structures with increasing non linearity, b) existence of non-unique antecedents, and c) non-smoothness (lack of a sufficient number of continuous derivatives) of the functions used in recurrences. Cause c) is especially noticeable between the levels of continuously differentiable functions and continuous but only piecewise continuously differentiable ones. For example, rounding off the corners in piecewise linear recurrences results often in the disappearance of stochasticity and the appearance of complete orderiliness. Since a correct description of natural dynamic processes should be inert (structurally stable) with respect to such subtleties, results obtained from non-smooth functions should be viewed with skepticism, unless there are sound natural (i.e. physical, biological, etc.) reasons to do otherwise. For this reason (cf. Introduction), the study of non-smooth recurrences was not attempted in this monograph.

A systematic study of first and second order recurrences would be greatly facilitated if solutions of certain practical and theoretical problems were available. A short (non-exhaustive) list of these problems is given below :

1) When a germ of an invariant curve is described by a convergent Mc Laurin series, what constitutes a good estimate of the radius of convergence.

2) What limits the continuation of an invariant curve germ, and how does an eventual deterioration of smoothness take place.

3) What are the necessary (or sufficient) conditions of existence and uniqueness of global invariant curves.

4) When the existence and uniqueness of a global invariant curve is known, how does its smoothness depend on a smooth variation of parameters.

5) How does one construct a germ of an invariant curve starting from a non-singular initial point.

6) When a second order recurrence admits non-unique antecedents, what rules govern a systematic determination of global antecedent invariant curves (especially when the latter originate at cycles).

7) How does one construct efficiently a numerical algorithm for the determination of a two-dimensional mean mass density (generalized solution of the Perron-Frobenius functional equation).

8) How does one construct efficiently a particular solution of a non homogeneous linear functional equation.

9) How does one construct an efficient numerical algorithm for the determination of cycles which favours a) cycles with at least one $|\lambda| \simeq 1$ (or $\varphi \simeq 0$ in the case of centres), b) cycles with at least one $|\lambda| \gtrsim 10$, c) cycles of order $k \gtrsim 20$.

10) How does one construct a symbolic algebra computer program permitting the determination of dominant terms of a recurrence without triggering a combinatorial explosion : a) when $\lambda = 1$, b) when $\lambda = -1$, and c) when $\varphi = 2\pi r/k$.

11) How does one construct a pair of involutes with some specified global properties, for example a given sum.

12) Is there an iterative version of the Birkhoff method of constructing formal k-th iterates.

Answers to problems 2) to 7) should of course be given in terms of the relevant properties of the functions f and g appearing in the recurrence (6-2).

# REFERENCES


A

1 R.L. Adler, A.G. Konheim, M.H. Mc Andrew, "Topological entropy", Trans. Amer. Math. Soc., 114 (1965), p. 309-319.

2 A.A. Andronov, "Les cycles limites de Poincaré et la théorie des oscillations autoentretenues", C.R. Acad. Sc. Paris, 18 (1929), p. 559.

3 A.A. Andronov, L.S. Pontryagin, "Inert systems", Dokl. Akad. Nauk S.S.S.R., 14 (1937), p. 247, and [A4], p. 183-187.

4 A.A. Andronov, "Collected papers", Izd. Akad. Nauk, U.S.S.R., Moscow (1956).

5 A.A. Andronov, A.A. Witt, S.E. Khaikin, "Theory of oscillations", 2nd edition, Fizmatgiz, Moscow (1959).

6 A.A. Andronov, E.A. Leontovich, I.I. Gordon, A.G. Maier, "Theory of bifurcations of dynamic systems in the phase plane", Nauka, Moscow (1967).

7 V.I. Arnold, "Ordinary differential equations", Nauka, Moscow (1975) (see in particular § 6, p. 39-40, and the particular problems discussed in D. Gale, H. Nikaido, "The Jacobian matrix and global univalence of mappings", Math. Annalen, 159 (1965), p. 81-93).

B

1 Ch. Babbage, "An essay towards the calculus of functions", Philos. Trans. of London (1815), p. 389-423.

2 J.R. Beddington, C.A. Free, J.H. Lawton, "Dynamic complexity in predator-prey models framed in difference equations", Nature, 255 (1975), p. 58-60.

3 J.R. Beddington, C.A. Free, J.H. Lawton, "Characteristics of successful natural enemies in models of biological control of insect pests", Nature, 273 (1978), p. 513-519.

4 G.D. Birkoff, "Collected Mathematical Papers", vol. 1 and 2, AMS Publications (1950), or Dover Publications (1968).

5 U.T. Bödewadt, "Zur Iteration reeller Funktionen", Mathematische Zeitschrift, 49 (1943), p. 497-516.

6 E. Borel, "Lectures on divergent series", Gauthier-Villars, Paris (1901).

7 R. Bowen, "Entropy for group endomorphisms and homogeneous spaces", Trans. Amer. Math. Soc., 153 (1971), p. 401-414.

C

1 J. Cameron, "I don't trust any economists today", Fortune (Sept. 11, 1978), p. 30-32.

2 B.V. Chirikov, "Research concerning the theory of non-linear resonance and stochasticity", C.E.R.N. Report : TRANS. 71-40 (1971), see also F.M. Izrailev, B.V. Chirikov, "Stochasticity of the simplest dynamic model with separated phase space", Institute of Nuclear Physics, Acad. of Sc. U.S.S.R., Sibirian Division, Novosibirsk, Preprint n° 191 (1968).

3 B.V. Chirikov, F.M. Izrailev, "Some numerical experiments with a non linear mapping : stochastic component", paper in Publication, n° 229, "Transformations ponctuelles et leurs application", C.N.R.S., Paris (1976), p. 409-428.

4 A.R. Cigala, "On a criterion of instability", Annali di Matematica, 11 (3), (1904), p. 67.

5 R. Clerc, Ch.Hartman, C. Mira, "Transition 'order to chaos' in a predator-prey model in the form of a recurrence", Informatica 77, Bled, Yugoslavia (1977), p. 3-116.

6 R.L. Clerc, Ch. Hartmann, "Bifurcation mechanism of a second order recurrence leading to the appearance of a particular strange attractor". Proc. 7-th ICNO, Academia, Prague (1978), Vol. I, p. 199-204.

7 R. Clerc, Ch. Hartmann, "Some properties of certain one-dimensional singularities of second order autonomous recurrence", C.R. Acad. Sc. Paris, 289 A (1979), p. 31-34.

8 R. Clerc, Ch. Hartmann, "Bifurcations of an isolated closed invariant curve of a second order recurrence". C.R. Acad. Sc. Paris, (1980),

9 J. Couot, "Computer simulation of invariant measures of a discrete dynamic system possessing chaotic regimes", Proc. Informatica - 78, Bled, Yugoslavia (1977), p. 3-207.

10 J. Couot, C. Gillot, "Equations fonctionnelles des densités de mesures invariantes par un endomorphisme de [0,1] et simulation numérique", Equadiff., Firenze (1978), p. 139-147.

11 J. Couot, "Invariant measures and topological entropy of complex stationary states of dynamic systems defined by piecewise monotonic endomorphisms of [0,1]", Proc. ICNO-78, Academia, Prague (1978), p. 205-210.

12 J. Couot, C. Gillot, G. Gillot, "Some numerical simulations of invariant measure densities", Report UER-Math., University of Toulouse III, (1979).

13 R. Courant, D. Hilbert, "Methoden der mathematischen Physik", Springer, Berlin (1937).

14 J.H. Curry, "On the structure of the Hénon attractor", Report of the National Center for Atmospheric Research, Boulder, Colorado (1978).

D

1 P.M. Diamond, "Analytic invariants of mappings of two variables", J. Math. Anal. and Appl., 27 (1969), p. 601-608.

2 E.D. Domar, "Essays on the theory of economic growth", New-York,(1957).

F

1 P. Fatou, Bull. Soc. Math. France, 47 (1919), p. 161-271 ; 48 (1920), p. 33-94 and 208-314.

2 J. Ford, "The transition from analytic dynamics to statistical mechanics", Paper in "Advances in chemical physics", I. Prigogine, S.A. Rice (editors), 24 (1973), J. Wiley and Sons, New-York, N.Y.,(1973).

3 R. Fricke, F. Klein, "Lectures on the theory of automorphic functions", Teubner Verlag, Stuttgart, Vol. 1 (1887) and Vol. 2 (1912).

G

1 G. Gandolfo, "Mathematical methods and models in economic dynamics", North Holland Publishing Co., Amsterdam (1971).

2 A. Gervois, M.L. Mekta, "Broken linear transformations", J. Math. Phys. 18 (1977), p. 1476-1479.

3 C. Gillot, private communication (1978).

4 A. Giraud, "Application des récurrences à l'étude de certains systèmes de commande", Thesis, Université de Toulouse (1969).

5 R.M. Goodwin, "The non linear accelerator and the persistence of business cycles", Econometrica, 19, 1, (1951), p. 1-17.

6 S. Grossmann, S. Thomae, "Invariant distributions and stationary correlation functions", Z. Naturforsch. 32a (1977), p. 1353-63.

7 Z. Grossman, I. Gumowski, "Random number generation and complexity in certain dynamic systems", Proc. Informatica-75, Bled, Yugoslavia, (1975), p. 3.4.

8 I. Gumowski, C. Mira, "Sur une solution particulière de l'équation de Schröder", C.R. Acad. Sc. Paris, 259 (Gr. 1), (1964), p. 2952-2954 and 259 (Gr. 1), (1964), p. 4476-4479.

9 I. Gumowski, C. Mira, "Optimisation in control theory and practice", Cambridge Univ. Press, (1968).

10 I. Gumowski, C. Mira, "Sensitivity problems related to certain bifurcations in non linear recurrence relations", Automatica, 5 (1969), p. 303-317.

11 I. Gumowski, C. Mira, "Boundaries of stochasticity domains in Hamiltonian systems", Proc. 8-th Int. Conf. on High-Energy Accelerators, Cern, Genève (1971), p. 374.

12 I. Gumowski, "Stochastic effects in longitudinal phase space", Proc. 9-th Int. Conf. on High-Energy Accelerators, Stanford, Ca., (1974), p. 439.

13 I. Gumowski, C. Mira, "A predator-prey model in the form of a recurrence. Problem of complex dynamics", First World Conference on Mathematics at the Service of Man", Barcelona, (July 1977).

14 I. Gumowski, R. Thibault, "Dynamic systems with a singular parametric resonance", Equadiff, Firenze, (1978), p. 91-98.

15 I. Gumowski, Ch. Hartmann, "Chaotic regimes of a class of discrete dynamic systems", Proc. Informatica-78, Bled, Yugoslavia, p. 3-204.

16 I. Gumowski, "Invariant curves of a second order autonomous recurrence possessing a non-classical singularity", Proc. 8-th ICNO, Akademia, Prague, (1978), p. 275-280.

17 W. Guzzardi Jr., "The down to earth economics", Fortune, (31 dec. 1978), p. 72-79.

H

1 J. Hadamard, "Sur les transformations ponctuelles", Bull. Soc. Math. France, 34 (1906), p. 349-363.

2 P.R. Halmos, "Lectures on Ergodic Theory", Chelsea Publ. Co., New-York (1956).

3 Ch. Hartman, Ch. Icard, "Search for invariants in a chaotic predator-prey regime", Proc. IMACS Congress, 1979, Sorrento (1979).

4 R.F. Harrod, "Towards a dynamic economics", London (1948).

5 C. Hayashi, Y. Ueda, H. Kawakami, "Transformation theory as applied to the solutions of non linear differential equations of the second order", Int. J. Non Linear Mech., 4 (1969), p. 235-255.

6 C. Hayashi, Y. Ueda, "Behaviour of solutions for certain types of non linear differential equations of the second order", Proc. 6-th ICNO, Poznań (1972), Polish Scientific Publishers, Warsaw (1973), p. 341-351.

7 M. Hénon, "A two dimensional mapping with a strange attractor", Commun. Math. Phys. 50 (1976), p. 69-77.

8 J. Herschel, "Consideration of various points of analysis", Philos. Trans. of London, 2 (1814), p. 440-468.

I

1 Isaacs, "Iterates of functional order", Canadian J. Math. 2 (1950), p. 409-416.

J

1 P. Johnson, A. Sklar, "Recurrence and dispersion of Čebyšev polynomials", J. Math. Anal. and Appl., 54 (3) (1976), p. 750-771.

2 G. Julia, "Rational substitutions", C.R. Acad. Sc. Paris, 164 (1917), p. 1098-1100, 166 (1918), p. 61-64, p. 153-156 and p. 599-601, 168 (1919), p. 147-149, 173 (1921), p. 690-693.

K

1 L.V. Kantorovich, "Functional analysis and applied mathematics", Usp. Mat. Nauk, 3 (1948), p. 89-185.

2 L.V. Kantorovich, G.P. Akilov, "Functional analysis", Nauka, Moscow (1977).

3 D.E. Knuth, "The art of computer programming", Section : "Random numbers", Addison-Wesley Co., New York, N.Y.,(1969).

4 V.I. Krylov, "Approximate computation of integrals", Figmatgiz, Moscow (1959).

5 G. Königs, "Recherches sur les intégrales de certaines équations fonctionnelles", Ann. Sci. Ecole Norm. Sup. 3 (1884), Supp. 3-41.

6 M. Kuczma, "Functional equations in a single variable", PWN, (Polish Scient. Publ.), Warsaw (1968).

7 M. Kuczma, "Fractional iteration of differentiable functions", Annales Polonici Mathematici, 22 (1969), p. 217-227.

8 M. Kuczma, "Uniqueness of solutions of one-variable functional equations", p. 187-199 in 'Differential equations with a deviating argument', Naukova Dumka, Kiev (1977).

9 H.E. Kyburg, "The logical foundations of statistical inference", D. Reidel Publishing Co., Dordrecht, Holland (1974).

L

1 L.D. Landau, E.M. Lifshitz, "Statistical Physics", Addison-Wesley, Reading, Mass. (1958), p. IX; Nauka, Moscow (1964), p. 11.

2 L.J. Laslett, E. Mc Millan, J. Moser, Courant Institute Report NYU - 1480 - 101. New-York University, N.Y.,(July 1968).

3 A. Lasota, J.A. Yorke, "On the existence of invariant measures for transformations with strictly turbulent trajectories", Bull. Acad. Polonaise des Sc., Série Sc. Math. Astr. et Phys., 25 (3), (1977), p. 233-238.
(N.B. "Turbulent trajectories" as used above have no relation with turbulent trajectories in a physically existing fluid).

4 S. Lattès, "Sur les équations fonctionnelles qui définissent une courbe ou une surface invariante par une transformation", Annali di Matematica, Série 3, 13 (1906), p. 1-69.

5 S. Lattès, "The reduced form of point-transformations in the domain of a fixed point", Bull. Soc. Math. France, 39 (1911), p. 309-345.

6 T. Levi-Civita, "On some instability criteria", Annali di Matematica, 5 (3), (1901), p. 221-305.

7 T.Y. Li, J.A. Yorke, "Ergodic transformations from an interval into itself", Trans. Am. Math. Soc., 235 (1978), p. 183-192.

8 A.M. Liapunov, "Collected Papers", U.S.S.R. Acad. of Science, Moscow (1948).

9 E.N. Lorenz, "Deterministic non periodic flow", J. Atmospheric Sc., 20 (1963), p. 130-141.

10 E.N. Lorenz, "The problem of deducing the climate from the governing equations", Tellus, 16 (1964), p. 1-11.

M

1 I.G. Malkin, "Some problems of the theory of non linear oscillations", Goztekhizdat, Moscow (1956).

2 R. May, "Simple mathematical models with very complicated dynamics", Nature, 261 (1976), p. 459-467.

3 V.N. Melekhin, "Theory of nonlinear difference equations and resonance instability in phase oscillations in a microtron and of oscillations of rays in open resonators", Soviet Physics JETP, 34(4) (1972), p. 702-708.

4 C. Mira, "Traversée d'un cas critique, pour une récurrence de deuxième ordre, sous l'effet d'une variation de paramètre", C.R. Acad. Sc. Paris, 268A (1969), p. 621-624.

5 C. Mira, "Systèmes à dynamique complexe et bifurcations de type 'boîtes emboitées'. Cas des récurrences d'ordre 1 déterminées par une fonction à un seul extrémum", RAIRO-Automatique, 12(1), (1978), p. 63-94, 12(2), (1978), p. 171-190.

6 C. Mira, "Dynamique complexe engendrée par une équation différentielle d'ordre trois", in Proc. Equadiff-78, R. Conti, G. Sentini, G. Villari, editors, Firenze (1978), p. 25-36.

7 Misiurewicz, W. Szlenk, "Entropy of piecewise monotone mappings", Soc. Math. de France, Astérisque, 50 (1977), p. 299-310.

8 A. Mitropolsky, A.M. Samoilenko, editors, Proc. 11-th Mathematical (Summer) School, Inst. of Math., AN Ukr. SSR, Kiev (1976).

9 A.S. Monin, "On the nature of turbulence", Uspekhi Mat. Nauk, 125(1), (1978), p. 97-122.

10 P. Montel, "Leçons sur les récurrences et applications", Gauthier-Villars, Paris (1957).

11 P.J. Myrberg, "On automorphic functions", Ann. Ecole Norm. Sup. 68 (1951), p. 383-424.

12 P.J. Myrberg, "Iteration von Quadratwurzeloperationen", Ann. Acad. Sc. Fenn. 259A (1958), p. 1-10.

13 P.J. Myrberg, "Iteration der reellen Polynome zweiten Grades", I, II, III, Ann. Acad. Sc. Fenn. 256A (1958), p. 1-10, 268A (1959), p. 1-10, 336A (1963), p. 1-10.

14 P.J. Myrberg, "Sur l'itération des polynomes réels quadratiques", J. Math. Pures Appl. 41(9) (1962), p. 339-351.

15 P.J. Myrberg, "Iteration der Polynome mit reellen Koeffizienten", Ann. Acad. Sc. Fenn., 374 AI (1965), p. 1-18.

N

1 In. I. Neimark, "Point-transformation methods in the theory of non linear oscillations", Radiofizika 1(1) (1958), p. 41-44.

O

1 D.S. Ornstein, "Ergodic theory, randomness and dynamical systems", Yale University Press, New Haven, Conn., (1974).

P

1 G. Pianigiani, "Existence of invariant measures for piecewise continuous transformations", private communication, submitted to Annales Polon. Math.

2 E. Picard, "Leçons sur quelques équations fonctionnelles", Gauthier-Villars, Paris (1950).

3 J. Pierpont, "Functions of a complex variable", Ginn. and Co., Boston, Mass. (1914).

4 S. Pincherle, "On the complete iteration of $x^2-2$". Rend. della Reale Accademia dei Lincei, 29 (1920), 1st sem., p. 329-333.

5 L. Pun, "Initial conditioned solutions of a second-order non linear conservative differential equation with a periodically varying coefficient", J. Franklin Inst. 295 (1973), p. 193-216.

R

1 Ph. Rabinowitz, "Numerical methods for nonlinear algebraic equations", Num. Analysis Acta, Gordon and Breach, New York, N.Y.,(1970)

2 W.E. Ricker, J. Fish. Res. Bd. Can., 11 (1954), p. 569-623.

3 O.E. Rössler, "Continuous chaos", paper in "Synergetics", H. Haken, ed. Springer (1977), p. 184-197.

4 O.E. Rössler, "Continuous chaos. Four prototype equations", Proc. of the International Congress "Bifurcation theory and applications in scientific disciplines", New-York Academy of Sciences (1978).

5 D. Ruelle, Springer Lecture Notes in Math., 565 (1976), p. 146.

6 D. Ruelle, "Applications conservant une mesure absolument continue par rapport à dx sur 0,1 ", Commun. Math. Phys. 55 (1977), p. 47-51.

S

1 B. Saltzman, "Finite amplitude free convection as an initial value problem", (I), J. Atmos. Sci., 19 (1962), p. 329-341.

2 P.A. Samuelson, "Foundations of economic analysis", Harward University Press, Cambridge, Mass., (1974).

3 G. Sansone, R. Conti, "Non linear differential equations", Pergamon Press, Oxford (1964).

4 E. Schröder, "Über iterierte Funktionen", Math. Ann. 2 (1871), p. 296-322.

5 V.V. Shakhildian, L.N. Beliustina, "Phase synchronization", Izd. Sviaz, Moscow (1975), p. 97-106.

6 A.N. Sharkovski, "A necessary and sufficient condition for the convergence of a one-dimensional iteration", Ukrainskii Mat. Jurnal, 12 (1960), p. 484-489.

7 D. Singer, "Stable orbits and bifurcation of maps of the interval", SIAM J. Appl. Math., 35 (1978), p. 260-267.

8 P. Souriac, "Recurrences with analytically known solutions", Internal Report, Dynamic Systems Research Group, University of Toulouse 3, March 1980.

9 G. Szekeres, "Fractional iteration of exponentially growing functions". J. Austr. Math. Soc., 2 (1962), part 3, p. 301-320.

T

1 G. Targonski, private communication (1978).

2 J. Tinbergen, H.C. Bos, "Mathematical models of economic growth", Gauthier-Villars, Paris, n° 28.

U

1 S.M. Ulam, "A collection of Mathematical problems", Interscience Tracts in Pure and Applied Mathematics, 8 (1960).

V

1 V. Volterra, "On functions depending on other functions", Atti della Reale Accademia dei Lincei, 3(2) (1887), p. 153-158.

2 V. Volterra, "Lectures on line-functions", Gauthier-Villars, Paris (1913).

3 V. Volterra, J. Pérès, "General theory of functionals", Gauthier-Villars, Paris (1936).

X

1 Various authors, see for example : "Sensitivity methods in control theory", Pergamon Press (1966), or "Sensitivity of automatic systems", Nauka, Moscow (1968).

Z

1 E. Zehnder, "Homoclinic points near elliptic fixed points", Comm. Pure and Appl. Math., 26 (1973), p. 131-182.

# INDEX